Springer Optimization and Its Applications

VOLUME 105

Aims and Scope
Optimization has been expanding in all directions at an astonishing rate during the last few decades. New algorithmic and theoretical techniques have been developed, the diffusion into other disciplines has proceeded at a rapid pace, and our knowledge of all aspects of the field has grown even more profound. At the same time, one of the most striking trends in optimization is the constantly increasing emphasis on the interdisciplinary nature of the field. Optimization has been a basic tool in all areas of applied mathematics, engineering, medicine, economics, and other sciences.

The series *Springer Optimization and Its Applications* publishes undergraduate and graduate textbooks, monographs and state-of-the-art expository work that focus on algorithms for solving optimization problems and also study applications involving such problems. Some of the topics covered include nonlinear optimization (convex and nonconvex), network flow problems, stochastic optimization, optimal control, discrete optimization, multi-objective programming, description of software packages, approximation techniques and heuristic approaches.

More information about this series at http://www.springer.com/series/7393

Giorgio Fasano • János D. Pintér

Editors

Optimized Packings
with Applications

 Springer

Editors
Giorgio Fasano
Thales Alenia Space Italia S.p.A.
Turin, Italy
giorgio.fasano@thalesaleniaspace.com
https://www.thalesgroup.com/en/worldwide/spac

János D. Pintér
Pintér Consulting Services, Inc.
Halifax, Nova Scotia, Canada
janos.d.pinter@gmail.com
www.pinterconsulting.com

ISSN 1931-6828   ISSN 1931-6836   (electronic)
Springer Optimization and Its Applications
ISBN 978-3-319-18898-0   ISBN 978-3-319-18899-7   (eBook)
DOI 10.1007/978-3-319-18899-7

Library of Congress Control Number: 2015944198

Mathematics Subject Classification (2010): 05B40, 37N40, 65K05, 90Bxx, 90-08, 90C11, 90C26, 90C29, 90C30

Springer Cham Heidelberg New York Dordrecht London

Printed on acid-free paper

Springer International Publishing AG Switzerland is part of Springer Science+Business Media (www.springer.com)

Preface

Optimized object packing (OOP) studies are aimed at finding the best possible non-overlapping arrangement of a given set of objects in a container (or a set of containers). This very general modeling paradigm can be specified in great many ways, thereby leading to interesting—and, as a rule, challenging—optimization models. OOPs can be important components, e.g., in cutting, covering, layout design, loading, scheduling, and supply chain management studies. Arguably, OOP is among the most significant application areas of operations research. Let us remark additionally that the study of atomic or molecular conformations, spherical point arrangements, the design of experiments and other related areas in computational physics, chemistry, biology, and numerical analysis are closely related to the OOP subject.

While the depth and quality of the decisions required to find high-quality OOPs is increasing, we have also witnessed significant and continuing progress regarding both theoretical advances and ready-to-use tools for actual OOP applications. Theoretical advances, scientific innovation, and algorithmic development are supported and enhanced by today's state-of-the-art computational modeling and optimization environments. Until quite recently, the numerical optimization approaches to tackle OOPs were essentially limited to handle convex (linear or nonlinear, continuous) optimization problems and linearly structured combinatorial and mixed integer-continuous optimization problems. The consideration of integer decision variables in more flexible nonlinear modeling frameworks gives rise to even harder combinatorial and mixed integer-continuous optimization problems. The solution of such computational challenges is becoming increasingly more viable.

In addition to the long-time theoretical interest directed towards OOPs, there is a strong practical motivation to solve various real-world packing problems. Our aim has been to offer a selection of efficient exact and heuristic algorithmic approaches and practical case studies related to the broadly interpreted subject of OOP. The contributing authors are well-recognized researchers and practitioners working (also) in the area of OOP-related modeling and optimization. Next we provide an overview of the contributed chapters (ordered on the basis of the family name of their first authors).

Chapter 1, titled "Using a Bin Packing Approach for Stowing Hazardous Containers into Containerships," has been authored by Daniela Ambrosino and Anna Sciomachen. They address the problem of determining stowage plans for containers loaded into a ship. This is the so-called master bay plan problem (MBPP). The MBPP consists in determining how to stow a set of containers—split into groups according to their size, type, class of weight, and destination—into a set of available slots (locations either on the deck or in the hold area) of predetermined bays of a container ship. Context-dependent structural and operational constraints, related both to the containers and to the ship, have to be satisfied by the MBPP. As an important variant of the MBPP, in this chapter the stowage of hazardous containers is considered. The need for stowing dangerous goods implies additional constraints concerning the safety of the entire cargo, since dangerous goods (categorized into different types) have to be stowed away from certain other goods. This variant of the MBPP is handled on the basis of its relationship with the bin packing problem, in which the packed items are containers and the bins are sections of the ship available for the stowage of hazardous (as well as other) containers. Following a step-by-step procedure for properly loading all containers on board, Ambrosino and Sciomachen show how the segregation rules derived from the International Maritime Dangerous Goods Code affect the available slots of the bins. The chapter reports a real-life case study solved by using the commercial software package CPLEX.

Chapter 2, titled "Dynamic Packing with Side Constraints for Datacenter Resource Management," has been written by Sophie Demassey, Fabien Hermenier, and Vincent Kherbache. Datacenter Resource Management (DRM) requires the assignment of virtual machines (VMs) with dynamically changing resource demands to physical machines with dynamically changing available capacities. The changes occurring at runtime invalidate the currently given assignments, thereby necessitating their updates (adjustments). The assignments are also subject to changing restrictions that express various datacenter user requirements. Within this context, the chapter surveys the application of vector packing (called the VM reassignment problem) providing insight into its dynamic and heterogeneous nature. The study advocates flexibility to answer the issues highlighted above, and presents BtrPlace, an open source resource manager based on the discipline of Constraint Programming. BtrPlace offers a flexible and scalable solution procedure as illustrated by sizeable numerical examples. The authors' experiments show that BtrPlace can effectively manage thousands of web applications running on thousands of physical machines.

Chapter 3, titled "Packing Optimization of Free-Form Objects in Engineering Design," has been authored by Georges M. Fadel and Margaret M. Wiecek. OOPs arising in the engineering design context—often referred to as layout optimization problems—require the determination of the arrangement of given subsystems or components within some enclosure (area or volume), to achieve a given set of objectives in the presence of spatial and/or performance constraints. As a rule, such optimization problems are challenging, due to their highly multimodal structure. In addition, the problems are often described by models that may not have closed-form analytical representations, and/or may require the use of computationally expensive evaluation procedures. The time needed to resolve object intersection calculations can increase exponentially with the number of objects to be packed, while the space available for the placement of these components becomes increasingly scarce. The chapter reviews the results of a multi-year research effort, specifically targeting the development of computational tools for automotive engineering design. The packing problems discussed are represented by single- or multi-objective optimization problems. The solution approaches reviewed rely on evolutionary algorithms, due to the level of complexity that precludes the use of sufficiently effective exact methods.

Chapter 4, titled "A Modeling-Based Approach for Non-standard Packing Problems," has been written by Giorgio Fasano. The chapter is focused on packing tetris-like items orthogonally, with the possibility of rotations into a convex domain, in the optional presence of additional constraints. Mixed Integer Linear Programming (MILP) and Mixed Integer Nonlinear Programming (MINLP) model versions, previously studied by the author, are reviewed. An efficient formulation of the objective function, aimed at maximizing the loaded cargo, is given as an MILP model. The MINLP model has been developed to address the relevant feasibility sub-problem: its purpose is to improve approximate solutions, as an intermediate step of a heuristic process. A space-indexed model is also introduced and the problem of approximating polygons by means of tetris-like items is studied. In both cases an MILP formulation has been adopted. Finally, a heuristic approach is proposed to provide effective solutions in practical applications.

Chapter 5, titled "CAST: A Successful Project in Support of the International Space Station Logistics," has been authored by Giorgio Fasano, Claudia Lavopa, Davide Negri, and Maria Chiara Vola. The International Space Station (ISS) is one of the most challenging currently active space programs: this program requires the handling of demanding logistic issues, mainly in relation to on-orbit maintenance and resource resupply. To serve the ISS, a fleet of launchers and vehicles is made available by the space agencies involved. An overall traffic plan schedules the recurrent upload and download interventions between the Earth and the ISS orbit. The European Space Agency (ESA) contributed annually to the ISS logistics from 2008 to 2014, by accomplishing five Automated Transfer Vehicle (ATV) missions. Within the related cargo accommodation context, in addition to tight balancing conditions, difficult packing issues arose: these had to be solved under conditions of strict deadlines and possible last minute changes. The Cargo Accommodation Support Tool (CAST) is a dedicated optimization framework funded by ESA and developed by Thales Alenia Space to create the ATV cargo accommodation plan. The chapter first describes the ATV loading problem. The basic concept of CAST is then reviewed, highlighting the advantages of the methodology adopted, both in terms of solution quality and time savings. Current extensions and possible future enhancements are also discussed.

Chapter 6, titled "Cutting and Packing Problems with Placement Constraints," has been written by Andreas Fischer and Guntram Scheithauer. In real-life cutting and packing problems additional placement constraints are often present.

For instance, defective regions of some raw material cannot become part of the end products. More generally, due to varying quality requirements, certain products may contain (material) parts of lower quality, while this is not allowed for some other products. The chapter considers one- and two-dimensional rectangular cutting and packing problems, in which items of given types have to be cut from (or placed on) a given raw material in such a way that optimizes the value of a context-specific objective function. In the one-dimensional (1D) case, it is assumed that for each item type the allocation intervals (segments of the raw material) are given, so that all items of the same type must be contained by one of these allocation intervals. The authors also consider problems in which the length of the 1D items could vary within known tolerances. In the two-dimensional (2D) case, rectangular items of different types have to be cut from a large rectangle. Here the authors investigate guillotine cutting plans under the condition that defective rectangular regions are not allowed to be part of the manufactured products. For these scenarios they present solution strategies which rely on the branch-and-bound principle or on dynamic programming. Based on the properties of the corresponding objective functions, they discuss possibilities to reduce computational complexity. This includes the definition of appropriate sets of potential allocation (cut) points which have to be inspected to obtain an optimal solution. Applying dominance considerations, the set of such allocation points can be kept small. In particular, the computational complexity becomes independent of the unit of measure of the input data. Possible generalizations of the solution strategy are also discussed.

Chapter 7, titled "A Container Loading Problem MILP-Based Heuristics Solved by CPLEX: An Experimental Analysis," has been authored by Stefano Gliozzi, Alessandro Castellazzo, and Giorgio Fasano. They consider a standard container loading model form: placing smaller boxes orthogonally (generally with the possibility of rotations) into a larger box, to maximize the loaded volume. Although this problem is NP-hard, a number of algorithms can handle it with high numerical efficiency. The task becomes even more challenging when additional conditions with an overall impact have to be taken into account. In such cases, a modeling-based global scope approach is advocated, e.g., when considering load balancing requirements. Mixed Integer Linear Programming (MILP) models relevant to the container loading problem including possible extensions are available in the literature. An MILP model, presented in Chap. 4 of this book, is taken as a basis. The chapter discusses some important computational aspects of the container loading problem in its classical form (i.e., without additional conditions). An *ad hoc* heuristics, derived from the above-mentioned overall approach, is also outlined. Next, the use of CPLEX as an MILP optimizer is considered. Case studies concerning the solution of the MILP model *tout court* for smaller model instances are reported first. Outcomes relevant to the *ad hoc* heuristics are shown next, in relation to a number of more difficult instances. Examples of container loading problems, involving additional balancing conditions, are also presented.

Chapter 8, titled "Automatic Design of Optimal LED Street Lights," has been written by Balázs L. Lévai and Balázs Bánhelyi. The authors discuss the issue of light pollution—i.e., the unnecessary lighting of outdoor areas—which has negative consequences, e.g., by disturbing wild life, not to mention energy conservation aspects. Based on its capabilities, light-emitting diode (LED) technology offers an efficient solution to this problem. LEDs have many advantages over incandescent light sources including lower energy consumption, longer lifetime, improved physical robustness, smaller size, and faster switching. Many cities in developed countries have LED street lights. Designing the orientation of LEDs in street lights is a nontrivial problem, however, since the use of multiple LED packages is required to replace a single incandescent light bulb. Specifically, the positional angles of LEDs in lamps have to be determined to produce an even light distribution over the target surface. Determining the set of best angles is a global optimization (GO) problem, induced by the underlying task of target area covering problems. The authors present an automatic design approach to find suitable LED configurations for street lights, including an embedded light pattern computation technique to evaluate these configurations. The resulting GO problems are solved (heuristically) using a genetic algorithm. In order to speed up the design process, a possible way of parallelization focused on the light pattern computation module is also discussed.

Chapter 9, titled "Approximate Packing: Integer Programming Models, Valid Inequalities and Nesting," has been authored by Igor Litvinchev, Luis Infante, and Lucero Ozuna. They suggest the use of a regular grid to approximate the container to be loaded. This way, the object packing problem is reduced to assigning objects to nodes of the grid, subject to non-overlapping constraints. This approximate packing problem is then formulated as a large-scale linear binary optimization problem. Different model formulations to express the non-overlapping constraints are presented and compared, and valid inequalities are proposed to strengthen the formulations. This general approach is applied first to the packing of circular and L-shaped objects into a rectangular container. Circular objects are defined in the general sense, as a set of points that are located at the same (not necessary Euclidean) distance from a given point. Different objects—including ellipses, rhombuses, rectangles, and octagons—can be handled by simply changing the definition of the norm used to define the distance concept. Nesting objects inside one another is also considered when appropriate, in the context of certain applications. Numerical results are presented to demonstrate the efficiency of the proposed approach: the optimization problems are solved using CPLEX.

Chapter 10, titled "Exploiting Packing Components in General-Purpose Integer Programming Solvers," has been written by Jakub Mareček. The author discusses the task of packing boxes into a large box; this task is often only a part of a more complex problem. As an example, in furniture supply chain applications, one needs to decide which trucks to use to transport furniture between production sites and distribution centers or stores: obviously, one has to search for packings that guarantee that all delivery items fit into the available trucks. Such problems are often formulated and solved using general-purpose integer programming solvers. This chapter studies the problem of identifying a compact formulation of the packing component in a general instance of integer linear programming. The space-indexed approach advocated is based on exploiting the problem structure and a reformulation using the adaptive discretization proposed by Allen, Burke, and Mareček, and then solving the extended reformulation. The solvers tested were CPLEX, Gurobi, and SCIP, with CLP as the linear programming solver. Results related to solving model instances with up to 10,000,000 boxes are reported.

Chapter 11, titled "Robust Designs for Circle Coverings of a Square," has been authored by Mihály Csaba Markót. The chapter investigates coverings of a square by a set of uniform size circles of optimized (minimal) radius, when uncertainties are present regarding the actual locations of the circles. This model statement is related to deploying sensors or other kinds of observation units with possible uncertainties regarding their actual deployments. Application examples include scenarios when the deployment has to be made remotely (e.g., from the air) into a potentially dangerous environment, or into a location with unknown terrain, or it is influenced by the weather conditions. The goal of the study is to produce coverings that are optimal in terms of a minimal radius, and are also robust in the following sense: wherever the circles are actually placed within a given uncertainty region, the end result is still guaranteed to be a covering. Markót investigates three special uncertainty regions: first he proves that for uniform circular uncertainty regions the optimal robust covering can be created from the exact optimal covering without uncertainties, provided that the exact covering configuration is feasible for the robust scenario. For uncertainty regions given by line segments and by general convex polygons, he proposes a bi-level optimization method combining a complete and rigorous global search and a derivative free black-box search. Numerical examples illustrate the efficiency of the suggested approach.

Chapter 12, titled "Batching-Based Approaches for Optimized Packing of Jobs in the Spatial Scheduling Problem," has been written by Sudharshana Srinivasan, J. Paul Brooks, and Jill Hardin Wilson. Spatial scheduling problems (SSPs) involve the non-overlapping arrangement of jobs within a limited physical workspace in such a manner that some scheduling objective is optimized. In the context of shipbuilding and other large-scale manufacturing industries, the jobs typically occupy large areas, requiring that the same contiguous units of space be assigned throughout the duration of their processing time. This adds an additional level of complexity to the corresponding scheduling problem. Since solving large-scale problem instances by using exact methods becomes computationally intractable, there is a need to develop efficient alternative strategies to provide near-optimal solutions. Much of the literature focuses on minimizing the makespan of the schedule. The authors propose two heuristic methods to minimize the sum of completion times. The approach is based on grouping jobs into batches and then applying a scheduling heuristic to these batches. It is shown that grouping jobs earlier in the schedule can result in poor performance when the jobs have large differences in processing times. The authors provide bounds on the performance of the algorithms, and present computational results comparing the solutions to the optimal objective obtained from the integer programming formulation for SSP. For a smaller number of jobs, both algorithms produce comparable solutions. For instances with a larger number of jobs and a higher variability in spatial dimensions, the efficient area-based model outperforms the iterative model, both in terms of solution quality and run time.

Chapter 13, titled "Optimized Object Packings Using Quasi-Phi-Functions," has been authored by Yuriy Stoyan, Tatiana Romanova, Alexander Pankratov, and Andrey Chugay. The authors here further develop the main conceptual tool—called phi-functions—of their previous related studies. New quasi-phi-functions are defined and used for the analytical description of relations of geometric objects placed in a container taking into account their continuous rotations, translations, and distance constraints. These new functions are substantially simpler to use than phi-functions for certain types of objects. In particular, quasi-phi-functions are derived for certain two- and three-dimensional (2D and 3D) objects. The authors formulate a generic optimal packing problem and introduce its exact mathematical model as a continuous nonlinear programming problem, using quasi-phi-functions. Next, they propose a general solution strategy that includes the construction of feasible starting points; the generation of nonlinear sub-problems of a smaller dimension and smaller number of constraints; and the search for local extrema of the problem using sub-problems. To show the advantages of quasi-phi-functions, two packing problems are considered which have a broad spectrum of industrial applications. The first of these is the packing of a given collection of ellipses into a rectangular container of minimal area taking into account distance constraints. The second problem is the packing of a given collection of 3D objects—including cuboids, spheres, spherical cylinders, and spherical cones—into a cuboid container of minimal height. The authors developed efficient optimization algorithms to obtain locally optimal object packings. The algorithms are applied to solve several hard model instances, including both known and new test cases.

Chapter 14, titled "Graph Coloring Models and Metaheuristics for Packing Applications," has been written by Nicolas Zufferey. He considers and discusses the link between graph coloring and packings. In the classical graph coloring problem, a color has to be assigned to each vertex of a given graph. If two vertices are connected with an edge, then their colors have to be different. The goal is to color the graph with the smallest number of colors. Next, he considers the packing problem of loading items into containers: for each item, one has to decide the container assigned. Since by assumption certain pairs of items are incompatible, they cannot be loaded in the same container. The goal is then to load all the items in a minimum number of containers. Although the correspondence between these two problems is obvious (a vertex corresponds to an item, a color corresponds to a container, and a connecting edge represents an incompatibility), there is no apparent bridge between the packing and the graph coloring literatures. Several packing problems are formulated and solved applying graph coloring models and methods, and metaheuristics.

The broad range of OOP models, solution strategies, and applications discussed and presented by the contributing authors to this volume clearly illustrate the relevance of the subject. This book will be useful for researchers and practitioners in the field of OOP and numerous related fields. It will be useful also for graduate and post-graduate students to broaden their horizon, by studying real-world applications and challenging problems that they will meet in their professional work. Researchers and practitioners working in mathematical modeling, engineering design, operations research, mathematical programming, and optimization will benefit from the case studies presented. This book also offers extensive literature links for further studies: hence, it can be used as a reference source to assist researchers and practitioners in developing new OOP and related applications.

Turin, Italy   Giorgio Fasano
Halifax, NS, Canada   János D. Pintér
March 2015

Acknowledgements

First of all, we wish to thank all contributing authors for their high-quality research work and diligent efforts to make possible the timely completion and publication of this volume.

One of the editors (GF) thanks also Piero Messidoro, and Annamaria Piras of Thales Alenia Space Italia S.p.A. for their support of the research and development activities related to modeling and optimization in a range of OOP and other space engineering applications. GF also wishes to thank Jane Alice Evans for her commitment and support of his work.

We have been glad to cooperate with Razia Amzad and her colleagues at Springer Science + Business Media on this book project, from its initial discussions to its completion.

Contents

Biography

Giorgio Fasano is a researcher and practitioner with nearly three decades of experience in the field of optimization. He works at Thales Alenia Space (Italy), managing and technically supporting major space engineering projects. He holds a M.Sc. in Mathematics, and he is a Fellow of the Institute of Mathematics and its Applications (IMA, UK), with the designations of Chartered Mathematician (IMA, UK) and Chartered Scientist (Science Council, UK). He is co-editor, with János D. Pintér, of Modelling and Optimization in Space Engineering, published by Springer (2013). He is also the author of a numerous publications on optimization, including the recent book titled Solving Non-standard Packing Problems by Global Optimization and Heuristics (SpringerBriefs in Optimization, 2014). His professional interests are related to operations research, mathematical programming, global optimization, optimized packings, and optimal control.

János D. Pintér is a researcher and practitioner with four decades of work experience. His professional interests are primarily related to model, algorithm and software development for nonlinear optimization, including a range of applications. He holds M.Sc. (ELTE, Budapest), Ph.D. (Moscow State University), and D.Sc. (Hungarian Academy of Sciences) degrees in Applied Mathematics / Operations Research / Optimization. He is the author of 4 books, and editor / co-editor of 5 other books; further research monographs and edited volumes are in progress. He is also the author or co-author of more than 200 articles, book chapters, proceedings contributions, book reviews, and technical reports. He serves or served on the editorial board of several scientific journals, and he also served as an officer of international professional societies affiliated with INFORMS and EURO. Dr. Pintér also runs PCS Inc., a consulting company. In this capacity, he is the principal developer of several nonlinear optimization software product versions: these products are in use around the world. For further details, please visit www.pinterconsulting.com.

Chapter 1
Using a Bin Packing Approach for Stowing Hazardous Containers into Containerships

Daniela Ambrosino and Anna Sciomachen

Abstract This chapter addresses the problem of determining stowage plans for containers into a ship, which is the so-called master bay plan problem (MBPP). As a novel issue and variant of MBPP, in the present work we consider the stowage of hazardous containers that follows the principles included in the segregation table of the International Maritime Dangerous Goods (IMDG) Code. Formally, the MBPP consists in determining how to stow a set of n containers, split into different groups, according to their size, type, class of weight and destinations, into a set of m available slots, that are locations either on the deck or in the stow, of predetermined bays of a containership. Some structural and operational constraints, related to both the containers and the ship, have to be satisfied. The need of stowing dangerous goods implies to take into account additional constraints to be verified in each slot concerning the safety of the whole cargo, for which dangerous goods are categorized into different types and forced to be stowed away from incompatible ones. We face such variant of MBPP on the basis of its relationship with the bin packing problem, where items are containers and the bins are sections of the ship available for the stowage of hazardous containers. In particular, following a step by step procedure for properly loading all containers on board, we show how the segregation rules derived from the IMDG Code impact on the available slots of the bins. A real life case study is reported.

Keywords Hazardous containers • International Maritime Dangerous Goods Code • Master bay plan problem • Bin packing • Combinatorial optimization

1.1 Introduction

Nowadays, mainly due to the increase in the shipping business and the phenomenon of naval gigantism, the sea is more and more becoming the main commercial channel. Following this trend, a still increasing number of works have been recently

D. Ambrosino (✉) • A. Sciomachen
Department of Economics and Business Studies, University of Genoa, Genoa, Italy
e-mail: ambrosin@economia.unige.it; sciomach@economia.unige.it

© Springer International Publishing Switzerland 2015
G. Fasano, J.D. Pintér (eds.), *Optimized Packings with Applications*, Springer
Optimization and Its Applications 105, DOI 10.1007/978-3-319-18899-7_1

1

proposed in the literature focusing on the performances of maritime terminals, whose activities are pivotal functions for operating supply chains efficiently. A recent overview of relevant literature about maritime terminal operations is provided in Stahlbock and Voss [1].

In this context, it is not surprising that container handling problems, and particularly the container loading aspects, have been dealt with frequently in the operations research literature (see, e.g., [2, 3], for surveys).

In this chapter, we focus our analysis on the quay and ships activities; more precisely, we devote our attention to the problem of determining stowage plans for containers into a ship, which is the so-called master bay plan problem (MBPP). Readers can find a detailed description of MBPP together with its main constraints in Ambrosino et al. [4]. MBPP is an NP-Hard problem [5], and a number of heuristics have been developed for efficiently facing this problem, usually applied to large size instances. Some heuristic methods for MBPP are compared in Ambrosino et al. [6].

Formally, the MBPP consists in determining how to stow a set of n containers of different size, type, class of weight and destinations, into a set of m available slots, that are locations either on the deck or in the stow, of predetermined bays of a containership. Some structural and operational constraints, related to both the containers and the ship, have to be satisfied. The aim is the operational efficiency of a port, depending on the loading and unloading containers' operations, and the minimization of the time that a ship is at the berth. It is also required to prevent damages to the goods, the ship, its crew and its equipment and the marine environment.

Regarding this, note that up to 8 % of the containers to be loaded into a ship consists of hazardous containers, that is containers carrying dangerous goods, such as solids, liquids, or gases, that can harm people, other living organisms, property, or the environment.

As a novel issue and variant of MBPP, in the present chapter we consider the stowage of hazardous containers that follows the principles included in the segregation table of the International Maritime Dangerous Goods (IMDG) Code, as it will be explained in the next section. In particular, the need of stowing dangerous goods implies to take into account additional constraints to be verified in each slot concerning the safety of the whole cargo, for which dangerous goods are categorized into different types and forced to be stowed away from incompatible ones. Note that, according to the ship certificate, hazardous containers can be stowed only in some slots in the hold of the ship.

Usually the MBPP involves loading decisions at a port which should take into account the possible loading operations at the next ports in the ship route; this means that stowing plans are determined for each port considering the sequence of ports that must be visited by the ship. Only few papers deal with the placement of containers into a containership on a multi-port journey. For instance, Imai et al. [7] present a unified approach for taking into account the route planning problem from both the liner and the terminal manager point of view. Different mathematical programming models are presented and evaluated throughout an extensive computational experimentation in Ambrosino et al. [8]. Delgado et al. [9] present a constraint programming approach for dealing with multi-port routes, focusing the attention on the loading problem at each departing port. Here, we are involved with the loading process of both standard and hazardous containers at a terminal: the stowage plan is defined for loading the containers that in a given port must be loaded and shipped to the different ports visited by the ship; note that the loading plan is not really affected by what happens in the next ports.

In particular, we present a methodological approach for facing the proposed MBPP with hazardous containers based on its relation with the bin packing problem, where items are containers and the bin is a slot of the ship. Relations between MBPP and BBP have been previously presented in Sciomachen and Tanfani [10] and in Zhang et al. [11], where the authors used the same similarity for packing containers into single ship bays. Sciomachen and Tanfani [12] extended the connection between MBPP and 3D-BPP proposed in the previous work by considering the loading pattern for maximizing the productivity of the quay operations at a maritime terminal thus balancing the crane work load. Recently, De Queiroz and Miyazawa [13] focus on the load balancing problem. For a review and classification of cutting and packing problems, the reader can refer to Wäscher et al. [14].

The way in which the international conventions about maritime transportation of dangerous goods impacts on the available slots of the ship, that is the bin, is explained in detail in Sect. 1.3. After the presentation of a real sized case study, reported in Sect. 1.4, in the last section of the chapter we derive some conclusions and outlines for future works.

1.2 International Regulations for Maritime Transport of Dangerous Goods

Today, the international law related to the maritime transport of dangerous goods issue includes many international treaties and codes. All of them have been written under the supervision of the International Maritime Organization (IMO). Note that, as agency of the United Nations, IMO sets internationally valid standards for safety, security and environmental performance of international shipping. Its aim is to create a high level playing-field so that ship operators can't address their financial interests by simply cutting costs and reducing safety, security and environmental performances.

The first Convention to mention is the International Convention for Safe Containers (CSC), entered into force in 1972 after the rapid increase in the use of freight containers for the consignment of goods by sea and the development of specialized container ships, seen in the 1960s. So IMO, in co-operation with the Economic Commission for Europe, developed the Convention which had two goals. The first one is to assure a high level of safety of human life in the transport and handling of containers by providing test procedures and related strength requirements.

The second goal is to facilitate the international transport of containers by providing uniform international safety regulations, equally applicable to all modes of surface transport. In this way, proliferation of divergent national safety regulations can be avoided.

Very important is also the International Convention for the Prevention of Pollution from Ships (MARPOL), adopted on 1973 and entered into force on 1983. This is the main international regulation related to the prevention of marine pollution caused by ships and due to accidental or operational causes; the first aim of MARPOL is preventing and/or minimizing pollution of the marine environment. Strictly related to the topic of the present chapter, the most important part is the Annex III, which contains general requirements for packing, marking, labelling, documentation, stowage, quantity limitations, exceptions and notifications, in case of substances carried in packaged form.

The SOLAS Convention is the most important treaty concerning the safety of merchant ships. The first version was written in 1914, after the Titanic disaster, but the last and official version was adopted in 1974. The main aim of the SOLAS Convention is to specify minimum standards for construction, equipment and operation of ships. Flag States are responsible for ensuring that ships under their flag comply with those requirements, and as a proof the Convention prescribes a number of certificates which ships and operators have to provide. For the purpose of this work, we have to focus on Chap. 7, which provides regulations about: (a) carriage of dangerous goods in packaged form; (b) construction and equipment of ships carrying dangerous liquid chemicals in bulk; (c) construction and equipment of ships carrying liquefied gases in bulk and gas carriers; (d) special requirements for the carriage of packaged irradiated nuclear fuel, plutonium and high-level radioactive wastes on board ships. Note that this chapter makes mandatory the International Maritime Dangerous Goods Code (IMDG Code), developed by IMO.

The IMDG Code has been edited as a uniform international reference for the transport of dangerous goods by sea, covering such matters as packing, container traffic and stowage, with particular reference to the segregation of incompatible substances. Since its adoption by the fourth IMO Assembly in 1965, the IMDG Code has been modified many times to be up-to-date with the ever-changing needs of industry. Amendments which do not affect the principles upon which the Code is based may be adopted by the MSC (Maritime Security Council), allowing IMO to respond to transport developments in reasonable time. The Code classifies dangerous goods into different classes, with the purpose of underlining, defining and describing main characteristics and properties of the substances, material and articles which would fall within each class or division. General provisions for each class or division are given. Individual dangerous goods are listed in the Dangerous Goods List, with the class and any specific requirements. In particular, all substances and articles subject to the provisions of this Code are assigned to one of the classes 1–9 according to the hazard (or the most predominant of the hazards) they present. These nine classes are reported in Fig. 1.1.

Above all the aspects faced by the IMDG Code, we have to underline the contents of Chap. 7, that is the segregation principles. Those are the guidelines which have to

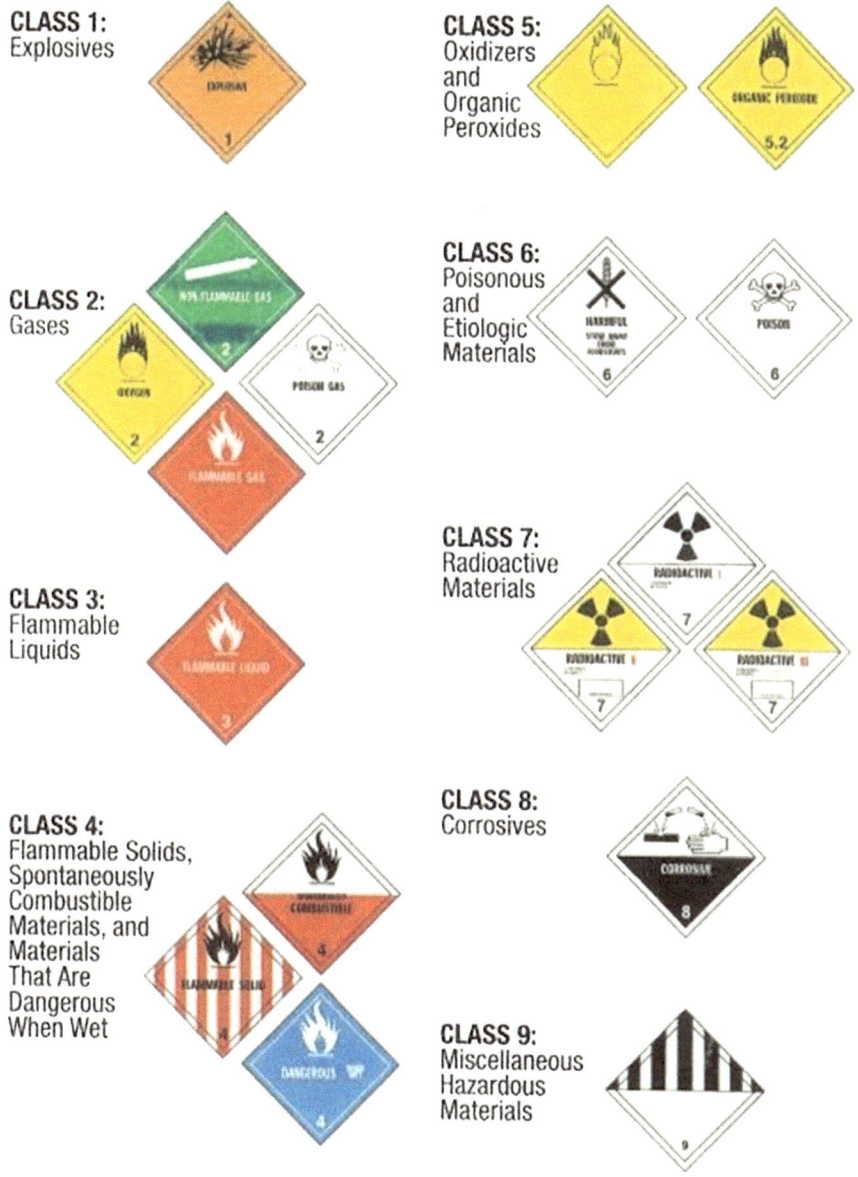

Fig. 1.1 The nine classes of dangerous goods

be followed by operators and carriers, in order to assure safety and security in every step of the transportation chain.

In fact, for their chemical properties, many substances are incompatible, continuously, because they could react mutually bringing to damages due even to

CLASS		1.1 1.2 1.5	1.3 1.6	1.4	2.1	2.2	2.3	3	4.1	4.2	4.3	5.1	5.2	6.1	6.2	7	8	9
Explosives	1.1, 1.2, 1.5	*	*	*	4	2	2	4	4	4	4	4	4	2	4	2	4	X
Explosives	1.3, 1.6	*	*	*	4	2	2	4	3	3	4	4	4	2	4	2	2	X
Explosives	1.4	*	*	*	2	1	1	2	2	2	2	2	2	X	4	2	2	X
Flammable Gases	2.1	4	4	2	X	X	X	2	1	2	X	2	2	X	4	2	1	X
Non-toxic, Non-flammable Gases	2.2	2	2	1	X	X	X	1	X	1	X	X	1	X	2	1	X	X
Toxic Gases	2.3	2	2	1	X	X	X	2	X	2	X	X	2	X	2	1	X	X
Flammable Liquids	3	4	4	2	2	1	2	X	X	2	1	2	2	X	3	2	X	X
Flammable Solids	4.1	4	3	2	1	X	X	X	X	1	X	1	2	X	3	2	1	X
Substances liable to sponateous combustion	4.2	4	3	2	2	1	2	2	1	X	1	2	2	1	3	2	1	X
Substances which, in contact with water, emit flammable gases	4.3	4	4	2	X	X	X	1	X	1	X	2	2	X	2	2	1	X
Oxidizing Substances (agents)	5.1	4	4	2	2	X	X	2	1	2	2	X	2	1	3	1	2	X
Organic Peroxides	5.2	4	4	2	2	1	2	2	2	2	2	2	X	1	3	2	2	X
Toxic Substances	6.1	2	2	X	X	X	X	X	X	1	X	1	1	X	1	X	X	X
Infectious Substances	6.2	4	4	4	4	2	2	3	3	3	2	3	3	1	X	3	3	X
Radioactive Materials	7	2	2	2	2	1	1	2	2	2	2	1	2	X	3	X	2	X
Corrisive Substances	8	4	2	2	1	X	X	X	1	1	1	2	2	X	3	2	X	X
Miscellaneous Dangerous Substances and Articles	9	X	X	X	X	X	X	X	X	X	X	X	X	X	X	X	X	X

Fig. 1.2 The segregation table

explosions, production of noxious or mortal gases and so on. For these reasons, a minimum distance has to be kept among these substances.

For this purpose, the Code provides a number of segregation rules, based on the properties of substances grouped in Classes and Divisions and listed into the Dangerous Goods List. Each relation between Classes is listed into the segregation table, reported in Fig. 1.2. Into the segregation table it is possible to identify specific segregation principles that must be followed for the stowage of every substance, if the cargo includes other harmful substances which are incompatible with the first. Furthermore, the IMDG Code provides different rules in relation to the type of cargo containers used, which could be open-top containers or standard closed containers.

In particular, the following four segregation principles are the most meaningful ones in terms of definition of stowage plans:

1. "Away from";
2. "Separated from";
3. "Separated by a complete compartment from";
4. "Separated longitudinally by an intervening complete compartment or hold from."

These principles will be investigated in more detail in the next section devoted to the definition of stowage planning problems and its related rules.

1.3 The Stowage Planning Problem and the Rules for Dangerous Goods

To give an idea of how a stowage plan is defined, let us consider the basic structure of a containership and its sections, depicted in Fig. 1.3 [4]; it consists of a given number of locations, which generally have a standard size of 8 feet (8′) in height, 8′ in largeness and 20′ in depth, corresponding to one TEU (Twenty Equivalent Unit). Each location is identified by three indices, namely bay, row and tier, each one consisting of two numbers that give its position with respect to the three dimensions.

Note that the address number of the ship locations depends on the numerical system adopted by each maritime company. Generally, each 20′ bay is numbered with an odd number, i.e. bay 01, 03, 05, etc., while two contiguous odd bays conventionally originate one even bay, used for the stowage of 40′ containers, i.e. bay 02 = bay 01 + bay 03 (see Fig. 1.3). As far as the row index, the ship locations have an even number if they are located on the left side, i.e. row 02, 04, 06, and an odd number if they are located on the right side, i.e. row 01, 03, 05, etc. Finally, for the tier index, the levels are numbered from the bottom of the hold to the top with even number, i.e. tier 02, 04, 06, etc., while in the upper deck possible numbers are 82, 84, 86, etc. Note that the tier numbers allow to distinguish in the final stowage plan the containers stowed in the hold from those in the upper deck.

In this chapter, we refer to the connection between MBPP and the 3D-BPP presented in Sciomachen and Tanfani [10], in which the exact branch-and-bound algorithm proposed by Martello et al. [15] is used for solving 3D-BPP instances. More precisely, we consider the MBPP as a three-dimensional orthogonal bin packing problem.

Formally, given a set of n rectangular-shaped items, each one characterized by width w_j, height h_j, and depth d_j, $j = 1, \ldots, n$, and a set of three-dimensional bins, having width W, height H, and depth D, 3D-BPP consists of orthogonally packing all items into the bins. As in most cutting and packing problems [16] we assume

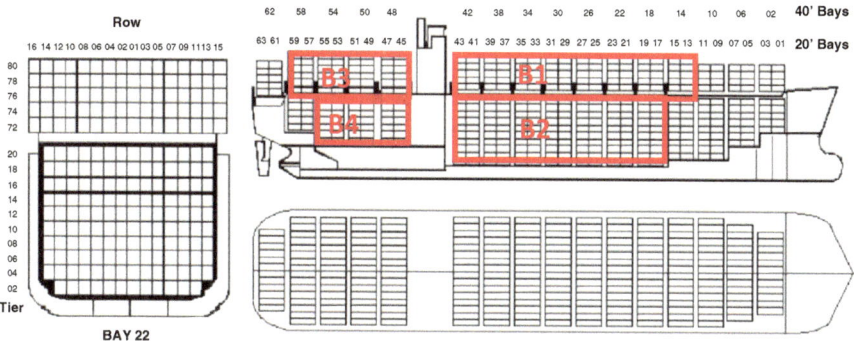

Fig. 1.3 Sections of a standard containership

that the considered bins are sufficient in number and size for containing all items, and the objective is either to minimize the number of bins or maximize the values of the loaded items or minimize the loading time.

The connection between 3D-BPP and MBPP implies that items are containers and the ship is the bin; however, note that the shape of a ship is different from a standard six-face solid that is utilized as the bin in 3D-BPP. Therefore, as in Sciomachen and Tanfani [10], in this chapter we assume the ship to be the bin and split it into different regular sections in order to be able to consider the above and below deck spaces, the bow and the stern as separate components. In this way, each section of the ship has a parallelepiped shape. In particular, for the purpose of the present work let us assume that four sections, i.e. bins, are considered for stowing dangerous containers; these bins, namely B1, B2, B3 and B4, are highlighted in Fig. 1.3. Further, note that all containers to be loaded, representing the items, are standard in size that is either $20'$ or $40'$ in length.

It is worth mentioning that we do not consider those slots in bays, rows and tiers where it is not possible to stow hazardous containers for safety reason, which are usually the most external bays and lowest tiers. Further, note that many maritime companies inhibit for stowage the whole external bays and those closest to the machineries and cabins of the crew; these are bays 43 and 45 in Fig. 1.3.

Each one of sections B_i, $i = 1, \ldots, 4$, can be hence considered as a bin and filled by following the main frame of the exact branch-and-bound algorithm for the 3D-BPP proposed by Martello et al. [15]; in that algorithm, it is assumed that items cannot be rotated, and are packed with each edge parallel to the corresponding edge of the bin. These assumptions are applicable to MBPP too. In particular, they are required for the definition of stowage plans since containers have to be stowed only in one orthogonal direction, one above the other in a stack.

In Sciomachen and Tanfani [10] the authors adapted the above-mentioned enumerative algorithm for 3D-BPP for finding feasible solutions for MBPP.

Here our goal is to show how the segregation rules derived from the IMDG Code impact on the available slots of the considered bins. In particular, we determine stowage plans filling simultaneously each one of the four bins, in such a way to satisfy the main structural constraints of the problem related to both the containers and the ship and the IMDG Code rules described in Sect. 1.2. In particular, having in mind the main segregation principles for dangerous goods presented in Sect. 1.2, let us define them in terms of stowage rules to be satisfied for loading the items, that is the containers, in the bin, that, for example referring to Fig. 1.3, could be the portion of the ship consisting of bays from 43 to 17 in the hold and of bays from 43 to 13 in the deck, that is bins B2 and B1, respectively. Further, let us focus on the segregation principles 2–4, concerning stowage rules for containerized items. Note that we always refer to closed containers.

- Principle 2: *Separated from*.
  This principle means that dangerous containers can never be put in the same stack (vertical line), unless they are separated by a deck, while can be stowed horizontally separated by one container space. Under deck this distances is not

necessary if there is a bulkhead; for example, given two hazardous containers that have to respect the separation principle 2, referring to Fig. 1.3, if one container is stowed in bay 23, row 05 and tier 02, the other one can be stowed in the same row, same tier and bay 25 thanks to bulkhead.

Figure 1.4 shows the implementation of this principle with respect to the available slots for stowing hazardous containers in the considered bin, both in the deck and in the hold, that is either B1 or B2, according to the longitudinal and cross sections of the ship.

In Figs. 1.4, 1.5, and 1.6 the slot coloured light represents a location where a hazardous container has been already stowed, while the slots coloured dark are those locations that are consequently forbidden for stowing other hazardous containers.

- Principle 3: *Separated by a complete compartment from.*

This principle means that dangerous containers can never be put in the same stack (vertical line), same hold or above the same hold. Thus, containers in the hold must be separated by a bulkhead (see Fig. 1.5); for example, given two hazardous containers that have to respect the separation principle 3, referring to Fig. 1.3, if one container is stowed in bay 41, row 05 and tier 02, the other one cannot be stowed in any location (both of the deck and the hold) in the bays 41 and 43 (i.e., there is a bulkhead separating bay 41 and bay 39 in the hold).

Containers on the deck must be separated by one container space along the bay direction (longitudinally: fore and aft) and two container space along the row direction (athwartships: port and strawboard side).

Again, given two hazardous containers that have to respect the separation principle 3, referring to Fig. 1.3, if one container is stowed in bay 41, row 05

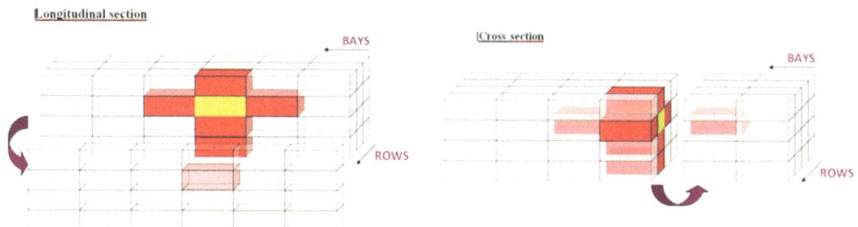

Fig. 1.4 Implementation of the second segregation principle for dangerous goods

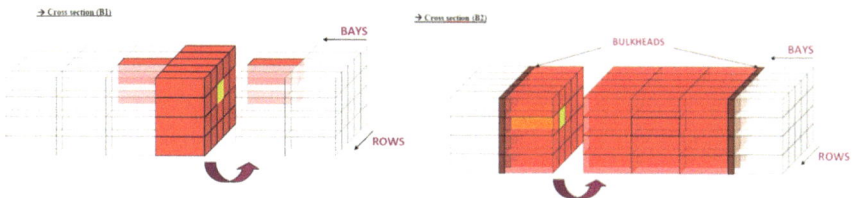

Fig. 1.5 Implementation of the third segregation principle in the deck (B1) and the hold (B2)

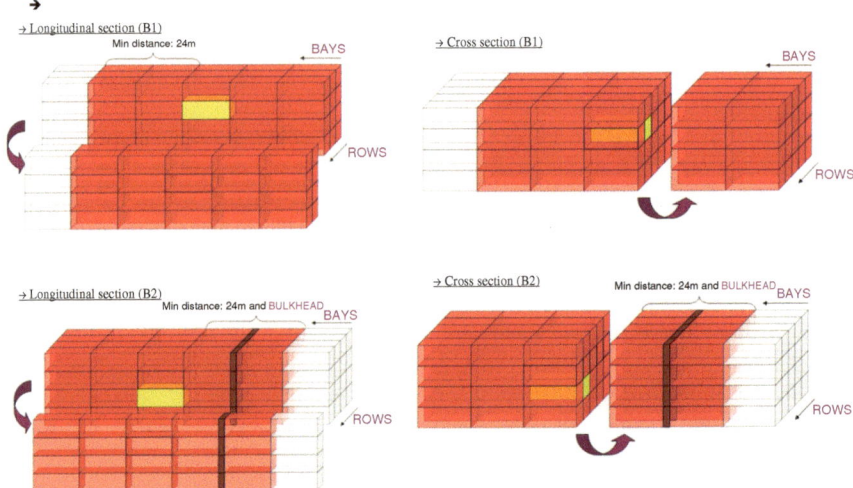

Fig. 1.6 Implementation of the fourth segregation principle in the deck (B1) and the hold (B2)

and tier 72, the other one cannot be stowed in any hold locations of bays 41 and 43, and in deck locations of rows 03, 01, 07, 09 of bays 43 and 39.

The implementation of this segregation principle is depicted in Fig. 1.5 in the cross sections; referring to Fig. 1.3, bin B1 and B2 for the deck and hold, respectively, are considered.

- Principle 4: *Separated longitudinally by an intervening complete compartment or hold from.*
  This principle requires that a minimum distance of two bays (24 m), including a complete compartment, must be maintained longitudinally between two containers that have to respect principle 4.

  For example, given two hazardous containers that have to respect the separation principle 4, referring to Fig. 1.3, if one container is stowed in any location belonging to bay 39, the other one cannot be stowed in any location (both of the deck and of the hold) belonging to bays 43, 41, 37 and 35.

  The implementation of this segregation principle, for the deck (B1) and the hold (B2), is depicted in Fig. 1.6.

  Note that the above requirements apply to the segregation of hazardous containers carried on board of containerships, either on decks or in holds, and compartments of other types of ship, provided that these cargo spaces are properly fitted to give a permanent stowage of the containers during transport.

Let us now see how the above segregation rules can be included in the loading pattern of the items in the bins. For the sake of simplicity, let us explain the proposed procedure focusing on bin B1 in the deck and B2 in the hold (see Fig. 1.3).

Note that the referring 3D-BPP algorithm proposed by Martello et al. [15] starts to position the biggest and the heavier items from the back left bottom corner of a bin and sequentially fills it in a vertical pattern, that is items are stacked one above the other, until the maximum height of the bin is reached; successively, the bin is filled width-wise and finally following a transversal pattern. Consequently, the weight of all packed items is concentrated near the origin of the axes, where they are positioned. Readers can easily understand that this loading pattern applied to the stowage planning can seriously compromise the cross and horizontal stability of the ship. In fact, during navigation and after any loading/unloading operation, it is required that the weight on the right side of the ship must be equal, within a given tolerance T1, to the weight on the left side of the ship (cross equilibrium constraint), and that the weight on the stern must be equal, within a given tolerance T2, to the weight on the bow (horizontal equilibrium constraint). The tolerance values T1 and T2 vary depending on the TEU capacity of the ship. For a detailed description of the ship stability constraints, readers is referred to Ambrosino et al. [4].

Further, destination constraints, which suggests loading first those containers having as destination the final port in the ship route and consequently load last those containers to be unloaded first, are violated by this loading pattern. Finally, loading the largest items first violates the size constraint, forcing the $40'$ containers to be stowed under the $20'$ ones. Note, in fact, that here items do not have the same size; that is, we consider both $20'$ and $40'$ containers. Further, bins associated with different sections of the ship can have different size too.

To remedy this situation, following the bay assignment procedure for a multiport route proposed by Ambrosino et al. [17], we first split the set of b bays of the ship according to the number p of ports to be visited by the ship and the number of containers to be shipped in each port. More precisely, let C_d, $d = 1, \ldots, p$, be the set of containers having port d as destination and t_d be the number of TEUs of set C_d. Note that value t_d allows us to define the minimum number of bays required to load all containers having destination d; in fact, remind that we assume that bins are large enough to load all items. Similarly, let $C_{d(h)} \subseteq C_d$ and $t_{d(h)}$ be, respectively, the subset of hazardous containers destined to port d and the corresponding TEUs. Once the number of bays necessary to stow containers of set C_d, $d = 1, \ldots, p$ is defined, we start from the central bay $b/2$ of the ship and assign to it the first port to be visited by the ship; then, alternatively, from the left and right side of the central bay, we assign bay $(b/2) + 1$ and $(b/2) - 1$ to the next port, and so on, according to the number of bays needed to stow all containers of the corresponding destination. If there is no incompatibility, that is if $C_{d(h)} = \varnothing$, the proposed bay assignment is accepted; otherwise, if $C_{d(h)} \neq \varnothing$ we have to check possible incompatibilities between classes of hazardous containers according to the segregation principles of the IMDG Code described above, such that incompatible containers could not be stowed in contiguous bays if they have to satisfy the segregation principles 3 and 4 (see Figs. 1.5 and 1.6). In particular, if a pair of containers, say c_1, c_2, belonging $C_{d(h)}$ are incompatible according to principles 3 or 4 we have to reassign one of the two to another bay, provided that the minimum distance between the bays satisfies the corresponding segregation rule. More precisely, suppose that c_1 and c_2 belong to $C_{d(h)}$ and more than one bay must be selected for stowing all containers of C_d; in case of the segregation principle 3, only for the deck locations (see Fig. 1.5), that is bin B1, two bays must be chosen (i.e., $(b/2) + 1$ for loading c_1 and $(b/2) - 1$ for c_2), while two more spaced bays are required in case of the segregation principles 4 for the deck and 3 and 4 for the hold (see Figs. 1.4, 1.5 and 1.6). If there are no bays available, we can either switch two contiguous full bays destined to different destinations or put one of the hazardous containers in a different bay, thus respecting the hazardous rule but violating the destination one.

As an example, suppose that two sets of containers C_{d1} and C_{d2} have to be stowed in the hold of the ship, corresponding to bin B2 of Fig. 1.3, for destination d_1 and d_2, respectively; following the bay assignment procedure described above, bays 30, 38 and 22 are assigned to C_{d1} and bays 42, 34, 26 and 22 are assigned to C_{d2}. Let two hazardous containers c_1 and c_2 of class 6.2 and 2.1 be loaded in bin B2 for being shipped to d_1. Note that the segregation rule for c_1 and c_2 requires satisfying principle 4. We see that in this case it is necessary to reassign bay 38 to C_{d2} and bay 42 to C_{d1}, in such a way that there are more than two bays between c_1 and c_2, loading the first in bay 30 and the last in bay 42.

Finally, if the pair of containers c_1 and c_2 belonging $C_{d(h)}$ are incompatible according to principle 2, we can assign them to the same bay but we have to provide the minimum distance between them required by the segregation rule (see Fig. 1.4).

Note that in both cases, that is either $C_{d(h)} = \varnothing$ or $C_{d(h)} \neq \varnothing$, the proposed bay assignment procedure balances the weight of the containers throughout the horizontal section of the ship, thus satisfying the given tolerance limit T2.

Knowing the set of containers to stow in each bay of the ship, we then start the loading process of each bin independently, assigning containers belonging to C_d, d, $d = 1, \ldots, p$, to the corresponding bay; bins corresponding to hold locations are loaded first. Note that considering loading pattern within each bin for single bay guarantees the horizontal stability of the ship verified by the previous bay assignment procedure. Further, note that executing in parallel the loading operations, either in different bins, like B4 and B2, or in different sufficiently spaced bays, like bays 41 and 30, allows us to minimize the total loading time of the ship, as it is shown in Sciomachen and Tanfani [12].

Finally, since the weight and size of a container located in a tier cannot be greater than those of a container located below it in the same row and bay, the containers assigned to a given bay are sorted in an increasing order of their size and in decreasing order of their weight, such that 20′ and heavier containers are loaded first, thus satisfying both the size and weight constraints, imposing that heavier containers cannot be put on a lighter one.

As a last step, in order to satisfy the cross stability constraint, we have to modify the origin of the axes of the 3D-BPP algorithm, as it starts to position the items from the left bottom corner to the bin, which is the origin, following a vertical pattern. Therefore, for each pair of even bays in the bin, we fix the origin considering first as x axis the depth, that is the lowest tier, the smallest bay and the highest even row; then, we consider the width as y axis, coming from the left side to the center of the bin continuing to the end of the tier, and finally the height as z axis, that is moving in a higher tier.

Note that this loading pattern is used if in the bay assigned to destination d $C_{d(h)} = \varnothing$; otherwise, since the less restrictive rule derived from the IMDG Code requires that a minimum distance of one slot in all direction has to be considered between a pair of dangerous containers, we split the corresponding bay in the bin into two parts along the cross section of the ship. In this way, the odd rows of the bay are included in one sub-bin, while the even rows form the other, thus separating the incompatible containers. Both sub-bins are then loaded starting from the lowest tier, the smallest bay and the highest odd row and the smallest even one, respectively, thus balancing the total weight of the loaded containers between the left and right side of the ship in the considered bay.

1.4 A Case Study

Let us detail the loading procedure for stowing containers into a containership described above with a simple case study, related to a containership leaving the port of Genoa, Italy, in which some hazardous containers have to be loaded. The ship has to visit four ports: Singapore, Hong Kong, Shanghai, Kaohsiung, shipping, respectively, 95, 175, 169 and 104 containers.

In each bay of the ship it is possible to stow up to 250 TEUs; therefore, to each destination the bay assignment procedure assigns two even bays and the corresponding odd bay. The bays of the ship go from 02 to 78; then, the central odd bay, that is bay 38, and the related even bays 37 and 39, is assigned to the first destination, that is Singapore. Successively, bay 42 is assigned to Hong Kong, while bay 34, corresponding to bay $(b/2) - 1$, is assigned to Shanghai; finally, bay $(b/2) + 2$ that is bays 46, with 45 and 47, is assigned to Kaohsiung. Let us focus on the stowage planning of this bay, since this last destination is the only one having hazardous containers to be shipped to. This bay, reported in Fig. 1.7, has 16 rows and 15 tiers; two bins are identified in it: bin B1, corresponding to the 6 tiers on the deck and without the external rows, and bin B2, corresponding to the regular shape of the hold, consisting of the first 6 tiers and the inner 12 rows.

To Kaohsiung we have to send 70 20′ containers and 34 40′ ones. Without loss of information from the loading procedure point of view, let us assume that the 20′ containers are named from $c1$ to $c70$, while the 40′ containers are named from $c71$ to $c104$. Further, among the 20′ containers, 30 are light, 38 are medium and 2 are heavy, with respect to their class of weight, while among those of 40′ let us assume that the first 29 containers are the heavy ones and the last 5 containers are the medium ones.

In order to see how different a stowage plan is when hazardous containers have to be loaded, first suppose that none of these containers contains dangerous goods.

The first step of the loading patterns is to sort the containers in an increasing order of their size and in decreasing order of their weight. The resulting sorted list is reported in Table 1.1, where each row corresponds to an ordered sequence of equivalent containers to load.

Then, we start to fill the hold of the ship, corresponding to bin B2.

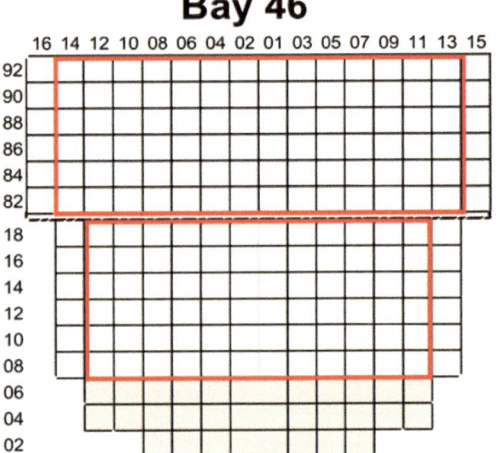

Fig. 1.7 The bay to be loaded

Table 1.1 The loading order of the containers to be shipped form Genoa to Kaohsiung

Load first	c69, c70
:	c31, c32, c33, c34, c35, c36, c37, c38, c39, c40, c41, c42, c43, c44, c45, c46, c47, c48, c49, c50, c51, c52, c53, c54, c55, c56, c57, c58, c59, c60, c61, c62, c63, c64, c65, c66, c67, c68
:	c1, c2, c3, c4, c5, c6, c7, c8, c9, c10, c11, c12, c13, c14, c15, c16, c17, c18, c19, c20, c21, c22, c23, c24, c25, c26, c27, c28, c29, c30
:	c71, c72, c73, c74, c75, c76, c77, c78, c79, c80, c81, c82, c83, c84, c85, c86, c87, c88, c89, c90, c91, c92, c93, c94, c95, c96, c97, c98, c99
Load last	c100, c101, c102, c103, c104

The resulting stowage configuration for bays 45 and 47 is reported in Fig. 1.8. Note that rows 1–11 are filled with $40'$ containers, thus corresponding to bay 46. As readers can easily note, this stowage plan allows the stowage of all containers in one bin.

Let us now assume, as it is the real case, that containers c31 and c49, having the same class of size and weight, are hazardous containers of the class 3 and 2.1, respectively. According to the segregation table reported in Fig. 1.2, this implies that they have to satisfy the second segregation principle (see Fig. 1.5), requiring, for the hold, one container space or a bulkhead and not in the same row. Consequently, we can see that the solution shown in Fig. 1.8 is not anymore feasible, since containers c31 and c49 are put in the same row (10), tiers 08 and 14, respectively, of bay 45. Thus, following the procedure presented in Sect. 1.3 for the loading pattern when hazardous containers requiring to respect principle 2 are given, we have to split the corresponding bin in the hold into two parts, separating the odd rows from the even ones; then, we partition the containers in the bins distributing them homogeneously with respect to the ordering sequence reported in Table 1.1, providing that one of the

	12	10	8	6	4	2	Bay 45 1	3	5	7	9	11
18	c11	c12	c13	c14	c15		c101	c102	c103	c104		
16	c5	c6	c7	c8	c9	c10	c95	c96	c97	c98	c99	c100
14	c48	c49	c1	c2	c3	c4	c89	c90	c91	c92	c93	c94
12	c42	c43	c44	c45	c46	c47	c83	c84	c85	c86	c87	c88
10	c36	c37	c38	c39	c40	c41	c77	c78	c79	c80	c81	c82
08	c69	c31	c32	c33	c34	c35	c71	c72	c73	c74	c75	c76

12	10	8	6	4	2	Bay 47 1	3	5	7	9	11
c26	c27	c28	c29	c30		c101	c102	c103	c104		
c20	c21	c22	c23	c24	c25	c95	c96	c97	c98	c99	c100
c67	c68	c16	c17	c18	c19	c89	c90	c91	c92	c93	c94
c61	c62	c63	c64	c65	c66	c83	c84	c85	c86	c87	c88
c55	c56	c57	c58	c59	c60	c77	c78	c79	c80	c81	c82
c70	c50	c51	c52	c53	c54	c71	c72	c73	c74	c75	c76

Fig. 1.8 The stowage plan obtained by the 3D-BPP loading pattern

	12	10	8	6	4	2	Bay 45 1	3	5	7	9	11
18	c6	c7	c8	C86	C87		c14	c15		c103	c104	
16	c3	c4	c5	C83	C84	C85	c11	c12	c13	c100	c101	c102
14	c39	c1	c2	C80	C81	C82	c49	c9	c10	c97	c98	c99
12	c36	c37	c38	C77	C78	C79	c46	c47	c48	c94	c95	c96
10	c33	c34	c35	C74	C75	C76	c43	c44	c45	c91	c92	c93
08	c69	c31	c32	c71	c72	c73	c40	c41	c42	c88	c89	c90

12	10	8	6	4	2	Bay 47 1	3	5	7	9	11
c21	c22	c23	c86	c88		c29	c30		c103	c104	
c18	c19	c20	c83	c84	c85	c26	c27	c28	c100	c101	c102
c58	c16	c17	c80	c81	c82	c68	c24	c25	c97	c98	c99
c55	c56	c57	c77	c78	c79	c65	c66	c67	c94	c95	c96
c52	c53	c54	cC74	c75	c76	c62	c63	c64	c91	c92	c93
c70	c50	c51	c71	c72	c73	c59	c60	c61	c88	c89	c90

Fig. 1.9 The stowage plan when dangerous containers are given

two dangerous container, for instance c31, is assigned to one sub-bin and container c49 to the other. Finally, in each sub-bin the same loading pattern as before is used. The resulting stowage plan is reported in Fig. 1.9, where, as before, 40' containers, depicted in both bays 45 and 47, are located in bay 46.

Note that there is at least one space distance between containers c31 and c49, and that the weight and size constraints are satisfied. Further, the cross stability constraint, requiring for the considered ship a tolerance value of $T1 = 100$ tons, is satisfied too. Finally, also in this case we are able to stow all containers in one bins, thus optimize the space occupancy in the ship.

However, in case of hazardous containers it is not always possible to follow the loading pattern suggested by an optimal 3D-BPP algorithm and find a feasible solution. In particular, the entire bay assignment procedure can become much more complex when hazardous containers need to respect the segregation principles 3 or 4. In fact, in such cases, it is not always possible to assign destination to bays, since often bays are not enough to respect segregation principles, requiring a minimum separation of two bays. For instance, in the present example if container c49 had been of class 6.2 instead of class 2.1, we would have to satisfy the third segregation principle (see Fig. 1.2). As a consequence, since a pair of odd bays is sufficient for stowing the containers of each destination, either container c31 or c49 should be placed in one of the bays destined to Singapore or Shanghai, that is in bay 33, 35, 37 or 39. Note that also stowing one of the two containers in the bin above the hold, that is in the deck, is inhibited. The serious drawback of the resulting stowage plan is that at the port, say Singapore, visited by the ship before Kaohsiung it is necessary to perform additional loading/unloading operations, which are the so-called unproductive moves, considered one of the most penalizing handling operations in the analysis of the performance indices of a maritime terminals, since impact on the overall berthing time of a ship.

For a better validation from a computational point of view of the procedure described in Sect. 1.3, small instances of the MBPP, similar in size to the above case study, have been generated, comparing the solutions with those obtained by solving the problem with hazardous constraints for respecting the segregation principles. As a main remark we can observe that the solutions are similar in terms of loading time of the bins but differ in the CPU time. More precisely, on average all instances are solved up to optimality by using a commercial software CPLEX 12.5 on a PC on a pc Intel(R) Core i5 CPU M520, 2,40 GHz Ram 6 GB in about 129 s, while few seconds are required by the proposed procedure.

The main negative impact of the presence of dangerous goods on the resulting stowage plans is a greater number of stacks (and sometimes bays) devoted to the stowage of containers having the same destination; this fact can impact also on the workload balance among the quay cranes and on the total loading time. Consequently, the performances of the maritime terminal can be affected too.

Finally, it is important to remark that hazardous containers cannot be unloaded in a port not corresponding to their destination due to the necessity of authority permissions. Thus, they cannot be unloaded for permitting other loading/unloading operations: all unproductive movements regarding this kind of containers must be executed on board.

1.5   Conclusions and Outlines for Future Works

In this chapter we have approached the problem of stowing containers into a containership (MBPP), in which some hazardous ones need to be loaded on board. We followed the relation between MBPP and 3D-BPP and have shown how the segregation rules for dangerous goods force to change the loading pattern.

We will go further in the direction of the present research considering both loading and unloading operations at each port visited by the ship.

Further, in order to manage efficiently all the requirements for stowing hazardous containers due to the segregation rules, it will be necessary to develop a new heuristic procedure. In fact, as remarked in the analysis of the above case study, it will be necessary another strategy for loading hazardous containers, particularly when the third and the fourth segregations principles have to be satisfied. One idea will be to investigate the possibility of relaxing the destination constraints for the hazardous containers and assigning them to the most profitable bays with respect to the minimization of the unproductive moves in each port visited by the ship.

Acknowledgements The present work has been partially supported by the project "Analysis and development of mathematical models for stowage plans with hazardous containers, in accordance with international maritime regulations" within the 2012 research funds of the University of Genoa. The authors wish to thank their friends and colleagues Giorgia Boi and Monica Brignardello for their valuable guide in understanding and applying the main principles of the international maritime law.

References

1. Stahlbock, R., Voss, S.: Operations research at container terminal: a literature update. OR Spectr. **30**, 1–52 (2008)
2. Bortfeldt, A., Wäscher, G.: Constraints in container loading – a state-of-the-art review. Eur. J. Oper. Res. **229**, 1–20 (2013)
3. Lehnfeld, J., Knust, S.: Loading, unloading and premarshalling of stacks in storage areas: survey and classification. Eur. J. Oper. Res. **239**, 297–312 (2014)
4. Ambrosino, D., Sciomachen, A., Tanfani, E.: Stowing a containership: the Master Bay Plan problem. Transport. Res. A **38**, 81–99 (2004)
5. Avriel, M., Penn, M., Shpirer, N.: Container ship stowage problem: complexity and connection to the colouring of circle graphs. Discret. Appl. Math. **103**, 271–279 (2000)
6. Ambrosino, D., Anghinolfi, D., Paolucci, M., Sciomachen, A.: An experimental comparison of different metaheuristics for the Master Bay Plan Problem. In: Festa, P. (ed.) Experimental Algorithms. Lecture Notes in Computer Science, pp. 314–325. Springer, Berlin (2010)
7. Imai, A., Sasaki, K., Nishimura, E., Papadimitriou, S.: Multi-objective simultaneous stowage and load planning for a container ship with container rehandle in yard stacks. Eur. J. Oper. Res. **171**, 373–389 (2006)
8. Ambrosino, D., Anghinolfi, D., Paolucci, M., Sciomachen, A.: Experimental evaluation of mixed integer programming models for the multi-port master bay plan problem. Flex. Serv. Manuf. J. (2013). doi:10.1007/s10696-013-9185-4
9. Delgado, A., Jensen, R.M., Janstrup, K., Rose, T.H., Andersen, K.H.: A constraint programming model for fast optimal stowage of container vessel bays. Eur. J. Oper. Res. **220**(1), 251–261 (2012)
10. Sciomachen, A., Tanfani, E.: The master bay plan problem: a resolution method based on its connection to the three-dimensional bin packing problem. IMA, J. Manage. Math. **14**(3), 251–269 (2003)
11. Zhang, W.-Y., Lin, Y., Jj, Z.-S.: Model and algorithm for container ship stowage planning based on bin packing problem. J. Mar. Sci. Appl. **4**(3), 30–36 (2005)
12. Sciomachen, A., Tanfani, E.: A 3DD packing approach for optimising stowage plans and terminal productivity. Eur. J. Oper. Res. **183**(3), 1433–1446 (2007)
13. De Queiroz, T.A., Miyazawa, F.: Two-dimensional strip packing problem with load balancing, load bearing and multi-drop constraints. Int. J. Prod. Econ. **145**, 511–530 (2013)
14. Wäscher, G., Haussne, H., Schumann, H.: An improved typology of cutting and packing problems. Eur. J. Oper. Res. **183**(3), 1109–1130 (2007)

15. Martello, S., Pisinger, D., Vigo, D.: The three-dimensional bin packing problem. Oper. Res. **48**(2), 256–267 (2000)
16. Oliveira, J.F., Wäscher, G.: Cutting and packing (editorial). Eur. J. Oper. Res. **183**(3), 1106–1108 (2007)
17. Ambrosino, D., Sciomachen, A., Tanfani, E.: A decomposition heuristics for the container ship stowage problem. J. Heuristics **12**, 211–233 (2006)

Chapter 2
Dynamic Packing with Side Constraints for Datacenter Resource Management

Sophie Demassey, Fabien Hermenier, and Vincent Kherbache

Abstract Resource management in datacenters involves assigning virtual machines with changing resource demands to physical machines with changing capacities. Recurrently, the changes invalidate the assignment and the resource manager recomputes it at runtime. The assignment is also subject to changing restrictions expressing a variety of user requirements. The present chapter surveys this application of vector packing—called the VM reassignment problem—with an insight into its dynamic and heterogeneous nature. We advocate flexibility to answer these issues and present BtrPlace, a flexible and scalable heuristic solution based on Constraint Programming.

Keywords Datacenter resource management • Vector packing • Dynamic side constraints • Constraint programming

2.1 Introduction

A datacenter is an infrastructure hosting computing machines. They supply different resources (CPU, RAM, etc.) in limited amount to execute software applications submitted by clients. Thanks to virtualization, a single physical machine (PM) can simultaneously run multiple application components, each embedded in a virtual machine (VM), if their total demand in each resource does not exceed the PM capacity, i.e. the amount of resource supplied by the PM.

S. Demassey (✉)
MINES ParisTech, PSL - Research University, CMA, CS 10207 rue Claude Daunesse, 06904 Sophia Antipolis Cedex, France
e-mail: sophie.demassey@mines-paristech.fr

F. Hermenier
University Nice Sophia Antipolis, CNRS, I3S, UMR 7271, Sophia Antipolis, France
e-mail: fabien.hermenier@unice.fr

V. Kherbache
INRIA, Sophia Antipolis, France
e-mail: vincent.kherbache@inria.fr

© Springer International Publishing Switzerland 2015   19
G. Fasano, J.D. Pintér (eds.), *Optimized Packings with Applications*, Springer
Optimization and Its Applications 105, DOI 10.1007/978-3-319-18899-7_2

A datacenter is a dynamic system since both demands and capacities vary over time: continuously, VMs are submitted, stopped, or resized according to the application needs; continuously, PMs are upgraded, powered on to support load spikes, or halted for maintenance purpose or due to a failure; continuously, execution rules are stipulated by the users of the datacenter—both the operators and the clients—for performance or security purpose. A datacenter is also a market place between the operators, who expect a maximal use of their resources at minimal operation cost, and the clients, who negotiate quality of service (QoS) contracts.

The resource manager of a datacenter is responsible for provisioning the submitted workload continuously. It assigns and reassigns VMs to PMs according to the current resource and user requirements so as to optimize QoS, operation costs and resource usage. The problem is a dynamic variant of vector packing with heterogeneous side constraints [18]: *dynamic* since the manager reoptimizes the problem at runtime, and *heterogeneous* since side constraints express a variety of user requirements and preferences.

Datacenters are commonplace nowadays with the advent of cloud computing. As their size keeps growing (to up to thousands of PMs in large IT companies) they necessitate more automation in resource management. Resource managers with advanced optimization abilities are, however, far from ubiquitous, as the dynamic and heterogeneous nature of the problem remains one major issue.

In this chapter, our first aim is to review this application of vector packing—which we call the VM REASSIGNMENT PROBLEM—with an insight into its two characteristics: dynamicity and heterogeneity. About dynamicity, we further discuss the induced problem of scheduling the reassignment actions. About heterogeneity, we survey some user requirements and preferences met in practical and seminal works. We provide generic formulations of these side constraints which may apply to many other practical applications of packing.

Our second aim is to illustrate the need for *flexibility* in optimization tools to address such characteristics. We present BtrPlace [7] our implementation of a flexible resource manager for virtualized datacenters. BtrPlace relies on Constraint Programming to provide dynamic reassignment and easy customization abilities while yet ensuring performance and scalability.

The chapter is structured as follows: in Sect. 2.2, we discuss the concepts of dynamicity, heterogeneity, and flexibility in the context of resource management. Section 2.3 formalizes the core packing problem and variants of the literature, then describes the induced scheduling problem. Section 2.4 catalogues typical user requirements. Sections 2.5 and 2.6 are devoted to BtrPlace and show empirical evaluations. Section 2.7 presents our conclusions and future research directions.

2.2   Flexible Resource Management

Datacenter resource management exhibits two facets: *dynamicity* and *heterogeneity*. This section describes how resource managers should accordingly offer *configurability* and *extensibility*. *Flexibility* refers to the combination of these two attributes.

2.2.1   Configurable Managers for Dynamic Datacenters

The infrastructure and the workload of a datacenter are highly volatile. They change, at variable pace and at variable intensity, as the user activities change and as failures occur. For example, the operators renew PMs in batches every month, they upgrade the PMs overnight, a hardware failure occurs about every day or week [11], the clients submit new applications every hour, and the load of service applications (such as websites) varies in minutes with spikes occurring at morning and off-peaks during weekends. These changes give the VM reassignment problem its dynamic nature and impact it in different ways:

Repair The problem is not to compute a new assignment but to repair a corrupted one. When changes invalidate the current assignment (e.g., when the new resource demand of a VM suddenly exceeds its current host capacity), the resource manager must compute a new valid assignment, then plan the appropriate reconfiguration actions: powering PMs on and off, launching and migrating VMs either live or off by cloning. These actions affect the performance of the applications during a significant time (e.g., about 10 s to halt or migrate live a VM of 1 GB RAM [17]). They also incur extra operation costs due to energy consumption and hardware usage. Hence, the resource manager should minimize the effects of the reconfiguration when computing a new assignment.

Reactivity Since the changes cannot be predicted accurately (e.g., when and where the next hardware failure will occur) and since the applications run in degraded mode while their requirements are violated and during the reconfiguration, a resource manager must (1) operate at runtime, (2) compute solutions quickly, and (3) compute fast reconfiguration plans.

Elasticity In addition to computing an assignment, the resource manager may command the VM and PM states (e.g., launch, halt, sleep) to accommodate the requirements. For example, it may adjust the number of replicas of a service according to the datacenter load and the required degree of fault tolerance. In these settings, the numbers and sizes of the VMs and PMs become new variables of the problem.

Structural Changes Finally, the changes affect not only the numeric values (the resource requirements) but also the logical constraints (the user requirements) of the problem. From one execution to another, the resource manager is then likely to solve a new optimization problem, not just a new instance of the problem.

Configurability is a required attribute of autonomic resource managers to address structural changes. A configurable resource manager takes as input the current assignment and the new user and resource requirements to merge them into its internal optimization model. If the current assignment violates at least one requirement, it then solves the model. For usability, the manager must offer a high-level interface to specify the new requirements. For reactivity, the internal reformulation of these requirements is expected to be fast. For robustness, the structural changes of the model should not deteriorate the performance of the solution algorithm.

2.2.2 Extensible Managers for Heterogeneous Datacenters

The infrastructure and the usage make each datacenter unique. The development of an universal resource manager remains utopian. Furthermore, each resource manager must deal with the heterogeneity inherent to its own datacenter.

Infrastructure and Workload The design of a datacenter depends on its function (e.g., for private business, internet service, or cloud computing). The size is a major characteristic as it varies from ten PMs gathered in a room to thousands of PMs geographically distributed. Resource management in such distinct environments refers to distinct problems and requires distinct solutions. Though, any resource manager must be scalable at some extent to support the probable growth of its infrastructure.

Within a datacenter, resources and machines come with a great diversity. Different types of resources are either provided by the PMs (e.g., CPU, RAM, disk storage, network interfaces) or shared by groups of PMs (e.g., licenses). Different PMs supply different types of resources and have different capacities. Furthermore, the PMs are connected through a hierarchical network offering different classes of bandwidth and latency.

Similarly, the workload usually presents a great heterogeneity in sizes and shapes from one application to another. This heterogeneity prevents to rely on symmetry arguments to help solve the packing problem.

User Requirements and Preferences Operators—who own or manage the infrastructure—and clients—who submit or use the applications—have multiple needs in terms of resource allocation. Clients expect a reliable QoS to guarantee the optimal execution of their applications by contracting service level agreements (SLAs). SLAs describe low-level metrics (e.g., the resource demand) and logical conditions on the relative assignment of VMs to express different concerns (e.g., grouping communicating VMs on PMs close to each other for performance purpose). Since the client pays for his SLAs and is refunded when violations occur, the operator is willing to enforce QoS while reducing operation costs (e.g., minimizing the number of powered PMs). Operators have strict requirements too, either permanent (e.g., isolating management services on specific PMs for security purpose) or temporary (e.g., freeing PMs to prepare for a maintenance). Hence, a great variety of user requirements and preferences exist. The resource manager must handle a number of them simultaneously at each reassignment step.

Extensibility refers, for a software, to the ease to design and to implement new functionalities. Resource managers release new features to clients on a regular basis. For example, the widely used VMware vSphere and Amazon EC2 were updated to support additional requirements, such as VM-to-PM affinity [13] or dedicated PMs [2]. Extensible resource managers should enable operators to implement desired features. One approach of extensibility is to rely on a modular framework providing an extensible set of primitives to express each feature.

2.2.3 Related Works

Flexibility is a recent concern in datacenter resource management. Pioneer approaches focused on scalability issues and proposed ad-hoc approximation algorithms ignoring everything but CPU and memory requirements [6, 16, 17, 25, 27]. The increasing energy consumption and the rise of SLAs shifted the goal of resource management to compromise between power saving and QoS guarantee. Ad-hoc partially configurable algorithms have been proposed to support these models (e.g., [13, 20]). The extensibility of these algorithms is, however, not discussed and the experiments limited to datacenters with less than 50 PMs. In the context of the Roadef/EURO Challenge 2012 [23], Google described a VM reassignment problem with a fixed set of eight user constraints including five violation penalties to minimize. The dataset consisted of synthetic instances up to 5,000 PMs and 20 resources to solve in 5 min. The competing algorithms were evaluated with regard to their optimization performance, not their flexibility.

Approaches based on Constraint Programming address extensibility but their experiments are often limited to datacenters of ten PMs (e.g., [5]). BtrPlace and its former version Entropy [17, 19] use Constraint Programming with the aim to address flexibility together with scalability. Currently, BtrPlace is bundled with 16 high-level user requirements but users already developed their own. It also provides a simple configuration language to invoke these side constraints on the fly. BtrPlace computes solutions for simulated instances of 5,000 PMs in less than 1 min.

2.3 Problem Statement

This section describes the core VM REASSIGNMENT PROBLEM—without user requirements—as a multi-dimensional vector packing problem. It also presents several objective variants and the induced reconfiguration scheduling problem.

2.3.1 The VM Reassignment Problem

Definition 1. A datacenter consists of a set $\mathscr{P}$ of PMs (the bins) and a set $\mathscr{R}$ of resources (the dimensions). A workload consists of a set $\mathscr{V}$ of VMs (the items). Each PM $p \in \mathscr{P}$ provides a given amount of each resource $r \in \mathscr{R}$, called its *capacity* and denoted c_{pr}. Each VM $v \in \mathscr{V}$ requires to run a given amount of each resource $r \in \mathscr{R}$, called its *weight*, and denoted w_{vr}. A feasible configuration is an assignment M of the VMs in $\mathscr{V}$ to the PMs in $\mathscr{P}$ that satisfies the *resource requirements*:

$$\sum_{v \in M^{-1}(P)} w_{vr} \leq c_{pr} \quad \forall p \in \mathscr{P}, r \in \mathscr{R}.$$

Given a current *source configuration* $M_0 : \mathscr{V} \to \mathscr{P}$, new capacities $c \in \mathbb{N}^{\mathscr{P} \times \mathscr{R}}$ and new weights $w \in \mathbb{N}^{\mathscr{V} \times \mathscr{R}}$, the VM REASSIGNMENT PROBLEM is to find a *target configuration* $M : \mathscr{V} \to \mathscr{P}$ satisfying the new resource requirements while optimizing a given quantitative *performance goal* $f(M) \in \mathbb{R}$.

2.3.2 Performance Goals

The performance goal estimates the quality of the target configuration M and of the reconfiguration process to reach M from M_0, in terms of service to the clients and of financial and energy savings for the operators. Performance goals are context-bound but, by contrast to the user requirements, they generally do not vary over time.

A single performance goal is typically integrated with the model as a function to optimize. A weighted sum allows to merge multiple goals as one objective function. However, intensive experiments and practical knowledge are needed to calibrate the weights accurately. An alternative is to bound the function values of a given goal by means of a hard constraint and to reoptimize the problem with progressively tightened bounds. For goals expressing user preferences, the user requirements are modeled as soft constraints that trigger penalty costs possibly proportional to the degree of violation; the objective then is to minimize the sum of the penalties.

Because of the theoretic cost models, the large size of the instances, the computational complexity of the problem, and the allowed solution time, resource managers do not seek optimality when solving the VM reassignment problem in practice.

2.3.2.1 Scoring the Target Configuration

Workload Consolidation aims at gathering the workload into the minimum number of PMs to power off the unused PMs [6, 17, 25, 27]. With no user requirements, the model coincides with the actual Multi-Dimensional Vector Packing

Problem [15]. A more elaborated model, matching the Capacitated Facility Location Problem, estimates the energy consumption through a fixed cost for each active PM and an execution cost for each pair of VM-PM [12]. Consolidation policies enforce PMs to run at full load. This tends to multiply resource shortages, thus reconfigurations, when the workload is subject to load spikes.

Load Balancing is a performance-oriented policy. It spreads the VMs across the PMs to get a desired load rate on each PM. Such policy can be achieved by minimizing the maximum load over all PMs [1], the sum of the deviations from the desired load rate [24], or the sum of the penalty costs for exceeding a desired safety capacity [23].

SLA Protection refers to policies expressing client satisfaction. An example of global satisfaction is to maximize the number of running applications [24]. The problem maps then to the Multiple Knapsack Problem. Individual client demands—such those described in Sect. 2.4—may also be turned into soft constraints then integrated with the objective when their satisfaction is more desired than required [23].

2.3.2.2 Scoring the Reconfiguration Process

The actions to execute on VMs and PMs to reach the target configuration from the source configuration impact the performance of the datacenter: they provoke downtimes and significant delays, and incur direct operation costs. The reconfiguration score reflects this impact and often dominates the target configuration score in many applications [4, 6, 17, 25, 27]. In fact, performance goals based on reconfiguration scores limit the distance between a source and a target configuration, thus preserve the stability of configuration scores.

Local Changes For instance, the workload consolidation policy yields no energy savings if PMs are turned on and off too frequently. Therefore the performance goal should limit the number of PM state transitions rather than the number of powered PMs. Limiting the number of VM migrations is less trivial. To minimize changes, most works on consolidation [4, 6, 19] and load balancing [14] solve the violations locally—one at a time or altogether—by repairing only a minimal subset of assignments. The decomposition obviously hinders optimality but drastically reduces the problem size.

Migration Numbers An alternative is to focus on minimizing the reconfiguration impact due, in particular, to the VM migrations. In the Load Rebalancing problem [14], the goal is to minimize the number of migrations. In [1], a hard constraint enforces to move less than k VMs. In [23], the maximum number of migrations per application is minimized as well as the weighted sum of migrations between each pair of PMs to simulate network bandwidth conservation.

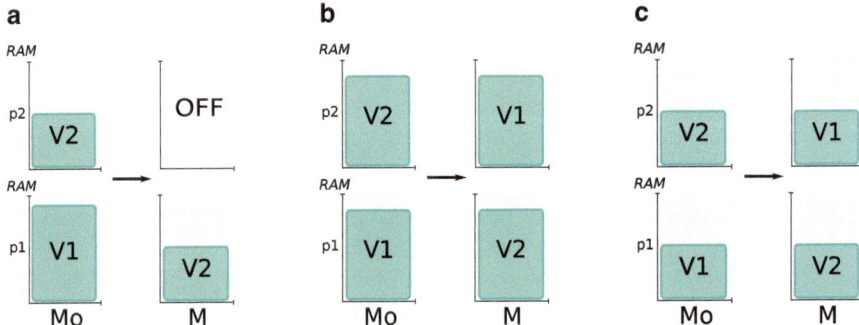

Fig. 2.1 Impacts of the resource (**a**, **b**) and user (**c**) requirements on the reconfiguration from M_0 to M: (**a**) requires to halt v_1 before migrating v_2, (**b**) prevents to migrate live both v_1 and v_2, (**c**) prevents to migrate live both v_1 and v_2 if they are in conflict

Action Durations In addition to the number of actions, [23] minimizes the sum of the predefined impacts of the actions. The action duration—including preparation and transfer times—is a relevant indicator of the impact. The duration can itself be evaluated as a function of the type of the action and of the size of the object. Several works consider this criterion in priority. They minimize either the duration of the whole reconfiguration process [21] or the sum of the completion times [17].

2.3.3 Scheduling the Reconfiguration Actions

Due to the resource limitations, the reconfiguration actions may have to be scheduled in a specific order. Figure 2.1a depicts such a situation: since PM p_1 supplies not enough RAM to run VMs v_1 and v_2 together, the resource manager must halt v_1 before starting the migration of v_2 to p_1.

Enabling live migrations makes the resource constraints still harder since a VM consumes resources on both the source and the target PMs during all the time of its live migration. As a result, a cycle of live migrations may cause a deadlock forbidding to reach the target configuration. Figure 2.1b illustrates this worst case: a cycle occurs between the live migrations of v_1 and v_2 since none of the two available PMs has enough RAM to colocate the two VMs at any time.

A common approach handles the reconfiguration scheduling problem separately after computing the target configuration [17, 27]. It requires to recompute both the reassignment and the scheduling if cycles occur. Furthermore, this two-phase approach disallows to consider the actual impact of the reconfiguration process within the reassignment problem.

As a workaround, [23] tightens up the resource constraints to ensure that all live migrations may happen simultaneously: the total requirement of both the newly assigned VMs and the previously hosted VMs must not exceed a PM capacity. This workaround discards feasible configurations by making permanent the temporary tighter resource requirements induced by live migrations. In addition, it permits to violate user requirements during the reconfiguration. Figure 2.1c considers the *conflict* constraint forbidding to colocate the two VMs v_1 and v_2. To enforce this constraint, one of the VMs must be stopped during the migration of the other one, then relaunched on its target PM. The impacts of these service interruptions and delays are not considered in the reassignment model of [23].

Bin et al. [5] consider the continuous satisfaction—including during the reconfiguration—of one user requirement in a particular use case. BtrPlace [7] generalizes this principle to any user requirements: by handling reassignment and scheduling as one global problem, it enforces the continuous satisfaction of the requirements [10] and controls the reconfiguration impact explicitly.

2.4 User-Defined Side Constraints

In this section, we present a catalog of packing side constraints issued from the literature and from practical user requirements. For each constraint, we discuss its main application contexts, cite some referring works, and introduce a mathematical set formulation using notations of Table 2.1.

spread(V) assigns all the VMs in V onto pairwise distinct PMs [7, 19].It is named *conflict* in [23], *VM-VM affinity* in [13] and *GroupAntiAffinity* in [22]. *Spread* is relevant to clients for fault-tolerance purpose by avoiding a single point of failure.

$$\mathrm{card}(M(V)) = \mathrm{card}(V).$$

gather(V) assigns all the VMs in V onto the same PM [7, 19]. It is named *VM-VM anti-affinity* in VMWare DRS[13]. *Gather* is relevant to clients for performance purpose by improving the intercommunication of a group of VMs.

$$\mathrm{card}(M(V)) = 1.$$

Table 2.1 Notations

$p \in \mathscr{P}$	Physical machines (PMs)
$v \in \mathscr{V}$	Virtual machines (VMs)
$r \in \mathscr{R}$	Resources
$c_{pr} \in \mathbb{N}$	Capacity of PM p in resource r
$w_{vr} \in \mathbb{N}$	Requirement of VM v in resource r
$M_0 \in \mathscr{P}^{\mathscr{V}}$	Source configuration
$M \in \mathscr{P}^{\mathscr{V}}$	Target configuration
$P \subseteq \mathscr{P}$	A set of PMs
$C^{\mathscr{P}} \subseteq 2^{\mathscr{P}}$	A set of sets of PMs
$V \subseteq \mathscr{V}$	A set of VMs
$C^{\mathscr{V}} \subseteq 2^{\mathscr{V}}$	A set of sets of VMs

ban(V, P) assigns all the VMs in V onto PMs not in P [7, 19]. It is named *VM-PM anti-affinity* in VMWare DRS[13]. *Ban* is relevant to operators for maintenance purpose—by freeing PMs before an upgrade—or for security purpose—by preventing client VMs to run on operator dedicated PMs.

$$M(V) \cap P = \emptyset.$$

fence(V, P) assigns all the VMs in V onto PMs in P [7, 19]. It is named *VM-PM affinity* in VMWare DRS [13]. *Fence* is relevant to operators for security purpose by partitioning VMs and PMs according to their compatibility.

$$M(V) \subseteq P.$$

mostlySpread(V, n) assigns all the VMs in V to at least $n \in \mathbb{N}$ PMs [7]. It is named *soft VM-VM affinity* in VMWare DRS [13]. *mostlySpread* is a soft version of *spread* when only a minimum number of distinct PMs is required.

$$\mathrm{card}(M(V)) \geq n.$$

quarantine(P) prevents the PMs in P to relocate their initial hosted VMs and to host new VMs [7, 13, 19]. *Quarantine* is relevant to operators for security purpose by isolating compromised PMs.

$$\forall p \in P, M^{-1}(p) = M_0^{-1}(p).$$

among($V, C^{\mathscr{P}}$) assigns all the VMs in V onto PMs belonging to a single group of $C^{\mathscr{P}}$ [7, 19, 28]. *Among* is relevant to clients and operators for performance purpose by running strongly communicant VMs on PMs with low network latency.

$$\exists P \in C^{\mathscr{P}}, M(v) \subseteq P.$$

root(V) prevents to reassign any VMs in V [7]. It is available as a property in [13]. *Root* is relevant to clients and operators for performance purpose by attaching VMs to some peculiar device.

$$\forall v \in V, M(v) = M_0(v).$$

split($C^{\mathscr{V}}$) prevents to collocate VMs belonging to two different groups in $C^{\mathscr{V}}$ (the groups are pairwise disjoint) [7]. It is available as the *dedicated instances* feature in Amazon EC2 [2]. *Split* is relevant to clients for security purpose by isolating groups of VMs from supposed malicious VMs.

$$\forall V_1 \in C^{\mathscr{V}}, \forall V_2 \in C^{\mathscr{V}} \setminus \{V_1\}, M(V_1) \cap M(V_2) = \emptyset.$$

$\texttt{splitAmong}(C^{\mathcal{V}}, C^{\mathcal{P}})$ assigns each group of VMs belonging to two different groups in $C^{\mathcal{V}}$ to distinct groups in $C^{\mathcal{P}}$ (the groups are pairwise disjoint) [7, 19]. It is a generalization of the *availability zones* in Amazon EC2 [2]. *SplitAmong* is relevant to clients for fault-tolerance purpose by isolating replicated VMs on dedicated PMs.

$$\forall V_1 \in C^{\mathcal{V}}, \forall V_2 \in C^{\mathcal{V}} \setminus \{V_1\}, \exists P_1, P_2 \in C^{\mathcal{P}}, M(V_1) \subseteq P_1, M(V_2) \subseteq P_2, \text{ and } P_1 \neq P_2.$$

$\texttt{maxOnline}(P, n)$ forces at most $n \in \mathbb{N}$ PMs in P to run [7]. *MaxOnline* is relevant to operators for performance purpose by restricting the number of running PMs due to license restrictions or cooling and powering limited capacities [9].

$$\text{card}(P \cap M(\mathcal{V})) \leq n.$$

$\texttt{capacity}(P, r, c)$ forces the total amount of resource r consumed on the PMs in P to be lower than $c \in \mathbb{N}$ [7]. *Capacity* is relevant to operators for performance purpose by restricting access to a shared resource, such as the number of Internet Protocol addresses.

$$\sum_{v \in M^{-1}(P)} w_{vr} \leq c.$$

$\texttt{spreadAmong}(V, C^{\mathcal{P}})$ assigns the VMs in V to at least $n \in \mathbb{N}$ groups of PMs among $C^{\mathcal{P}}$. It is named *spread* in [23]. *SpreadAmong* is relevant to clients for fault-tolerance purpose.

$$\text{card}\{P \in C^{\mathcal{P}} | M(V) \cap P \neq \emptyset\} \geq n.$$

$\texttt{dependency}(V_1, V_2, C)$ given a partition C of the set of the PMs, assigns the VMs in V_1 to elements of C that run at least one VM in V_2 [23]. *Dependency* is relevant to clients for performance purpose.

$$C(M(V_1)) \subseteq C(M(V_2)).$$

2.5 BtrPlace: A Flexible Resource Manager

In this section, we present BtrPlace, an open source resource manager based on Constraint Programming [7]. BtrPlace is the evolution of the former consolidation manager Entropy [17] with a focus on flexibility [18, 19].

2.5.1 Global Design

For regular users, BtrPlace is a configurable VM reassignment algorithm bundled with 16 placement constraints addressing security, performance, reliability, and fault-tolerant concerns. The algorithm takes as input (1) a description of the infrastructure extracted from a monitoring service and (2) a collection of resource and user requirements declared through an API and configuration scripts [19]. The algorithm first checks if the current infrastructure satisfies all the requirements. If not, it computes a new valid VM-to-PM assignment and a schedule of the reconfiguration actions. For advanced users, BtrPlace is an extensible VM reassignment algorithm where third-party developers can implement and integrate new constraints and extensions. BtrPlace is employed for different usages by companies and in research projects such as the Fit4Green European project [12] which addresses energy efficiency in datacenters.

2.5.2 Implementing Flexibility

The flexibility of BtrPlace results from the composability of its core Constraint Programming model through the use of global constraints [3]. In Constraint Programming, a combinatorial problem is modeled as *variables* taking their values in discrete sets called *domains* and *constraints* that represent the required relations between the variables. Each constraint provides a dedicated algorithm to identify and *filter* values in a variable domain that are inconsistent with regard to the relation and to the other variable domains. A *propagation algorithm* calls the filtering algorithms in turn until no more inconsistencies are detected. If a domain becomes empty, the problem is proved to be infeasible. If all domains are singletons, then they figure a solution. Otherwise, a decision tree is built. Successively at each node, a variable-value assignment to explore is selected—in a heuristic order called the *branching strategy*—and the propagation algorithm is recalled.

Flexibility is a strength of Constraint Programming compared to other paradigms like Mathematical Programming or SAT solving: A Constraint Programming model decomposes a problem in global constraints which are altogether processed by a generic algorithm. As a constraint may express any logical relation, a user can embed a part of the complexity of his problem in one constraint as soon as he can define a reasonably efficient filtering algorithm for it. Finally, any Constraint Programming solver supplies a—usually extensible—set of fundamental constraints which can be easily invoked through predicates to compose a model. BtrPlace relies on the Java open-source solver Choco [8].

For each call to the reconfiguration algorithm, BtrPlace generates the Constraint Programming model in two phases. The first phase generates the core model including decision variables for the VM assignments, the PM states, the starting times of the reconfiguration actions, and ad-hoc constraints of vector packing and scheduling. The second phase specializes the core model with the resource and user requirements. Each component attaches variables of the core model with new generated variables through some global constraints of the Choco API. The resulting model is solved heuristically by a truncated branch-a-bound.

The extensibility of BtrPlace has yet some limitations. It cannot infer the next VM states, or perform multiple actions on a same element during a reconfiguration. It also historically focuses on hard constraints. Problems that are vastly organized over soft constraints such as the one formulated for the Roadef challenge [23] could be supported by the framework of BtrPlace but are not currently implemented.

2.5.3 The Optimization Algorithm

The optimization problem—assignment and scheduling—is obviously NP-hard and is intractable for medium to large-size datacenters. The BtrPlace algorithm uses two heuristic strategies to accelerate the resolution.

The *filter optimization* limits the set of VMs to reassign [19]. Each constraint uses a dedicated algorithm—similar to the filtering algorithm—to check the viability of the current assignment. On failure, it computes a set of *candidate* VMs to migrate to resolve the conflict. For example, the spread constraint checks if the VMs to spread are already on distinct PMs. If not, it selects all collocated VMs. All other VMs are fixed to their initial PM in the model prior to its resolution.

Our second strategy relies on a truncated DFS branch-and-bound with a dedicated branching heuristic. The heuristic first focuses on the assignment variables in decreasing order of criticality: first the running VMs that are no longer hosted on a suitable PM, then all other running VMs, and finally the new VMs to launch. To minimize the number of migrations, the heuristic tries to assign a VM first to its current PM, then to the other possible PMs in random order.

2.6 Evaluation of BtrPlace

Highly available (HA) web applications are typical applications running on datacenters. Their architecture illustrates typical user requirements. They are usually composed of three tiers: one deserves static HTTP content, a second one handles the business logic and the last one manages data. To ensure performance and fault-tolerance, each tier is composed of replicated VMs to run on distinct PMs. The replicas of the last tier run databases that must synchronize themselves. To reduce the synchronization latency, they have to run on PMs close together.

To evaluate BtrPlace empirically in a realistic context, we generated workloads made of HA web applications. Each application uses between 6 and 30 VMs with at least two VMs per tier. The resource requirements of the VMs are defined according to one of 12 templates; all VMs in a same tier instantiate the same template.

Each template defines a demand in RAM ranging from 1 to 3 GB and a maximum CPU usage ranging from 30 to 60 uCPU. The CPU consumption varies at any time randomly between 20 and 90 % of its maximum usage. User requirements may be attached to a HA application by means of one *spread* constraint per tier (to model fault-tolerance) and one *among* constraint over the VMs of the third tier (to model synchronization latency).

To evaluate the scalability of BtrPlace, we considered a large datacenter of 5,000 PMs each providing 200 uCPU and 16 GB RAM. To evaluate the impact of the resource usage, we varied the consolidation ratio from 3 to 6 VMs on each PM in accordance with a common observation of real service-oriented datacenters [26]. This amounts to up to 1,700 applications running a total of 30,000 VMs and an overall resource usage varying from 36 to 73 %. For each consolidation ratio, we generated 50 instances for different source configurations.

We considered two scenarios of reconfiguration: LI simulates Load Increases and NR simulates a maintenance for Network Rewiring. In LI, the CPU demand of 10 % of the applications increases by 30 % (capped at 100 %): it increases the overall demand by an average of 5 %. In NR, 5 % of the PMs are randomly selected to be powered off for maintenance: it corresponds to the rate of rewiring observed in Google's datacenters at any moment [11].

BtrPlace ran on one core of an Intel Xeon X3440 at 2.53 GHz with 16 GB RAM running Linux 2.6.32-5-amd64 and Sun's JVM 1.8.0. We gave a time limit of 5 min and stopped BtrPlace at the first solution.

Impact of the Number of VMs Figure 2.2 shows that the solution time grows with the number of VMs, as expected by the complexity of the problem. However, it never exceeds 30 s in the NR case—which is almost the time to halt or to migrate one large VM. Instances of LI appear to be much harder as only one of the 50

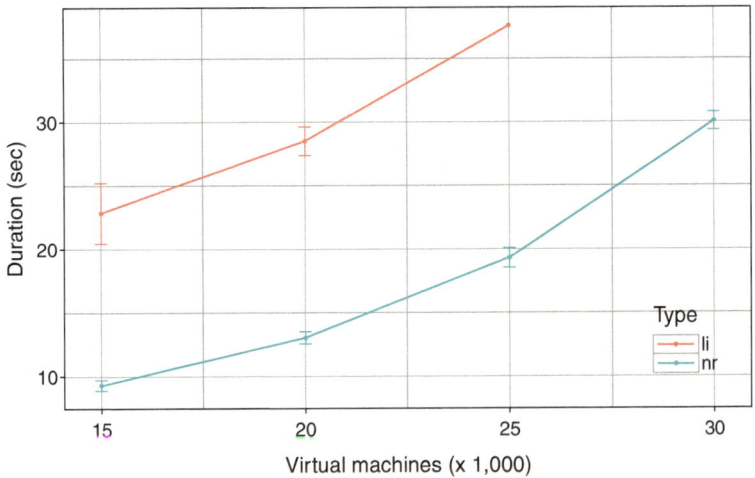

Fig. 2.2 Solution time according to the number of VMs

instances with 25,000 VMs was solved within the five allotted minutes and all
instances having 30,000 VMs hit the timeout. The difference comes from the *filter
optimization* that reduces the problem size more effectively in the NR case. In NR
the algorithm only reassigns VMs that have to be restarted after a failure, while in
LI all the VMs assigned to overloaded PMs are considered. It amounts to 1,500
VMs and 3,000 VMs, respectively, on average for the largest instances. The gap
of performance is also explained by the intrinsic difficulty of the LI case due to
the tighter resource constraints. The number of nodes explored grows exponentially
with the consolidation rate.

To address this scalability issue, we envisage three solutions. A first solution is to
provide a stronger filtering algorithm of the vector packing constraint since our current implementation is limited on purpose to reduce the memory consumption and
to speed up the resolution for large and easy instances. To preserve genericity, we
could automatically adapt the filtering level according to the instance characteristics.
A second solution is to rely on stronger branching strategies. The strategies must
remain instance-independent or, at least, auto-adaptive. The last solution proposed
in [19] automatically splits instances into independent sub-problems.

Impact of the User Constraints In the previous experiments, no applications were
constrained by the HA requirements (i.e., with *spread* and *among* constraints). In
Fig. 2.3, we vary the percentage of applications with HA requirements and compute
the solution time overhead on the solving process for scenarios NR and LI.

We observe that the overhead is acceptable in both cases as it never exceeds one
third of the total solution time. In the worst case, the average overhead is 11 s (34 %)
in NR and only 4 s (11 %) in LI. This low overhead demonstrates that the resource
constraints are dominant. We also observe that with 25,000 VMs, the total solution
time decreases when there is up to 66 % of the constraints in the NR case and 100 %
in the LI case. This phase transition reveals that the reduction of the search space
due to the side constraints may compensate the extra computation time.

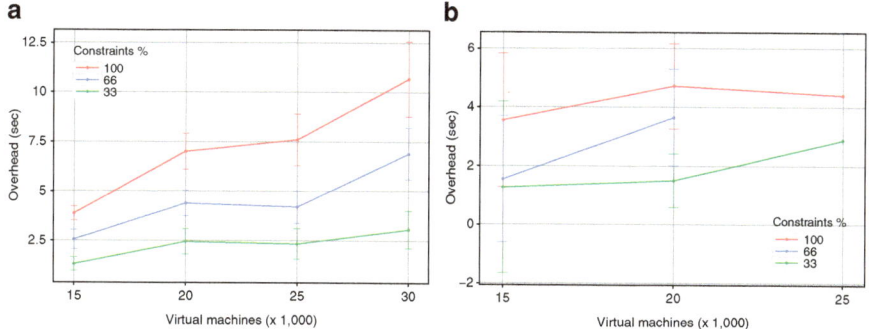

Fig. 2.3 Impact of the user constraints over the solution time. (**a**) NR; (**b**) LI

2.7   Conclusion and Future Works

The resource manager has the critical task to efficiently deploy the client applications throughout a datacenter. This involves to assign VMs to PMs and to revise the assignment recurrently as the environment changes. This optimization problem matches the multi-dimensional vector packing problem with various objectives and side constraints depending on the context. For the past years, many companies and researchers have proposed meaningful solutions for their own context.

In this chapter we advocated an unifying approach that would be able to embrace all these specificities. We emphasized the dynamic and heterogeneous nature of resource management and proposed flexibility as a solution. We assessed this approach through BtrPlace, a flexible and scalable solution based on Constraint Programming. While fully generic optimization algorithms may not always be faster than *ad hoc* solutions, our experiments showed that BtrPlace is effective to manage thousands of highly available web applications running on thousands of PMs.

In future works, we plan to keep improving BtrPlace in terms of performance and flexibility. Regarding performance, we will develop enhanced algorithmic components (partitioning, filtering, branching) to address the scalability issue when solving both large and difficult instances. Regarding configurability, our next step is to make the algorithm auto-adaptive. Similarly to the model, the solver will automatically invoke the right algorithmic components with respect to the instance characteristics. Finally regarding extensibility, we plan to add support for soft constraints—to manage user preferences in addition to user requirements—and for network concerns—by modeling new elements like topology, bandwidth, and latency. One main challenge will be to multiply the case studies to further assess the gain of flexibility.

Acknowledgements BtrPlace is available online under the terms of the LGPL license [7]. Instances used in the experiments are available at https://www.github.com/btrplace/workloads-tdsc. Experiments presented in this paper were carried out using the Grid'5000 experimental testbed (https://www.grid5000.fr), being developed by INRIA with support from CNRS, RENATER and several universities as well as other funding bodies.

References

1. Aggarwal, G., Motwani, R., Zhu, A.: The load rebalancing problem. In: 15th Annual ACM Symposium on Parallel Algorithms and Architectures (SPAA '03), pp. 258–265 (2003)
2. Amazon EC2: http://www.aws.amazon.com/ec2/
3. Beldiceanu, N., Carlsson, M., Rampon, J.X.: Global Constraint Catalog, 2nd edn. Technical report, Swedish Institute of Computer Science (2010). http://www.sofdem.github.io/gccat/
4. Beloglazov, A., Abawajy, J., Buyya, R.: Energy-aware resource allocation heuristics for efficient management of data centers for cloud computing. Future Gener. Comput. Syst. **28**(5), 755–768 (2012)
5. Bin, E., Biran, O., Boni, O., Hadad, E., Kolodner, E., Moatti, Y., Lorenz, D.: Guaranteeing high availability goals for virtual machine placement. In: International Conference on Distributed Computing Systems (2011)

6. Bobroff, N., Kochut, A., Beaty, K.: Dynamic placement of virtual machines for managing sla violations. In: 10th IFIP/IEEE International Symposium on Integrated Network Management (IM'07), pp. 119–128 (2007)
7. BtrPlace: An Open-source flexible virtual machine scheduler. http://www.btrplace.org/
8. Choco: An open source Java constraint programming library. http://www.choco.emn.fr
9. Citrix store. http://www.store.citrix.com
10. Dang, H.T., Hermenier, F.: Higher sla satisfaction in datacenters with continuous vm placement constraints. In: 9th ACM Workshop on Hot Topics in Dependable Systems (HotDep '13), pp. 1:1–1:6 (2013)
11. Dean, J.: Designs, lessons and advice from building large distributed systems. In: Keynote of the International Conference on Large-Scale Distributed Systems and Middleware (2009)
12. Dupont, C., Hermenier, F., Schulze, T., Basmadjian, R., Somov, A., Giuliani, G.: Plug4green: A flexible energy-aware VM manager to fit data centre particularities. Ad Hoc Netw. 25(B), pp. 505–519 (2015)
13. Epping, D., Frank, D.: VMware vSphere 4.1 HA and DRS technical deepdive, CreateSpace (2010)
14. Fukunaga, A.S.: Search spaces for min-perturbation repair. In: 15th International Conference on Principles and Practice of Constraint Programming (CP'09), pp. 383–390 (2009)
15. Garey, M.R., Graham, R.L., Johnson, D.S., Yao, A.C.C.: Resource constrained scheduling as generalized bin packing. J. Comb. Theory Ser. A 21(3), 257–298 (1976)
16. Hermenier, F., Loriant, N., Menaud, J.M.: Power management in grid computing with xen. In: International Conference on Frontiers of High Performance Computing and Networking (ISPA'06), pp. 407–416 (2006)
17. Hermenier, F., Lorca, X., Menaud, J.M., Muller, G., Lawall, J.: Entropy: A consolidation manager for clusters. In: ACM SIGPLAN/SIGOPS International Conference on Virtual Execution Environments. ACM, New York (2009)
18. Hermenier, F., Demassey, S., Lorca, X.: Bin repacking scheduling in virtualized datacenters. In: Principles and Practice of Constraint Programming (CP'11), pp. 27–41. Springer, Berlin (2011)
19. Hermenier, F., Lawall, J., Muller, G.: Btrplace: A flexible consolidation manager for highly available applications. IEEE Trans. Dependable Secure Comput. 10(5), 273–286 (2013)
20. Jung, G., Hiltunen, M.A., Joshi, K.R., Schlichting, R.D., Pu, C.: Mistral: Dynamically managing power, performance, and adaptation cost in cloud infrastructures. In: 30th IEEE International Conference on Distributed Computing Systems (ICDCS '10), pp. 62–73 (2010)
21. Nus, A., Raz, D.: Migration plans with minimum overall migration time. In: 2014 IEEE Network Operations and Management Symposium (NOMS), pp. 1–9 (2014)
22. Openstack cloud software: http://www.openstack.org/
23. Roadef/Euro challenge 2012: Machine reassignment (2012). http://www.challenge.roadef.org/2012/
24. Tang, C., Steinder, M., Spreitzer, M., Pacifici, G.: A scalable application placement controller for enterprise data centers. In: 16th ACM International Conference on World Wide Web, pp. 331–340 (2007)
25. Verma, A., Ahuja, P., Neogi, A.: pmapper: Power and migration cost aware application placement in virtualized systems. In: Middleware 2008. Lecture Notes in Computer Science, vol. 5346, pp. 243–264. Springer Berlin (2008)
26. Virtualization penetration rate in the enterprise. Technical report, Veeam Software (2011)
27. Wood, T., Tarasuk-Levin, G., Shenoy, P., Desnoyers, P., Cecchet, E., Corner, M.D.: Memory buddies: Exploiting page sharing for smart colocation in virtualized data centers. In: ACM SIGPLAN/SIGOPS International Conference on Virtual Execution Environments, pp. 31–40 (2009)
28. Zheng, J., Ng, T.S.E., Sripanidkulchai, K., Liu, Z.: COMMA: Coordinating the migration of multi-tier applications. In: 10th ACM SIGPLAN/SIGOPS International Conference on Virtual Execution Environments (VEE '14) (2014)

Chapter 3
Packing Optimization of Free-Form Objects in Engineering Design

Georges M. Fadel and Margaret M. Wiecek

Abstract Packing for engineering design involves the development and use of methods to determine the arrangement of a set of subsystems or components within some enclosure to achieve a set of objectives without violating spatial or performance constraints. Packing problems, also known as layout optimization problems are challenging because they are highly multimodal, are characterized by models that lack closed-form representations, and require expensive computational procedures. The time needed to resolve intersection calculations increases exponentially with the number of objects to be packed while the space available for the placement of these components becomes less and less available.

This paper presents a multiyear research effort targeting the development of computational tools for packing optimization problems which are encountered at different stages of engineering design with special interest in automotive design. Due to increasingly realistic engineering applications, the problems feature a rising level of complexity and therefore require optimization models and approaches with growing sophistication. To be relevant to automotive design, the packing problems account for the free shape of objects and consider either their compact packing within an envelope or their noncompact packing in the presence of multiple criteria used to evaluate system performance. The packing problems are represented by single or multiobjective optimization problems (MOPs) while the solution approaches rely on evolutionary algorithms due to the level of complexity that precludes development of effective exact methods.

Keywords Packaging • Configuration layout • Compact packing • Multiobjective optimization • Automotive design • Pareto solutions

G.M. Fadel • M.M. Wiecek (✉)
Department of Mechanical Engineering, Clemson University, Clemson, SC 29634, USA
e-mail: fgeorge@clemson.edu; wmalgor@clemson.edu

© Springer International Publishing Switzerland 2015
G. Fasano, J.D. Pintér (eds.), *Optimized Packings with Applications*, Springer
Optimization and Its Applications 105, DOI 10.1007/978-3-319-18899-7_3

3.1   Introduction

This paper focuses on free-form packing, i.e., the packing of objects in a container where the objects, the container, or both, have a nonregular shape. Three basic concerns have until recently precluded the study of free-form packing: First, the complexity of the problem has been such that even the most performing algorithms executed on the best computers could not give a satisfactory answer in a reasonable amount of time. Second, the descriptions of free-form objects were not amenable to efficient interference computation, and third, there was no theoretical interest in these cases since results were not mathematically provable to be the best. Nonetheless, engineering designers are most interested in this problem in that their job requires providing solutions that are at least feasible, and preferably optimal to some combination of criteria.

To the authors' knowledge, the groups of Cagan and his students [3, 45, 46, 61] and Teng and his students [47] are the only ones beside the authors' team focusing on layout optimization in 3D for realistic engineering problems. All three groups use heuristic methods to progress towards a solution, Cagan favors simulated annealing and pattern search whereas Teng and Fadel use genetic algorithms (GAs).

The problem of packing 3D free-form objects within a specific free-form envelope is the most general of the configuration design problems involving constraints, multiple criteria, and mixed discrete/continuous variables. Furthermore, the objectives and constraints are a mixture of linear, quadratic, nonlinear, multimodal, continuous or discontinuous functions that are often analytically unavailable.

The handling of such problems requires differentiating between both object and system. The object is an atomic solid that cannot be taken apart, whereas the system is an aggregation of objects or components that can move with respect to each other. The objective is to design the system, placing the objects inside the container or enclosure, in order to satisfy a multiplicity of constraints and maximize one or a set of criteria. Thus a description of the objects as an indivisible whole must be provided, and a description for the system as a relative or global positioning of the objects is sought. The criteria of interest are system level characteristics, such as volume, inertia, heat transfer, and maintainability.

The packing problem can be dealt with from three different perspectives. The first is motivated by geometric considerations and is based on the concept of the envelope. An outer shell or envelope is constructed for each object to be packed while its internal details are ignored. Additionally, an inner shell or envelope is constructed for an enclosure within which the objects are packed. While mathematical optimization is not used at this stage, the other two perspectives, known as compact packing and noncompact packing, rely heavily on it. The generic compact or noncompact packing problem may be formulated as a multiobjective optimization problem (MOP) of the form [29]:

Given:

- *The design space R^n and a global Cartesian coordinate system.*
- *A set of N objects defined by their shape, material, and positioned in space by six variables, i.e., three Cartesian coordinates (x, y, z) and three angles (α, β, γ) defining the orientation of the object with respect to a local Cartesian coordinate system.*
- *The vector $\mathbf{x}$ called the vector of design variables and composed of subvectors $\mathbf{x_i}$ where $\mathbf{x_i} = (x_i, y_i, z_i, \alpha_i, \beta_i, \gamma_i), i = 1, \ldots, N$.*
- *A set of m equalities, usually called functional constraints,*

$$\mathbf{h}(\mathbf{x}) = \mathbf{0} \tag{3.1}$$

positioning the objects with respect to a reference coordinate system, where $\mathbf{h}$ is a vector-valued function defined as $\mathbf{h} : R^{6N} \to R^m$.
- *A set of n inequalities*

$$\mathbf{g}(\mathbf{x}) \leq \mathbf{0} \tag{3.2}$$

used to identify the interior and exterior of the objects, and enable intersection and overlap calculations, where $\mathbf{g}$ is a vector-valued function defined as $\mathbf{g} : R^{6N} \to R^n$.
- *A set of lower and upper bounds*

$$\mathbf{x^L} \leq \mathbf{x} \leq \mathbf{x^U} \tag{3.3}$$

restricting the position variables.
- *A vector-valued objective function $\mathbf{F} : R^{6N} \to R^p$, evaluating the designs with respect to p scalar-valued objective functions*

$$\mathbf{F} = [f_1(\mathbf{x}), \ldots, f_p(\mathbf{x})] \tag{3.4}$$

Find:

- *A set of values for $\mathbf{x}$ optimizing the objective function $\mathbf{F}$*

Satisfying:

- *The feasibility constraints in the set X of feasible designs*

$$X = \{\mathbf{x} \in R^{6N} : \mathbf{g}(\mathbf{x}) \leq \mathbf{0}, \mathbf{h}(\mathbf{x}) = \mathbf{0}, \mathbf{x}^L \leq \mathbf{x} \leq \mathbf{x}^U\} \tag{3.5}$$

Although the envelope generation makes use of the position variables $\mathbf{x}$ and the constraints of the MOP, it is not driven by any objective function and so does not assume the form of an optimization problem. In compact packing, the MOP is reduced to a single-objective optimization problem (SOP) ($p = 1$) because the optimization of a measure of compactness is the only criterion of interest to designers. The task is then to find an optimal feasible solution x^* representing the optimal locations of all objects.

In noncompact packing, designers are interested in optimizing several objective functions evaluating the performance of packing, and the resulting optimization problem assumes the MOP form. Solving the MOP is understood as calculating its set of efficient or Pareto-optimal solutions [23]. Because MOPs typically have infinitely many Pareto-optimal solutions, approximation approaches are of importance in the selection of a solution method. Some noncompact packing problems that assume a bilevel formulation make use of the MOP or SOP at each level. Therefore, advanced methodologies are required for deriving their solutions.

The multiplicity and the type of the objective and constraint functions inherent in packing optimization problems make them most difficult to solve. Further, their level of complexity precludes development of traditional, exact optimization methods such as gradient-based algorithms. Consequently, evolutionary algorithms (EAs), and in particular GAs and simulated annealing, have been the only approaches effective in resolving these difficulties.

EAs are nature-inspired adaptive search techniques that can efficiently deal with problems having discreteness and multimodality in the search space. They do not require that the optimization problem assume a functional form and do not rely on the gradients of objective and constraint functions. GAs, one such class of EAs with a working principle based upon Darwin's theory of the survival of the fittest, can work with almost any kind of variable representations (discrete, integer, real) so long as suitable genetic variation operators are provided. GAs have therefore emerged as an attractive optimization tool for a wide class of SOPs and MOPs. Their effectiveness is also justified by the need to generate sets of Pareto-optimal solutions for packing problems formulated as MOPs.

The Simple Genetic Algorithm [26, 32], Evolution Strategies [43], Genitor (a steady-state GA) [56, 57], CHC (cross-elitist generation, heterogeneous recombination, cataclysmic mutation) [24], and Covariance Matrix Adaptation [30] are but a few of the notable single-objective optimization algorithms based on heuristic evolutionary methods.

Multiobjective genetic algorithms (MOGAs), and in a broader sense multiobjective evolutionary algorithms (MOEAs), have been very attractive to engineers because they operate on a set of solutions and not on a single one, and can therefore effectively obtain in a single run a set of nondominated solutions that are a good approximation of the (true) set of Pareto-optimal solutions. Various methods have been investigated, typically based on evolving the solutions towards nondomination using methods such as rank-based evaluation, and towards separating solutions using niching or alternative methods. Again, notable efforts in designing MOEAs include: Strength Pareto evolutionary algorithm (SPEA2) [63], Pareto-envelope based selection algorithm (PESA-II) [8], non-dominated sorting genetic algorithm (NSGA-II) [17], neighborhood cultivation genetic algorithm (NCGA) [55], intelligent multi-objective evolutionary algorithm (IMOEA) [31], MOEA [18], Omni-optimizer (OmniOpt) [16], fast Pareto genetic algorithm (FPGA) [25], and archive-based micro genetic algorithm (AMGA, AMGAII) [49, 50, 52]. A comprehensive survey of MOEAs can be found in [4] and a more recent one is [62].

This paper presents a multiyear research effort targeting the development of computational tools for solving packing optimization problems which are encountered within the different stages of engineering design with a special interest in automotive design. Due to increasingly realistic engineering applications, the problems feature a rising level of complexity and therefore require optimization models and approaches with growing sophistication. In the next three sections of this paper, the authors describe methodologies that have been developed according to the three packing perspectives discussed above. Modeling approaches and algorithms are discussed in each section.

In Sect. 3.2, methods for geometric representation of objects are described with a particular attention given to effective algorithms that convert computer-aided design (CAD) representations to formats used in the fast calculation of intersections or overlap between objects. These methods give a foundation for the development of models and algorithms for compact and noncompact packing that are described in Sects. 3.3 and 3.4, respectively. Applications in automotive design are also presented. The paper is concluded in Sect. 3.5.

The presented work has been accomplished at Clemson University over the last decade as an interdisciplinary research effort between Mechanical Engineering and Mathematical Sciences Departments. Since the resulting research papers have been published in engineering journals, the objective of this review paper is to highlight those aspects and results of the accomplished work which might be of interest to the operations research community.

3.2 Geometric Considerations

The representation of objects to pack is a critical first step in packing algorithms. When the shape is regular, be it prismatic, spherical or cylindrical or derivatives of such, identifying the volume occupied by the object is relatively straightforward once an origin and direction are specified. Much of the literature on compact packing is based on such shapes [22], with a plethora of algorithms and approaches to deal with the packing of such objects. If their shape or the shape of the enclosure in which they are to be placed is irregular, especially if nonconvex, then a detailed representation of the object is needed. This is the case of most mechanical components considered in the packing problems of interest to engineers such as placing the components under the hood of the car, or placing components inside a satellite. It is thus desired to compute a geometric representation of the space available to package the object as well as a geometric representation of the space occupied by it. For a single object, an envelope that could be nonconvex and that hides (encapsulates) all the detailed internal geometry of that object is desired. Once the geometry is available, an efficient evaluation of the overlap of two objects is needed. These aspects are detailed next.

3.2.1 Modeling

3.2.1.1 CAD

The complexity of engineering packing problems is due to the complex geometries encountered, and to the multiplicity of objectives and functional constraints that have to be met. Engineers represent objects using CAD software. The capabilities of current commercial packages such as Dassault CATIA, Siemens NX, PTC Creo, Solidworks, Autodesk Inventor, and many others provide the designer with very powerful tools that can be used to represent components at the level of detail needed to be able to manufacture them and virtually prototype the overall system. Geometric modeling techniques are used in these CAD packages and produce objects that are modeled exactly in the same geometries as the built artifact. In the vehicle design application, every component is represented in the CAD software, and the envelope in which they reside, be it the body of the vehicle, the trunk space or any entity is also represented accurately. The underhood, for instance, is represented with all the details required, including wheel wells, open bottom, and structural members. These data files are typically very large, and are difficult to manipulate with external software such as optimizers.

One major issue is to develop an application that is not dependent on a single or specific software package. It is therefore currently common to extract from the CAD data just the relevant information, and since geometric representations are often nonplanar surfaces and complex curves, a simpler, approximate representation of the needed shapes must be extracted to be able to solve the problem. The next sections detail approximation approaches addressing this issue.

3.2.1.2 Tessellation

Most CAD software can output the surface geometry of an object as an ".STL" file, a format originally developed for 3D Systems company, the first developer of 3D printers based on stereolithography. Known as the STereoLithography format [2], this is currently the format generally accepted as a standard for communicating with 3D printers. It describes the surface of an object as a collection of triangles in 3D space. This conversion from native CAD format to STL is called tessellation. Since the free-form surface is converted to a set of triangles, there is some loss of information in the approximation, and the user has to make a tradeoff between the number of triangles and the approximation accuracy. The CAD software typically uses some acceptable error tolerance in generating the tessellation, and this format is therefore ideal to be used in packaging optimization. There are two STL formats, one encodes the information in readable ASCII format, the other in binary format, which is a much more compact representation.

3.2.1.3 Voxelization

Voxels are the 3D equivalent of a pixel, which represents a unit of surface with a property, typically color, and is used to represent images. The voxel is a unit of volume, usually cuboidal with also some property. The choice of the size of the voxel affects the degree of accuracy of the approximation. Smaller voxels represent the object more accurately, but then increase the number of voxels needed to fill the volume of the object. Larger voxels introduce more error in the approximation, but would be more efficient in computing interferences. Note that various methods such as octree [39] representation address both these issues, but the performance of the interference checking algorithm would be affected by such a representation. The generation of voxels is typically not available in CAD software and is discussed next.

3.2.2 Algorithms

3.2.2.1 Surface Voxelization

The surface voxelization engine takes a binary STL file (CAD data in tessellated format) and generates the corresponding voxel data. The schematic of the procedure for surface voxelization is as follows [51, 53]:

- Compute the bounding box of the object by selecting three orthogonal directions and identifying the minimum and maximum coordinates in each direction thus forming a box.
- Select a discretization size and divide the box into cells or voxels of the specified size.
- For every voxel of the bounding box, perform a triangle-voxel overlap computation. If the facet intersects the voxel, mark the voxel as nonempty. The approach described in [40] is used to identify the intersections.

This approach identifies the voxels that are on the boundary of the object; the next task is to fill the object with voxels.

3.2.2.2 Volume Voxelization

Volume voxelization converts the surface voxel data to volume voxels using ray tracing [1, 53]. The rays are fired from the sides of a larger bounding box encapsulating the object and are stopped as soon as they touch the surface voxels. The surface voxels are not assumed to be illuminated by the rays. All the voxels that are not outside the object are assumed to be either inside or on the surface of the object thus constituting the volume of the object. The volume obtained from this process is almost always a superset of the actual volume. Since volume voxelization is used for the objects to be packed, this approximation also gives a conservative packing. The ray tracing algorithm also assumes that the objects are manifold (which also implies a closed volume).

3.2.2.3 Interference Checking

Once a geometric representation is selected, collision detection or overlap calculation is needed to ensure that no two objects occupy the same location. This is also known as interference checking. One can use a surface representation and come up with ways to efficiently identify interferences, or convert the object into a volumetric representation such as voxels and then use the regular voxels to identify interferences, or use an implicit equation that describes the object and then use mathematical techniques to identify interferences. There is again much work on interference checking with surface representations [37]. Many of these algorithms depend on a tessellated representation which was described earlier.

Some packing algorithms may require calculating the amount of interference (volume overlap) in order to use some gradient technique to minimize the overlap. These algorithms are described in the literature and work reasonably well when identifying the collision between two objects [41]. The first implementation of the configuration design optimization method (Sect. 3.4.1.1) and its extension (Sect. 3.4.1.2) for the two cases of noncompact packing used the software I-COLLIDE developed by Lin's group [6]. Note that these methods become computationally expensive when computing the collisions and amount of interference between multiple objects two at a time.

Collision detection for the case of compact packing is computationally challenging using tessellated objects since the objects have to be placed in contact with one another. With nonconvex objects having cavities and holes, it is extremely computationally expensive (and often impossible) to compute penetration depth and the direction of movement to reduce interference. The voxelization described earlier overcomes this limitation and facilitates the computation of collision detection and amount of interference. The voxel-based approach to collision detection is used for the both compact (Sect. 3.3.3) and the noncompact (Sects. 3.4.1.3 and 3.4.1.4) packing problems considered.

To detect whether two objects overlap, the physical coordinates (matrix indices) for the bottom-left-back corner of an object are determined. Once that location is known, the coordinates (indices) of all the voxels in the matrix are determined. Thus, for the two objects, the physical location of all the voxels is known. Based on the physical location, the relative index of all the voxels (as compared to the entire voxel grid) is determined. The global matrix is parsed to determine if any voxel is occupied by any two objects considered; if a voxel is occupied by the two objects, they overlap with each other, and the amount of overlap can be computed by counting the voxels that are occupied by both objects [51].

Once these geometric operations have been set up and the objects and container processed, the packing optimization can be carried out as presented in the next two sections.

3.3 Compact Packing

The compact packing problem (CPP) of interest to engineers consists in placing free-form objects with full rotational freedom inside an arbitrarily shaped enclosure such that the volume of the objects inside the enclosure or their number is maximized.

3.3.1 Modeling

The approach used to model and solve the CPP is inspired from the human packing of objects inside a container. The objects to be packed are placed sequentially, in an order to be determined, inside the container one after the other in a specified orientation and position. The CPP is therefore modeled as a single-objective combinatorial optimization problem. The feasible set contains permutations of objects defining packing sequences and the objective is to maximize the packing efficiency which is defined as a percent of the total volume of the container that is occupied by the packed objects. The CPP is known to be NP-complete [28].

3.3.1.1 Optimization Variables

The choice and representation of optimization variables is implied by the type of algorithm that is developed for the CPP. Since the first objective is to identify the order and orientation of placement of the objects, a genetic algorithm is used to identify possible sequences. The following description in Sect. 3.3.1 is extracted from [51].

There are two types of optimization variables: packing sequences and orientations. Let N denote the number of objects that are to be packed inside the container. A packing sequence for N objects is a permutation of the form $\pi = (\pi(1), \pi(2), \pi(3), \ldots, \pi(N))$, where $\pi(i)$ ($i = 1; 2; \ldots N$) denotes the index of an object. Also $\pi(i) \neq \pi(j)$ for $i \neq j$. The object with index $\pi(1)$ is packed first, the object with index $\pi(2)$ is packed next, and so on. The orientation of the objects is represented using a mixed representation comprising multi-parity bits and real numbers.

For the sequential placement of N distinct objects, the number of permutations is $N!$ which also represents the dimension of the sequence search space. If there are k_i

objects of type $i, 1 = 1, \ldots, M$, where M is the number of distinct types of objects, then the dimension s of the search space for the M distinct types is given by [51]:

$$s = \frac{(\sum_{k=1}^{M} k_i)!}{\prod_{i=1}^{M} (k_i!)} \tag{3.6}$$

The orientation variables cannot be represented with real numbers since they are circular entities (0° and 360° are the same). The genetic operators designed to work with real numbers (noncircular) cannot reflect the circular property of rotation. Also, for prismatic and free-form objects, the number of possible orientations differs, both for the orthogonal case and for the continuous case. The desired characteristics of a good representation are as follows [51]:

1. It should preserve the circular property of the rotation.
2. It should have minimal redundancy in the representation.
3. All orientations should be equally probable.
4. It should not impose any pseudo-ordering on the rotation variables.

The representation of the orientation variables depends upon the complexity of the objects and the desired rotational freedom. The representation is decomposed into multiple parts each of which captures some orientation aspect. The complexity of the representation increases with the increase in the complexity of the objects to be packed and the desired rotational freedom. For example, if only prismatic objects with orthogonal orientation are to be packed, only six possible orientations exist; whereas if a free-form object with full rotational freedom is to be packed, there are infinitely many possibilities. The representation of the orientation variables is designed to capture this variability and also adjust the dimension of the search space accordingly. All possible scenarios that can occur when representing the orientation variables are discussed next in the order of increasing dimension of the search space.

3.3.1.2   Scheme 1: Prismatic Objects with Orthogonal Orientations

This is the simplest of all the scenarios and is encountered in the orthogonal rectangular packing problem. Let the three dimensions of an object to be packed in 3D be $(l \times w \times h)$, then the six possible orientations (span along the x, y, and z coordinate directions) are 1. l-w-h, 2. l-h-w, 3. w-l-h, 4. w-h-l, 5. h-l-w, and 6. h-w-l. Any of the three edges (l, w, h) may be oriented along the x axis; either of the two remaining edges may be oriented along the y axis, and the remaining edge must then be oriented along the z axis.

A single bit with a parity of six is used for every prismatic object for which only orthogonal orientations are desired. This representation satisfies all the desirable characteristics mentioned above. Since there are six possible choices; for N objects, there are 6^N different combinations.

3.3.1.3   Scheme 2: Free-Form Objects with Orthogonal Orientations

Consider a cuboid with all three dimensions different, and with the faces marked as $(1, 2, 3, 4, 5, 6)$. In this particular case, there are a total of 24 distinct orientations for an object. The representation for Scheme 1 can be extended to accommodate this case by adding one extra bit for every free-form object with orthogonal orientation (the bit will have a parity of four). In this particular case, orientation is represented using two bits. Mutating the first bit (parity six) changes the bounding box of the object (large change), whereas mutating the second bit only changes the profile visible on every face of the bounding box (small change). With this scheme, there is no redundancy or pseudo-ordering and no explicit handling of the circular property is required. For N free-form objects, there are 24^N possible combinations.

3.3.1.4   Scheme 3: Free-Form Objects with Full Rotational Freedom

For free-form objects with full rotational freedom, a perturbation of θ, where $-45° \leq \theta \leq 45°$, can be added to the rotation of the objects in each coordinate direction. The perturbation does not represent the orientation but rather the difference in orientation. Three real variables are added for every object that has full rotational freedom. The dimension of the search space in this case is infinite. Below is an example of a complete chromosome for the case of three free-form objects with full rotational freedom:

- permutation: $(3, 2, 1)$,
- orientation: 6, 1, 5 (parity 6),
- facial orientation: 2, 4, 3 (parity 4),
- perturbation about x: $15°$, $23°$, $-12°$,
- perturbation about y: $21°$, $13°$, $-34°$,
- perturbation about z: $22°$, $35°$, $-1°$.

In this variable representation scheme the packing sequence is specified by the permutation $(3, 2, 1)$ implying that object 3 is packed first, then object 2, and then object 1. The next field, object orientation, is used if an object is prismatic, that is, the object has six unique orientations (e.g., a cuboid). The object orientation is represented by a number specifying which side of the object is placed in a specific plane. The parity is 6 meaning that this number can have six possible values which are all equally probable. In the example, all objects are prismatic and the numbers 6, 1, and 5 specify the orientations of the objects 3, 2, and 1, respectively. The third field, object facial orientation, has parity of 4 and specifies the rotation of that object by $90°$ while the side that has already been identified by the object prismatic orientation remains on the same plane. In the example, the objects 3, 2, and 1 have facial orientations 2, 4, and 3, respectively. If only the orthogonal orientations are allowed, then an arbitrary object can have 24 ($= 6 \times 4$) possible orientations that are represented by two numbers, the prismatic orientation and the facial orientation. If full rotational freedom of an object is desired, then three perturbations are added (one each about x, y, and z axis). In such a case, these three numbers are sufficient for capturing full rotational freedom and the two variables representing the orthogonal orientation are not needed. In the example, the objects 3, 2, and 1 are rotated 15°, 23°, and −12°, respectively about x.

While it is extremely difficult to get good packing with full rotational freedom, an additional difficulty comes from the fact that voxelization has to be performed whenever the orientation changes. A layout algorithm, that is presented in the next section, checks voxel overlap to identify if two objects overlap or if an object overlaps with the enclosure. The object is rotated as desired, and then placed in an enclosure. If the object has been rotated by some arbitrary angles, its voxels would not be aligned anymore with the enclosure voxels, and it would need to be revoxelized in the new orientation to compute overlap.

In the problems considered, the search space is discretized and few predefined orientations (e.g., 0°, 30°, 60°) are used. The additional orientations are represented as perturbations (and are therefore less than 90°). Thus the overall orientation is obtained by changing the prismatic and facial orientations and then introducing perturbations. Furthermore, since some enclosure surfaces are not orthogonal, patch aligned orientations are used because the alignment of some portion of the object surface to an enclosure surface actually helps obtaining a better packing efficiency.

Having detailed the different variable encodings needed for compact packing, the next topic is a description of the algorithms proposed.

3.3.2 Algorithms

The optimization task is to find an optimal packing sequence and optimal orientations of all the objects (or of as many objects as possible) resulting in the most compact packing. Given the packing sequence and orientations, an algorithm is required to place the objects inside the container and compute the packing efficiency. Thus, the overall solution approach needs two algorithms:

1. An optimization algorithm to generate an optimal packing sequence (or position in 3D space) and an optimal orientation of every object.
2. A layout algorithm to pack the objects according to the provided sequence, ensuring that objects neither collide or overlap with each other nor with the enclosure, to determine which objects can be placed inside the enclosure in the specified orientation in the remaining volume, and to compute the packing efficiency.

3.3.2.1 The Optimization Algorithm

The working principle of the proposed single-objective optimization algorithm is based on the steady-state GA [54]. The proposed algorithm also borrows concepts from several existing single-objective GAs.

Since the optimal solution to the CPP is a packing sequence of oriented objects, the GA requires a suitable encoding of the packing sequence and orientations. The algorithm needs crossover and mutation operators for each variable type in the chromosome. For permutation variables, an order-based crossover [44] is used and mutation is modeled using the swap operator [13]. For multiparity bits, one-point crossover [26] is used and mutation is modeled by applying bit flipping [13]. For real variables, simulated binary crossover [14] is employed and polynomial mutation [15] for real variables is used to maintain diversity in the population. A detailed description of all the genetic variation operators applied in the algorithm can be found in [5]. A set of rules is used to decode the chromosome. The pseudo-code of the modified GA is as follows.

```
   The optimization algorithm
1  Begin
2  Generate the initial population randomly.
3  Evaluate the initial population.
3  Repeat
4    Choose two random parents.
5    Create one offspring from the two parents using the genetic
       variation operators.
6    Evaluate the offspring solution.
7    Choose a solution randomly from the population.
8    Compare the offspring against the chosen solution;
       if the offspring has better packing efficiency, then replace
       the chosen solution with the offspring.
9    Compute the diversity in the population.
10   If the diversity in the population is lost, then store the
       best solution and regenerate the remaining population.
11 Until(100 % packing efficiency is reached or number of
       function evaluations is exhausted).
11 End
```

Thus, the proposed optimization algorithm is an elite preserving steady-state GA, which, contrary to typical GAs, incorporates an explicit diversity preserving mechanism. The algorithm does not have a very high selection pressure (it does not follow the best solution at every iteration) which increases the resiliency to premature convergence. The genetic variation operators used to create the offspring solution depend upon the solution representation. Since the proposed algorithm solves a SOP, the phenomenon of genetic drift drives the entire population towards a single point which often results in a loss of diversity. The algorithm therefore incorporates a diversity preservation operator which is computed in the variable space.

3.3.2.2 The Layout Algorithm

The layout algorithm receives the packing sequence and orientation for every object from the optimization algorithm and communicates with CAD algorithms to generate a packed configuration. In particular, the CAD algorithms process the 3D CAD data, perform geometric transformations, collision detection, interference evaluation, etc. The CAD algorithms are coupled with layout heuristics that consist of a set of rules specifying the movement of an object until a suitable location for it has been found. The placement of the objects is according to the bottom-left-back-fill (BLBF) heuristic inspired by the bottom-left (BL) strategy, that has been implemented in [22] and [38] for two-dimensional packing of rectangular objects, and the bottom-left-fill (BLF) strategy, that places the objects from the bottom left but also attempts to fill voids as proposed by Hopper and Turton [33]. In the BLBF heuristic, the placement of an object is started from the bottom-left-back position of the container. The object is moved until a suitable position has been found which does not overlap with either the container or the already placed objects. Such a strategy ensures that every object is placed at the bottom-left-back-most position available. The computational complexity of the BLBF heuristic varies as a cubic function of the grid resolution used for packing.

Following are the sequence of steps performed by the layout algorithm.

```
The layout algorithm
1 Receive the packing sequence and the orientation of every
  object from the optimization algorithm.
2 Construct the rotation matrix for every object.
3 Use geometric algorithms to rotate every object (in trian-
  gulated form).
4 Voxelize all the objects whose bounding box could fit inside
  the bounding box of the container.
5 Pick the objects in the order of the packing sequence and pack
  them using the BLBF heuristic.
6 Compute the packing efficiency based on the volume of the
  objects inside the container.
7 Report the packing efficiency to the optimization algorithm.
```

3.3.3 Applications

The models and algorithms presented in this section have been applied to the automotive design problems such as the packing of engine components and the packing of luggage in the trunk [48, 51]. Figures 3.1, 3.2, and 3.3 below show the implementation of the compact packing algorithm on three packing problems of increasing level of difficulty. In the first problem a set of 34 rectangular boxes that fit exactly inside of a rectangular container are packed into that container (Fig. 3.1). The objective of this exercise was to ensure that the algorithms can pack regular-shaped objects. The performance speed of the algorithm has not been compared

Fig. 3.1 34-box packing problem [51]

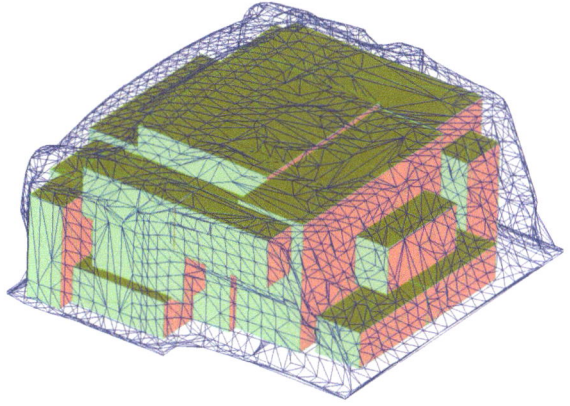

Fig. 3.2 SAE packing problem with a nonconvex trunk geometry [51]

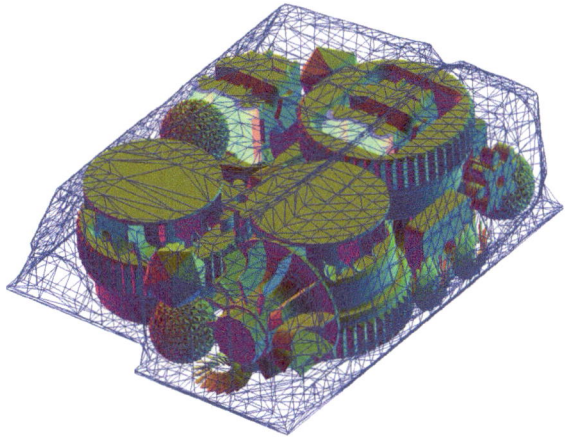

Fig. 3.3 Free-form packing with full rotational freedom in nonconvex trunk [51]

Table 3.1 Simulation results: 34-box packing problem

Iterations	Success rate	Time (s)
5,000	19	4
10,000	33	8
15,000	42	12
20,000	52	15

Table 3.2 Simulation results: SAE and free-form packing problems

Results	SAE	Free-form
Number of function evaluations	10,000	10,000
Grid resolution	$10 \times 10 \times 10$ (mm^3)	$10 \times 10 \times 10$ (mm^3)
Number of voxels	$124 \times 133 \times 55 = 907{,}060$	$124 \times 133 \times 55 = 907{,}060$
Best packing efficiency (BPE)	72.95 %	30.85 %
Number of objects corresponding to BPE	21	30
Median packing efficiency	69.74 %	28.9 %
Execution time (for single simulation)	1 h 8 min approx	4 h approx

with the algorithms existing in the literature since the main objective was to pack nonconvex shapes inside a nonconvex enclosure. The packing of rectangular boxes representing a set of suitcases of prescribed dimensions (SAE standard J1100 [42]) inside a nonconvex trunk space is shown in Fig. 3.2. The packing of any nonconvex shape inside the nonconvex trunk enclosure, as shown in Fig. 3.3, is also within the capabilities of this algorithm.

For the 34 box packing problem, the success rate is defined as the number of instances 100 % efficiency is obtained from 99 simulation runs starting with different random seeds for the GA. In this problem, 17 out of 34 boxes fit perfectly in the bin, making 100 % efficiency possible. The population size is equal to the number of objects to be packed. Crossover probability is set to 1.0 and mutation to 1/N where N is the number of objects to pack. The results obtained when running the algorithm on a 2 GB DDR2 667 MHz RAM and 2 GHz Intel Core 2 Duo Processor are provided in Table 3.1.

The results obtained for the SAE and free-form packing problems are presented in Table 3.2. Ninety nine simulation runs were performed for each problem starting with different random seeds for the GA. The total number of possible objects in the SAE problem is 38 while in the free-form packing problem it is 40. In the SAE problem, the objects represent suitcases and golf bags, not all of which fit in the trunk.

3.4 Noncompact Packing

Compact packing is driven by the single objective of compactness that is to be maximized by the tight placement of system components in an enclosure. In noncompact packing or configuration layout, the tightness of packing is not the only criterion of interest to designers and other criteria evaluating the system performance become of significance. Since the compactness criterion is accompanied by other criteria, the mathematical formulation of the overall problem assumes the form of an MOP to account for multiple objective functions. Occasionally the compactness criterion is replaced with another criterion that indirectly measures the tightness of packing by directly gauging another metric, namely the moment of inertia of the system. This results in components coming close to each other, but not necessarily maximizing the filling of the container's volume.

3.4.1 Modeling and Applications

In the four subsequent subsections MOP packing models of increasing levels of complexity are presented. They address packing problems in automotive design with a growing degree of realistic engineering applications. The models are improved in two directions: by integrating other physics-based processes that occur in the vehicle and interact with its configuration layout, or by advancing the optimization approaches to allow designers to exert distributed control during the design process.

3.4.1.1 Basic Packing of Vehicle Underhood

At the initial stage of the study the MOP is meant to model the packing of the vehicle underhood [28, 29]. The design variables represent the position and orientation of the components with respect to the Cartesian coordinate system. As described earlier, the information about the shape of a component is carried in the tessellated description of its surface.

The objective functions include compactness, balance, and maintainability. Compactness of a system is often measured by the volume of a box bounding the system or by the volume of the convex hull of the system. Both these approaches while mathematically reasonable may lead to misleading designs in that various configurations of the same volume are of different utility to the designer [29]. Instead, in this study, the inertia matrix norm is calculated for every component and the system inertia matrix norm, which is the sum of all component matrix norms, is used as the measure of compactness. The static system balance is achieved by bringing the system center of gravity to a target while the dynamic balance is measured by the system moment of inertia which is evaluated, as mentioned above, as a compactness measure. The maintainability of a component is understood as its accessibility or the amount of mechanical work to be done to access the component and remove from the system. This work is proportional to the component weight and

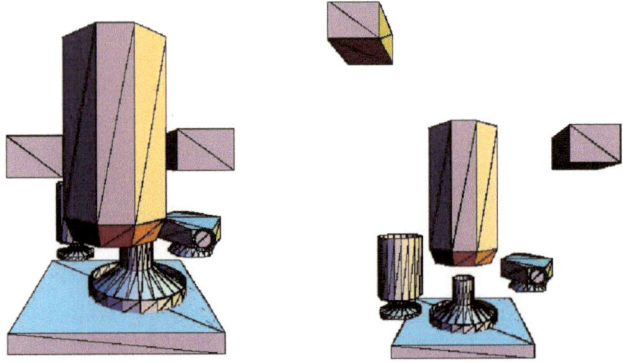

Fig. 3.4 Simplified underhood packing with the highest importance assigned to compactness (*left*) and with the highest importance assigned to maintainability from the front (*right*) [28, 29]

the number of other components that have to be removed to access the component of interest. The system maintainability is then the sum of the maintainability of all components.

The MOP model has two types of constraints: functional and interference (or overlap). The functional constraints (such as distance between components, mutual location) assume the form of geometric conditions that reduce the feasible set since the location of certain components is known a priori (e.g., the radiator is typically located in front of the engine). These constraints are embedded in the definition of the optimization variables and as such do not yield any equality or inequality constraints. The interference constraints prevent any overlap between any two components and between components and the enclosure. The interference is measured by the volume of the intersection of the overlapping components, which is supposed to be zero and therefore yields equality constraints. The functional and interference constraints are used in each model presented in Sect. 3.4.

In Fig. 3.4 the results obtained with this approach are displayed, specifically the tessellated representations of several components that fit under the hood. On the right side, the compactness metric is low, but the maintainability is best. In the one on the left, the compactness is best, and the maintainability is low. In both cases, the center of gravity is roughly at the same location.

3.4.1.2 Packing Considering Vehicle Dynamics or Heat Transfer

At the next stage the basic packing model of the underhood is upgraded by including an overall vehicle dynamic model which allows evaluation of the dynamic performance of the vehicle when the weight of the components in the various packing designs affects the vehicle's behavior [58, 59]. This specific problem identifies the placement of hybrid hydraulic components and an auxiliary power unit in a large truck.

Two types of design variables are used, the position and orientation variables as before and new dynamic input variables.

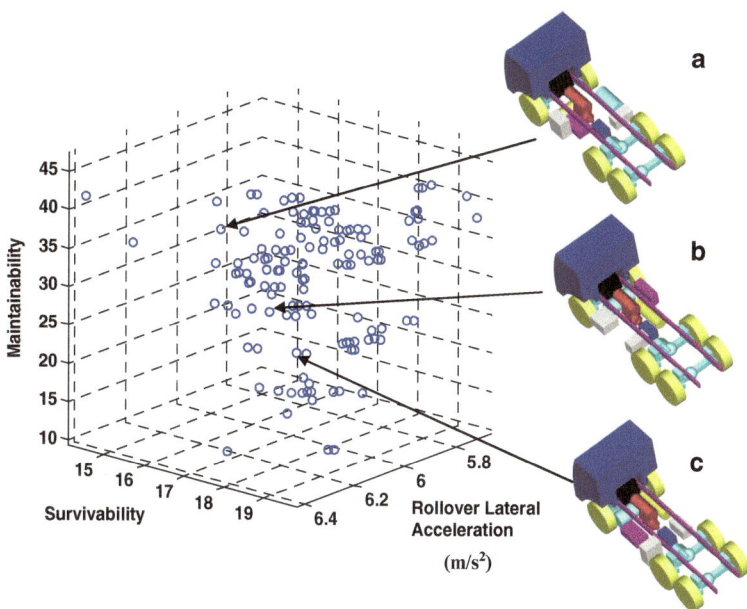

Fig. 3.5 Three vehicle configurations and their location in the nondominated set produced by the packing GA [58, 60]

The maintainability and the dynamic balance remain as the objective functions and are accompanied by a new criterion of vehicle survivability. The balance is measured with vehicle rollover propensity which is quantified by the vehicle lateral acceleration, a quantity provided by the dynamic model. The survivability characterizes the vehicle ability to survive attacks from explosives and bullets and is defined similarly to the maintainability. The overall survivability is the sum of the survivability of all components while each component survivability is quantified by the level of protection provided by the overlap with the other components.

The vehicle ground clearance inequality constraint reflected in the maximum angle of the slope the vehicle can climb is a new constraint added to the MOP model.

Because the complete dynamic analysis provided by the vehicle dynamic model requires expensive computations, the model is called only if the design being evaluated does not violate the interference and ground clearance constraints.

Figure 3.5 displays a set of nondominated points available to the designer in the three-dimensional objective space. Every point corresponds to a vehicle packing configuration and shows the performance of this configuration with respect to the maintainability, survivability, and rollover acceleration. Three specific configurations are depicted for demonstration. From among the computed configurations designers can select a preferred configuration that satisfies additional criteria not considered in the optimization.

Another upgrade of the basic packing problem, which is of interest to automotive manufacturers, is to take heat transfer issues into consideration when deciding on the packing configuration of the underhood. Because of the heat generated by the engine, some components have to be placed at some distance from the heat sources.

For this problem, the design variables are the locations of the components similarly to the previous case, but in addition, a temperature variable is applied to each component, and heat sources are imposed. The temperature distribution in the underhood is computed using computational fluid dynamics (CFD). The objective functions include the location of the center of gravity, maintainability, survivability, and additionally an objective related to the heat transfer under the hood. The additional objective is to minimize the root mean square value of the average temperatures under the hood while maintaining the temperature of critical components such as the battery below some critical threshold. The functional and interference constraints complete the model [27, 34, 35].

3.4.1.3 Packing with Morphing Components

In the traditional configuration design the shapes of components are fixed prior to the packing process during which only their position and orientation are optimized. At a subsequent stage of research, it is assumed that the shape of some objects changes, meaning that they morph while their shape and functional requirements are respected [20, 21]. The need to account for morphable components in a packing algorithm results from the lack of or limited communication between the layout designer and the component designer and the fact that the designs of the latter may not be optimal for the former. The ability of changing the shapes of components during packing may clearly lead to far better packing solutions than those resulting from the traditional approach. Even though integrating the component shape design into packing is challenging, it is essential for advancing the art of packing.

In the MOP model of this packing problem the position and orientation variables are accompanied by the shape variables of morphable components. A mass-spring physics system represented by a differential equation is used to model the movements of a morphing component whose expansions or contractions are similar to inflation and deflation of a balloon filled with air and are controlled by air pressure variations. The system consists of masses that are placed at the vertices of the tessellation triangles and of springs that connect the masses along the triangle edges.

The objective functions include again the dynamic balance measured by the vehicle moment of inertia and the maintainability. In that morphing components cannot be placed too low in the vehicle, the vehicle ground clearance, that acted as a constraint in the dynamic packing model, is now treated as a third objective function and therefore more significantly affects the optimization. The constraint functions include the functionality and interference constraints of the basic packing model (Sect. 3.4.1) and new constraints on the shape of morphable components in the form of inequality constraints bounding from below the volume of the components. The upper and lower bounds on the position and orientation of all components are also imposed.

The packing problem with one morphing component is already challenging due to the large number of design variables and expensive evaluation of objective and constraint functions. The attempts to solve this MOP in its original all-in-one (AiO) formulation encounter computational issues since the shape variables affect the size of a morphing object and its position, and the position of the surrounding objects results in a maximum volume attainable. If an increase of volume is desirable, all the components may have to be displaced to generate new nondominated solutions. In effect, the AiO formulation is not solvable and requires decomposition into solvable subproblems whose optimal solutions are coordinated to obtain an AiO optimal solution. The MOP is decomposed into a bilevel problem with a smaller MOP at the system (upper) level and a SOP at the component (lower) level.

At the system level, the position and orientation of components serve as the design variables, while the shape of the morphable component remains fixed. Though the objectives are identical to that in the AiO formulation, they are only considered as a function of the position and orientation of components. The optimization is performed subject to the functionality and interference constraints, which are functions of the position and orientation of components while the shape of the morphable component is kept constant, and subject to the bounds on the position and orientation of components.

At the component level, the design variables are the shape variables of the morphing component and the scalar-valued objective is to maximize its volume or reach a target volume. The optimization is performed subject to the functionality and interference constraints, which are functions of the morphing component shape while the position and orientation of all components are kept constant, and subject to the new inequality constraints bounding the morphing component volume from below [19–21].

The objective of the bilevel optimization problem is to find a design that is Pareto-optimal for the upper-level MOP and optimal for the lower-level SOP, and at the same time, Pareto-optimal for the AiO MOP.

Figure 3.6 depicts a CAD representation of a vehicle underhood to reflect the realism of the underhood packing problem. The effect of a morphing component on the packing is illustrated in Fig. 3.7. In the image on the left, the water container starts expanding to attempt to reach a specified volume and occupy the available space. The image on the right shows a bigger morphed water container which slightly affects the location of the other components.

3.4.1.4 Packing for Distributed Design

Engineering design of a complex system, that is composed of subsystems and components, requires interaction among several engineering disciplines (such as fluid dynamics, thermodynamics, structures, controls, and others) that are involved in the design process of the system. Because system and component designs are typically assigned to independent engineering teams with complementary

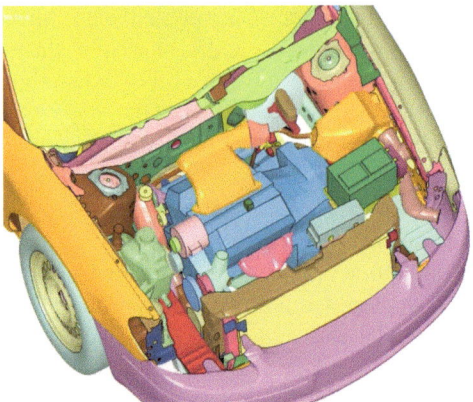

Fig. 3.6 CAD representation of the underhood [19, 20]

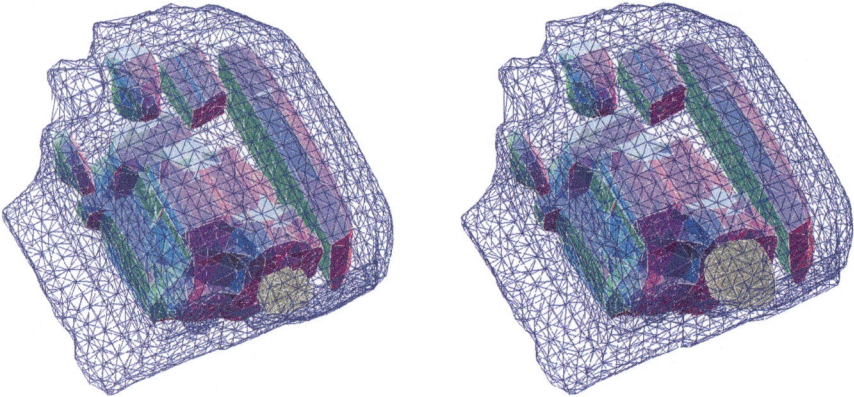

Fig. 3.7 Realistic underhood packing with morphing water container in initial position (*left*) and while expanding to a target volume and filling the available space (*right*) [19, 20]

background and expertise, packing for a distributed or decentralized design process needs to be studied. The distinction among designing teams is reflected in the corresponding MOPs which have different feasible sets and objective functions, belong to different disciplines, and require different solution algorithms. Multidisciplinary multiobjective design optimization provides models and methods to address this level of complexity in engineering design.

The multidisciplinary MOP (MDMOP) packing problem involves the optimization of the layout of components in a container and the simultaneous optimization of the design of one (or more) of these components. Each optimization is performed with respect to different objective functions while the design solutions of the component affect the optimal placement of all components in the enclosure. While each design problem is carried out by a different team because it requires highly specialized knowledge, the two problems display interactions due to the placement of the component within the container. Even that the MDMOP problem is still an MOP and its Pareto-optimal solutions could be computed once all of the problem data are known, in practice it is never directly solved due to the distributed character of the design process.

In this study, the MDMOP packing problem involves optimally placing six components (battery, engine, radiator, coolant reservoir, air filter, and brake booster) within the underhood of a hybrid electric vehicle, while one of the components, the battery, is being designed under demanding thermal criteria. The battery design depends on its functionality but also affects the optimal placement of all other components under the hood. Two design teams are involved in this packing problem: the vehicle level team responsible for packing the underhood and the component level team who designs the battery. The MDMOP is decomposed into a bilevel problem with the optimization of the layout of components within the underhood at the upper level and the optimization of the Lithium-ion (Li-ion) battery at the lower level [11, 12].

The upper-level MOP uses the design variables being the positions of all components and the shape variables of the battery while the objective functions include the compactness, maintainability, and survivability of the vehicle. The embedded functionality constraints, the interference equality constraints, and bounds on the position variables of all the components complete the formulation of this MOP.

The lower-level MOP models the battery design focusing on its optimal thermal behavior. The formulation uses the design variables particular to the battery thermal behavior and the battery shape variables. The multiple objective functions also model the thermal behavior and, due to their specific characteristics, allow the concept of Pareto-optimality to be strengthened to equitability. In effect, the set of Pareto-optimal solutions is reduced to the set of equitable solutions [36]. A coordinating equality constraint is added to the MOP at each level which gives each-level designers the possibility to modify the component shape to improve the design at their level while enforcing consistency with the other level.

The objective of the bilevel problem is to find a design that is Pareto-optimal for the upper-level MOP and equitable for the lower-level MOP, and at the same time Pareto-optimal for the AiO MDMOP.

In Figs. 3.8 and 3.9 the underhood configurations for different arrangements of the battery cells and different importance of optimization levels are illustrated. When the maximum importance is assigned to the vehicle level, the battery (the rectangular box) is placed on the left (Fig. 3.8). In contrast, when the maximum importance is assigned to the battery level, the battery is placed to the right in both cell arrangements (Fig. 3.9). Note that the other components also change places due to the different location of the battery.

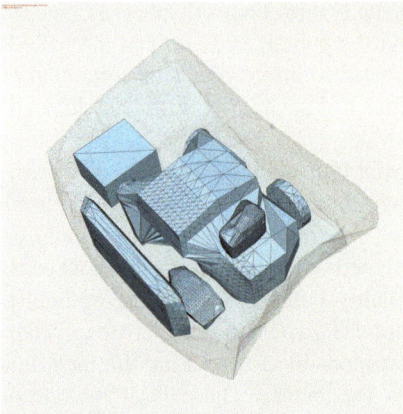

Fig. 3.8 Vehicle layout for the battery cells arranged to nine columns and eight rows, and the maximum importance given to the vehicle level [12]

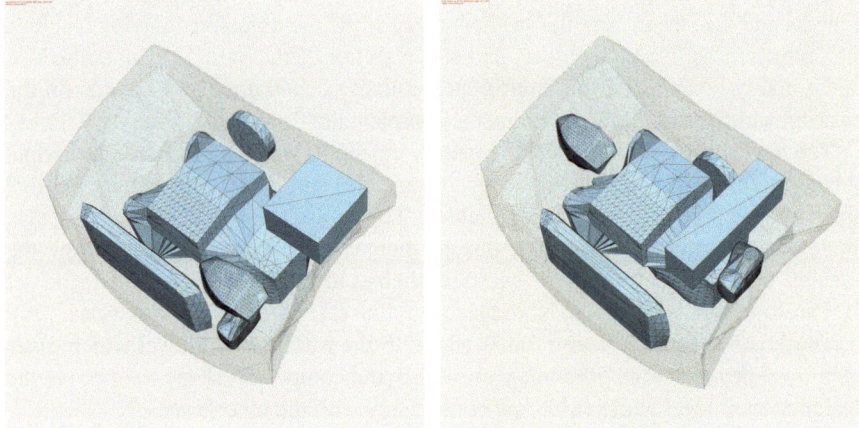

Fig. 3.9 Vehicle layout for the maximum importance given to the battery level: the battery cells arranged to nine columns and eight rows (*left*) and to 18 columns and four rows (*right*) [12]

3.4.2 Algorithms

As the noncompact packing problem evolved over the years to become a more effective methodology, the accompanying solution algorithms have been developed and adapted and better interference computation methods have been proposed and used.

Because of the nature of the packing problems which consist of placing objects inside an enclosure subject to multiple objectives, GAs have been an appealing tool to identify a set of nondominated solutions and eventually a set of Pareto-optimal solutions. Furthermore, nonexact algorithms have another important capability that deterministic algorithms lack. A deterministic or gradient-based optimization approach would start at a certain initial arrangement of objects and attempt to move them in a direction that optimizes some aggregate metric of the various objective functions. An object moving in the identified direction would encounter other objects, and the positions of some of those objects might need to be swapped. A deterministic method that enables such a swap has not been readily available. In contrast to exact algorithms, a GA can be programmed to operate on a chromosome representing an object identifier and its location. By having the GA operators change the identifier, objects are easily swapped and or moved to other locations in the design space. Typically, a higher mutation rate is imposed to ensure a more global exploration of the design space.

The following paragraphs describe the evolution of the algorithms developed to deal with the noncompact packing problem.

The Configuration Design Optimization Method is built using the library LibGA available on the web [7]. The algorithm is modified to work interactively and with clouds of solutions to ensure convergence to different areas of the nondominated or Pareto set. The interactivity is introduced to allow the designer to restart the process as the algorithm performs best in its first generations, and to intervene and modify the variable bounds to steer the optimizer towards some solution space [28, 29].

As the demands on efficiency increase with the use of simulations to study dynamic aspects of the problem, and then thermal aspects, the packing GA is rewritten to use NSGA-II [17], the most performing MOGA available at the time. A customized encoding method and novel GA operators are developed to help the GA explore and exploit the packing design space more efficiently [58, 60].

The next improvement considers the heat transfer calculations which are even more computationally costly than the dynamic simulation mentioned earlier. To address this problem, the packing optimization is carried out without the heat transfer objective initially, just considering the other objectives. One hundred nondominated solutions are obtained using the same packing algorithm as for the dynamic packing but with the in-house developed AMGA2, the Archive-based Micro GA originally tailored for the CPP [52]. The configurations obtained are then used to perform CFD in parallel on a supercomputer and to obtain temperature maps in the underhood. These temperature profiles are used to train a neural network to relate the position of the objects to their temperature, and to validate the neural network training. Once the neural network is obtained, it is then used with the packing optimization algorithm as an approximation for the heat transfer objective. The newly computed solutions are then validated by running the CFD again on them [27, 34, 35].

The attempts to solve the packing problem with morphing components as the AiO problem with the packing genetic algorithm described earlier were unsuccessful; the GA was unable to simultaneously place the components and modify the shape variable of the morphing object to occupy the available space. The bilevel formulation was then implemented with the packing algorithm based on NSGA-II at the system level, and the Matlab sequential quadratic programming (SQP) algorithm at the component level [20, 21].

The bilevel formulation for distributed design requires a solution strategy that deals with the distributed character of the design process. A multiobjective decomposition algorithm (MODA) [9, 12] is used for computing the Pareto-optimal solutions to the AiO MDMOP while only having access to the Pareto-optimal solutions to the MOP at each level. MODA performs distributed multiobjective optimization that is conducted by two entities working independently with distinct mathematical models. The limited information exchange between the entities is sufficient for the computation of the Pareto-optimal solutions to the AiO MDMOP. MODA is an exact algorithm making use of block coordinate descent, a Gauss–Seidel decomposition technique, and the method of multipliers. The convergence results are available in [9, 10]. In the distributed vehicle layout study, MODA employs the AMGA to solve the vehicle-level MOP and the Matlab SQP algorithm to solve the battery-level MOP. The SQP algorithm is tied to the battery model implemented in Matlab/Simulink [9].

Running the Packing GA without a morphing object takes about 1 min on a personal computer similar to the one used for compact packing (Sect. 3.3.3) using an archive of 10–50 individuals (nondominated solutions) and a population equal to the number of objects to pack (of an order of 8–10). Adding the morphing and optimization of the battery and the bilevel approach increases the computational time, which is approximately 5 min per nondominated solution. Typically, multiple runs are performed with different seeds for the AMGA, with the time multiplied by the number of runs that are performed in parallel to obtain multiple solution sets.

All five noncompact packing models presented in Sect. 3.4 are summarized in Table 3.3.

3.5 Conclusion

This paper reviews a multiyear interdisciplinary research program on the packing optimization of free-form objects. The engineering and science disciplines that have contributed to this research include structural design, automotive design, heat transfer, mathematical analysis and optimization, and computer science.

As mechanical designs such as automobiles, satellites, consumer products, airplanes increase in functionalities and complexity, issues of packaging become increasingly critical. Designers of hybrid vehicles, in particular, have been dealing with packaging issues, and the approaches and methodologies presented provide them with a tool to handle this type of problems.

The authors contend that the presented optimization models, algorithms, and results determine the state-of-the art in free-form packing. The models make direct use of single or multiobjective mathematical optimization problems. Because the algorithms are (meta)heuristic, they are neither backed up with mathematical proofs of correctness, nor are the results supported with proofs of optimality. As argued in the paper, exact algorithms do not seem to have the capability to work as solvers in the challenging engineering packing optimization. While development of new exact

Table 3.3 Summary of noncompact packing models presented

	Basic	+ Dynamic	+ Thermal	+ Morphing	Distributed
Dec. variables					
Position	✓	✓	✓	✓	✓
Orientation	✓	✓	✓	✓	
Temperature			✓		
Shape				✓	✓
Obj. functions					
Compactness	✓		MoI	MoI	MoI
Balance					
Static (CoG)	✓				
Dynamic	MoI	Acceleration	MoI	MoI	MoI
Heat			Temperature		Temperature
Maintainability	✓	✓	✓	✓	✓
Survivability		✓	✓		✓
Ground clearance			✓	✓	
Constraints					
Functional	Imbed	Imbed	Imbed	Imbed	Imbed
Interference	✓	✓	✓	✓	✓
Ground clearance		✓			
Shape				✓	✓

Note: Moment of inertia (denoted *MoI*) can be used as a compactness measure as well as a dynamic behavior measure

algorithms is highly desirable, the progress in effective (meta)heuristic algorithms for packing has made it possible to solve very difficult optimization problems once thought intractable.

Future research is likely to be directed toward more advanced packing problems such as packing with multiple morphable components, packing taking into consideration wiring, hoses, pipes, and handling the multiple objectives related to their placement. Those problems may be approached with hierarchical optimization to reflect the inherent hierarchy in the system (e.g., vehicle at the upper level and components at the lower level) or with distributed optimization to model multiple design teams responsible for designing the components of the system. The resulting optimization problems may become a collection of interacting MOPs or SOPs making up a multilevel or network structure and may require sophisticated decomposition and coordination strategies to compute optimal or Pareto-optimal solutions. It will then be of interest to conduct mathematical studies on those classes of optimization problems.

The authors believe that maintaining the interdisciplinary character of work by integrating engineering and science perspectives will continue leading to new significant accomplishments in future studies on packing optimization.

Acknowledgements This work was partially supported by the Automotive Research Center (ARC), a US Army Center of Excellence for modeling and simulation of ground vehicles, and by the National Science Foundation, grant number CMMI-1129969.

References

1. Amanatides, J., Woo, A.: A fast voxel traversal algorithm for ray tracing. In: Eurographics '87, pp. 3–10 (1987)
2. Burns, M.: Automated Fabrication. Prentice Hall, Upper Saddle River (1993)
3. Cagan, J., Degentesh, D., Yin, S.: A simulated annealing-based algorithm using hierarchical models for general three dimensional component layout. Comput. Aided Des. **30**(10), 781–790 (1998)
4. Coello Coello, C.A.: A comprehensive survey of evolutionary-based multiobjective optimization techniques. Knowl. Inf. Syst. **1**(3), 269–308 (1999)
5. Coello Coello, C.A., Van Veldhuizen, D.A., Lamonts, G.B.: Evolutionary Algorithms for Solving Multi-objective Problems. Springer, Berlin (2002)
6. Cohen, J., Lin, M., Manocha, D., Ponamgi, K.: I-COLLIDE: An interactive and exact collision detection system for large-scaled environments. In: Proceedings of ACM International 3D Graphics Conference (1995)
7. Corcoran, A.: LibGA: Library of GA routines written in C (1993). http://www.cs.cmu.edu/afs/cs/project/ai-repository/ai/areas/genetic/ga/systems/libga/
8. Corne, D.W., Knowles, J.D., Oates, M.J.: The Pareto envelope-based selection algorithm for multi-objective optimization. In: Parallel Problem Solving from Nature, pp. 839–848. Springer, Berlin (2000)
9. Dandurand, B.: Mathematical optimization for engineering design problems. Ph.D. thesis, Clemson University, Clemson (2013)
10. Dandurand, B., Wiecek, M.M.: Distributed computation of Pareto sets. SIAM J. Optim. (in print)
11. Dandurand, B., Guarneri, P., Fadel, G., Wiecek, M.M.: Equitable multiobjective optimization applied to the design of a hybrid electric vehicle battery. ASME J. Mech. Des. **135**(4), 041004 (2013)
12. Dandurand, B., Guarneri, P., Fadel, G., Wiecek, M.M.: Bilevel multiobjective packaging optimization for automotive design. Struct. Multidiscip. Optim. **50**(4), 663–682 (2014)
13. Deb, K.: Multi-objective Optimization Using Evolutionary Algorithms. Wiley, Chichester (2001)
14. Deb, K., Agrawal, R.B.: Simulated binary crossover for continuous search space. Complex Syst. **9**(2), 115–148 (1995)
15. Deb, K., Goyal, M.: A combined genetic adaptive search (GeneAS) for engineering design. Comput. Sci. Inf. **26**(4), 30–45 (1996)
16. Deb, K., Tiwari, S.: Omni-optimizer: A procedure for single and multi-objective optimization. In: Proceedings of the 3rd International Conference on Evolutionary Multi-criterion Optimization (EMO'2005). Lecture Notes on Computer Science, vol. 3410, pp. 41–65 (2005)
17. Deb, K., Agrawal, S., Pratap, A., Meyarivan, T.: A fast and elitist multi-objective genetic algorithm: NSGA-II. IEEE Trans. Evol. Comput. **6**(2), 182–197 (2002)
18. Deb, K., Mohan, M., Mishra, S.: Evaluating the ϵ-domination based multi-objective evolutionary algorithm for a quick computation of Pareto-optimal solutions. Evol. Comput. J. **13**(4), 501–525 (2005)
19. Dong, H.: Physics based shape morphing and packing for layout design. Ph.D. thesis, Clemson University, Clemson (2008)
20. Dong, H., Fadel, G., Guarneri, P.: Bi-level approach to vehicle component layout and shape morphing. ASME J. Mech. Des. **133**(4), 041008 (2011)

21. Tiwari, S., Dong, H., Fadel, G., Fenyes, P., Kloess, A.: A physically-based shape morphing algorithm for packing and layout applications. Int. J. Interact. Des. Manuf. **8**(3), 171–185 (2014)
22. Dowsland, K.A., Vaid, S., Dowsland, W.B.: An algorithm for polygon placement using a bottom-left strategy. Eur. J. Oper. Res. **141**, 371–381 (2002)
23. Ehrgott, M.: Multicriteria Optimization, 2nd edn. Springer, Berlin (2005)
24. Eshelman, L.J.: The CHC adaptive search algorithm: How to have safe search when engaging in nontraditional genetic recombination. In: Foundations of Genetic Algorithms 1 (FOGA-1), pp. 265–283 (1991)
25. Eskandari, H., Geiger, C.D., Lamont, G.B.: FastPGA: A dynamic population sizing approach for solving expensive multiobjective optimization problems. In: Evolutionary Multiobjective Optimization Conference (EMO-2007). Lecture Notes in Computer Science, vol. 4403, pp. 141–155. Springer, Berlin (2007)
26. Goldberg, D.E.: Genetic Algorithms in Search, Optimization and Machine Learning. Addison-Wesley Longman, Boston (1989)
27. Gondipalle, S.: CFD analysis of the underhood of a car for packaging considerations. Master's thesis, Clemson University, Clemson (2011)
28. Grignon, P.M.: Configuration design optimization method. Ph.D. thesis, Clemson University, Clemson (1999)
29. Grignon, P., Fadel, G.M.: A GA based configuration design optimization method. ASME J. Mech. Des. **126**(1), 6–15 (2004)
30. Hansen, N., Ostermeier, A.: Adapting arbitrary normal mutation distributions in evolution strategies: The covariance matrix adaption. In: Proceedings of the Third IEEE Interntational Conference on Evolutionary Computation, pp. 312–317. IEEE, New York (1996)
31. Ho, S.Y., Shu, L.S., Chen, J.H.: Intelligent evolutionary algorithms for large parameter optimization problems. IEEE Trans. Evol. Comput. **8**(6), 522–541 (2004)
32. Holland, J.: Genetic algorithms and adaptation. In: Adaptive Control of Ill-Defined Systems. NATO Conference Series (1984)
33. Hopper, E., Turton, B.C.H.: An empirical investigation of meta-heuristic and heuristic algorithms for a 2D packing problem. Eur. J. Oper. Res. **128**, 34–57 (2001)
34. Katragadda, R.T.: Predicting the thermal performance for the multiobjective vehicle underhood packing optimization. Master's thesis, Clemson University, Clemson (2012)
35. Katragadda, R.T., Gondipalle, S.R., Guarneri, P., Fadel, G.M.: Predicting the thermal performance for the multi-objective vehicle underhood packing optimization problem. In: Proceedings of ASME DETC 2012. Paper DETC2012-71098 (2012)
36. Kostreva, M.M., Ogryczak, W., Wierzbicki, A.: Equitable aggregations and multiple criteria analysis. Eur. J. Oper. Res. **158**(2), 362–377 (2004)
37. Lin, M., Gottshalk, S.: Collision detection between geometric models: A survey. In: Proceedings of IMA Conference on Mathematics of Surfaces (1998)
38. Liu, D., Teng, H.: An improved BL-algorithm for genetic algorithm of the orthogonal packing of rectangles. Eur. J. Oper. Res. **112**, 413–420 (1999)
39. Meagher, D.: Octree encoding: a new technique for the representation, manipulation and display of arbitrary 3-D objects by computer. Technical Report IPL-TR-80-111. Rensselaer Polytechnic Institute, Troy, NY (1980)
40. Moller, T.A.: Fast 3D triangle-BOC overlap testing. J. Graph. Tools **6**(1), 29–33 (2001)
41. Redon, S., Lin, M.: A fast method for local penetration depth computation. J. Graph. Tools **11**(2), 37–50 (2006)
42. SAE: SAE Standard J1100. Motor Vehicle Dimensions (2005)
43. Schwefel, H.-P.: Evolution and Optimum Seeking. Wiley, New York (1995)
44. Syswerda, G.: Schedule optimization using genetic algorithms. In: Davis, L. (ed.) Handbook of Genetic Algorithms 1, pp. 332–349. Van Nostrand Reinhold, New York (1991)
45. Szykman, S., Cagan, J.: A simulated annealing approach to three-dimensional component packing. ASME J. Mech. Des. **117**(2A), 308–314 (1995)

46. Szykman, S., Cagan, J.: Constrained three dimensional component layout using simulated annealing. ASME J. Mech. Des. **119**(1), 28–35 (1996)
47. Teng, H.F., Sun, S.L., Liu, D.Q., Li, Y.Z.: Layout optimization for the objects located within a rotating vessel – a three-dimensional packing problem with behavioral constraints. Comput. Oper. Res. **28**, 521–535 (2001)
48. Tiwari, S., Fadel, G., Fenyes, P.: A fast and efficient compact packing algorithm for free-form objects. In: ASME 2008 IDETC & CIE Conference, New York (2008)
49. Tiwari, S., Koch, P., Fadel, G.M., Deb, K.: AMGA: An archive-based micro genetic algorithm for multi-objective optimization. In: GECCO'08 Conference Proceedings (2008)
50. Tiwari, S., Fadel, G.M., Koch, P., Deb, K.: Performance assessment of the hybrid archive-based micro genetic algorithm (AMGA) on the CEC09 test problems. CEC09 MOEA Competition (2009). Nominated for best paper award
51. Tiwari, S., Fadel, G., Fenyes, P.: A fast and efficient compact packing algorithm for the SAE and ISO luggage packing problems. J. Comput. Inf. Sci. Eng. **10**(2), 021010 (2010)
52. Tiwari, S., Fadel, G.M., Deb, K.: AMGA2: Improving the performance of the archive-based micro genetic algorithm for multi-objective optimization. J. Eng. Optim. **43**(4), 377–401 (2011)
53. Tiwari, S., Fadel, G., Fenyes, P., Kloess, A.: An envelop generation algorithm for packing and layout applications. Int. J. Interact. Des. Manuf. **8**(3), 171–185 (2014)
54. Vavak, F., Fogarty, T.C.: Comparison of steady state and generational genetic algorithms for use in nonstationary environments. In: Proceedings of IEEE Conference on Evolutionary Computation, Nagoya, pp. 192–195 (1996)
55. Watanabe, S., Hiroyasu, T., Miki, M.: NCGA: Neighborhood cultivation genetic algorithm for multi-objective optimization problems. In: Proceedings of the Genetic and Evolutionary Computation Conference (GECCO 2002), pp. 458–465 (2002)
56. Whitley, D.: The GENITOR algorithm and selection pressure: Why rank-based allocation of reproductive trials is best. In: Proceedings of the 3rd International Conference on Genetic Algorithms, pp. 116–121 (1989)
57. Whitley, D.: Cellular genetic algorithms. In: 5th International Conference on Genetic Algorithms, p. 658 (1993)
58. Yi, M.: Packing optimization of engineering problems. Ph.D. thesis, Clemson University, Clemson (2005)
59. Yi, M., Blouin, V., Fadel, G.M.: Multi-objective configuration optimization with vehicle dynamics applied to midsize truck design. In: ASME 2003 International Design Engineering Technical Conferences & Computers and Information in Engineering Conference, Chicago (2003)
60. Yi, M., Fadel, G.M., Gantovnik, V.B.: Vehicle configuration design with a packing genetic algorithm. Int. J. Heavy Veh. Syst. **15**(2/3/4), 433–448 (2008)
61. Yin, S., Cagan, J.: An extended pattern search algorithm for three-dimensional component layout. ASME J. Mech. Des. **122**, 102–108 (2000)
62. Zhoua, A., Qu, B.-Y., Li, H., Zhaob, S.-Z., Suganthan, P.N., Zhang, Q.: Multiobjective evolutionary algorithms: A survey of the state of the art. Swarm Evol. Comput. **1**(1), 32–49 (2011)
63. Zitzler, E., Laumanns, M., Thiele, L.: SPEA2: Improving the strength Pareto evolutionary algorithm for multiobjective optimization. In: Proceedings of the EUROGEN 2001 Conference, pp. 95–100 (2001)

Chapter 4
A Modeling-Based Approach for Non-standard Packing Problems

Giorgio Fasano

Abstract This chapter examines the problem of packing tetris-like items, orthogonally, with the possibility of rotations, into a convex domain, in the presence of additional conditions. An MILP (Mixed Integer Linear Programming) and an MINLP (Mixed Integer Nonlinear Programming) models, previously studied by the author (Fasano, Solving Non-standard Packing Problems by Global Optimization and Heuristics. SpringerBriefs in Optimization, Springer Science + Business Media, New York, 2014), are surveyed. An efficient formulation of the objective function, aimed at maximizing the loaded cargo, is pointed out for the MILP model. The MINLP one, addressed to the relevant feasibility sub-problem, has been conceived to improve approximate solutions, as an intermediate step of a heuristic process. A space-indexed model is further introduced and the problem of approximating polygons by means of tetris-like items investigated. In both cases an MILP formulation has been adopted. An overall heuristic approach is proposed to provide effective solutions in practice. One chapter of this book focuses on the relevant computational aspects (Gliozzi et al., Container loading problem MIP-based heuristics solved by CPLEX: an experimental analysis. In: Fasano, G., Pintér, J.D. (eds.) Optimized Packings and Their Applications. Springer Optimization and Its Applications, Springer Science + Business Media, New York, 2015).

Keywords Tetris-like items • Orthogonal packing • Convex domain • Additional/balancing conditions • Mixed integer linear/nonlinear programming models • Global optimization (GO) • Efficient formulation • Feasibility sub-problem • Space-indexed/grid-based-position paradigms • Polygon approximation • Heuristics

G. Fasano (✉)
Thales Alenia Space Italia S.p.A., Turin, Italy
e-mail: giorgio.fasano@thalesaleniaspace.com

© Springer International Publishing Switzerland 2015
G. Fasano, J.D. Pintér (eds.), *Optimized Packings with Applications*, Springer Optimization and Its Applications 105, DOI 10.1007/978-3-319-18899-7_4

4.1   Introduction

This chapter summarizes and extends results descending from a long-lasting research effort aimed at solving complex three-dimensional packing problems arising in the space industry [1]. In this challenging context, the relevant issues could hardly be considered applying a standard typology. Quite often, indeed, the operational scenarios to deal with are characterized by the presence of tricky geometries and complex additional conditions that can even be of global impact, such as in the case of balancing.

Often irregularly shaped and of non-negligible dimensions, the objects involved cannot be realistically approximated in terms of single cuboids (i.e. rectangular parallelepipeds). Significant effort has therefore been addressed to allow for tetris-like items, i.e. objects consisting of clusters of mutually orthogonal (rectangular) parallelepipeds. Similarly, the domains (containers) to take account of are generally not box-shaped and often several internal volumes are not exploitable, since these correspond either to clearance/forbidden zones or actual holes. Additionally, separation planes (with no fixed position specified a priori) can partition the domain into sub-domains. Some items may be requested to assume pre-defined positions/orientations or are subject to placement restrictions, such as the requirement of having a given side parallel or orthogonal to a specified direction.

In order to cope with overall conditions such as balancing, when necessary in addition to those mentioned above, a Global Optimization (GO) based view is highly desirable. This is essentially based on a modeling philosophy, as opposed to a pure algorithmic one, consisting of sequential procedures limited to local search.

A number of modeling-based works are present in the literature, although these are usually restricted to the case of box-shaped items (e.g. [2–5]). On the other hand, very interesting studies consider strongly irregularly shaped objects, even though the adopted philosophy is mainly focused on local optimization [6–8].

This chapter emphasizes the solution of non-standard packing issues, in the context outlined above, by a GO approach. Mixed Integer Linear/Non-linear (MILP/MINLP) formulations have been conceived and a library of mathematical models set up. This supports ad hoc heuristics, implemented to obtain satisfactory, albeit probably sub-optimal (or at least non-optimal proven), solutions to a wide collection of real-world instances [1].

The general problem of placing tetris-like items orthogonally into a convex domain, without pair-wise intersection, so that the total volume loaded is maximized, is the main topic of this chapter.

Section 4.2 investigates a dedicated MILP model [1], specifically constructed to overcome the challenging computational difficulties that are typically associated with the problem in question, when formulated in terms of Mathematical Programming. It is, indeed, well known that, even when single parallelepipeds are involved (i.e., tetris-like items consisting of one component only), the relevant MILP models available in the specialist literature (e.g., [3, 4]) are very hard to solve. This holds also if a number of *valid inequalities* are purposely added. The model discussed in this section can be used to solve small-size instances, tout court. In addition, it can advantageously be adopted as a basic element of the above-mentioned heuristics that act recursively, following an overall *greedy* approach.

MINLP models (e.g., [2]) have been built up for the *feasibility* sub-problem, derived from the general one, when a set of items need to be loaded (without any possibility of rejection, provided that the instance is feasible) and no *objective* function is assigned. Moreover, they can be adopted [1] to improve approximate solutions where intersections between items are admitted, "minimizing" the overall overlap (actually this optimization target is attained only partially, through *surrogate* functions). An MINLP version, implemented for this specific case is summarized in Sect. 4.3.

An alternative formulation of the model reported in Sect. 4.2 (currently being looked into) is presented in Sect. 4.4. The relevant MILP model extends, in the case of tetris-like items and convex domains, previous formulations available in the literature, based on the discretization of the domain and often referred to as *space-indexed* or *grid-based-position* paradigms (e.g., [9, 10]). All models presented in Sects. 4.2, 4.3, and 4.4 are suitable for considering additional conditions, such as specific loading requirements or balancing. Nevertheless, these aspects, albeit frequent in a number of real-world applications, are not considered in this chapter and the reader is referred to [1] for an extensive discussion (except the *space-indexed* formulation). Section 4.5 introduces the generation of (two-dimensional) covering tetris-like items, providing outer approximation of polygons. The issue of simplifying the representation of complex objects in such a way is a very interesting optimization problem per se, especially considering its potential applications. The three-dimensional extension is not surveyed in this chapter (since it is quite straightforward). Section 4.6 proposes a novel heuristic approach, mainly based on the MILP model presented in Sect. 4.2.

An extensive experimental analysis has recently been carried out, concerning the MILP model presented in Sect. 4.2. One chapter of this book [11] reports and examines the computational results available to date, in depth, highlighting the advantages of the overall methodology suggested. Since this chapter is restricted to the computational aspects (assuming the relevant model as known) the present work serves also the scope of providing a topical framework. Fasano [1] offers an extensive bibliography, both on packing problems in general and on the more specific subjects considered here.

In order to state the general problem discussed in this chapter, the following definition is introduced.

A tetris-like item is a set of rectangular parallelepipeds positioned orthogonally, with respect to an (orthogonal) reference frame. This frame is called "local" and each parallelepiped is a " component".

Hereinafter, "tetris-like item" will usually be simply referred to as "item," if no ambiguity occurs; similarly, "rectangular parallelepipeds" are referred to as "parallelepipeds."

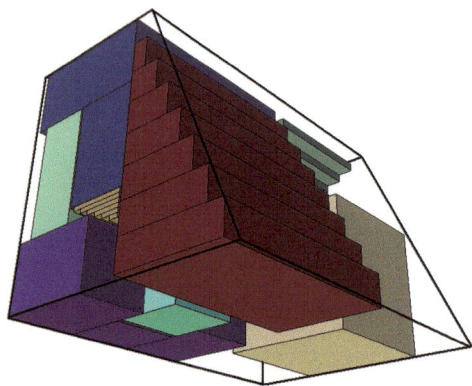

Fig. 4.1 Tetris-like item packing into a convex domain

A set I of N items, together with a domain D, consisting of a (bounded) convex polyhedron, is considered (see Fig. 4.1). This is associated with a given orthogonal reference frame, indicated in the following as the main frame. The general problem is to place items into D, maximizing the loaded volume, considering the following positioning rules:

- *each local reference frame has to be positioned orthogonally, with respect to the main one (orthogonality conditions);*
- *for each item, each component has to be contained within D (domain conditions);*
- *the components of different items cannot overlap (non-intersection conditions).*

4.2 Direct MILP Formulation

An MILP model for the general problem stated in Sect. 4.1 is described next, expanding on some aspects not pointed out in its previous discussion [1]. Recalling the basic concepts introduced there, the main orthogonal reference frame has origin O and axes w_β, $\beta \in \{1, 2, 3\} = B$. It is assumed, without loss of generality, that the whole domain D is entirely contained inside its first octant. Similarly, each local reference frame, associated with every item, is chosen so that all item components lie within its first octant. Its origin coordinates, with respect to the main reference frame, are denoted by $o_{\beta i}$. The set Ω of all (24 possible) orthogonal rotations, admissible for any local reference frame, with respect to the main one, is introduced.

The set of components of a generic item i is denoted by C_i. For each item i, the set E_{hi} of all (8) vertices associated with each of its components h is defined. An extension of this set is obtained by adding to E_{hi} the geometrical center of component h. This extended set is denoted by $\widehat{E}_{hi}$. For each item i and each possible orthogonal orientation $\omega \in \Omega$, the following binary (0–1) variables are introduced:

$\chi_i \in \{0, 1\}$, with $\chi_i = 1$ if item i is chosen; $\chi_i = 0$ otherwise;
$\vartheta_{\omega i} \in \{0, 1\}$, with $\vartheta_{\omega i} = 1$ if item i is chosen and it has the orthogonal orientation $\omega \in \Omega$; $\vartheta_{\omega i} = 0$ otherwise.

The *orthogonality* conditions can be expressed as follows:

$$\forall i \in I \quad \sum_{\omega \in \Omega} \vartheta_{\omega i} = \chi_i, \tag{4.1}$$

$$\forall \beta \in B, \forall i \in I, \forall h \in C_i, \forall \eta \in \widehat{E}_{hi} \tag{4.2}$$

$$w_{\beta \eta h i} = o_{\beta i} + \sum_{\omega \in \Omega} W_{\omega \beta \eta h i} \vartheta_{\omega i}.$$

Here $w_{\beta \eta h i}$ ($\forall \eta \in \widehat{E}_{hi}$) are the vertex coordinates of component h, with respect to the main reference frame, or its geometrical center ($\eta = 0$), relative to item i; $W_{\omega \beta \eta h i}$ are the projections on the axes w_β of the coordinate differences between points $\eta \in \widehat{E}_{hi}$ and the origin of the local reference frame, corresponding to orientation ω of item i.

The *domain* conditions are expressed as follows.

$$\forall \beta \in B, \forall i \in I, \forall h \in C_i, \forall \eta \in E_{hi} \tag{4.3}$$

$$w_{\beta \eta h i} = \sum_{\gamma \in V} V_{\beta \gamma} \lambda_{\gamma \eta h i},$$

$$\forall i \in I, \forall h \in C_i, \forall \eta \in E_{hi} \quad \sum_{\gamma \in V} \lambda_{\gamma \eta h i} = \chi_i \tag{4.4}$$

Here V is the set of vertices delimiting D, $V_{\beta \gamma}$ are their coordinates (with respect to the main reference frame) and $\lambda_{\gamma \eta h i}$ are non-negative variables. These conditions correspond to the well-known necessary and sufficient conditions for a point to belong to a convex domain.

The *non-intersection* conditions are represented by the constraints shown below, see [1] for more details:

$$\forall \beta \in B, \forall i, j \in I/i < j, \forall h \in C_i, \forall k \in C_j \tag{4.5-1}$$

$$w_{\beta 0 h i} - w_{\beta 0 k j} \geq \frac{1}{2} \sum_{\omega \in \Omega} \left(L_{\omega \beta h i} \vartheta_{\omega i} + L_{\omega \beta k j} \vartheta_{\omega j} \right) - D_\beta \left(1 - \sigma^+_{\beta h k i j} \right),$$

$$\forall \beta \in B, \forall i, j \in I/i < j, \forall h \in C_i, \forall k \in C_j \tag{4.5-2}$$

$$w_{\beta 0kj} - w_{\beta 0hi} \geq \frac{1}{2} \sum_{\omega \in \Omega} \left(L_{\omega \beta hi} \vartheta_{\omega i} + L_{\omega \beta kj} \vartheta_{\omega j} \right) - D_\beta \left(1 - \sigma_{\beta hkij}^- \right),$$

$$\forall i,j \in I/i < j, \forall h \in C_i, \forall k \in C_j \tag{4.6}$$

$$\sum_{\beta \in B} \left(\sigma_{\beta hkij}^+ + \sigma_{\beta hkij}^- \right) \geq \chi_i + \chi_j - 1,$$

$$\forall i,j \in I/i < j, \forall h \in C_i, \forall k \in C_j \tag{4.7-1}$$

$$\sum_{\beta \in B} \left(\sigma_{\beta hkij}^+ + \sigma_{\beta hkij}^- \right) \leq \chi_i,$$

$$\forall i,j \in I/i < j, \forall h \in C_i, \forall k \in C_j \tag{4.7-2}$$

$$\sum_{\beta \in B} \left(\sigma_{\beta hkij}^+ + \sigma_{\beta hkij}^- \right) \leq \chi_j.$$

Here the constants D_β are the sides (respectively parallel to the main reference frame axes) of the parallelepiped, of minimum dimensions, containing D; $w_{\beta 0hi}$ and $w_{\beta 0kj}$ are the center coordinates, with respect to the main reference frame, of components h and k of items i and j, respectively; $L_{\omega \beta hi}$ and $L_{\omega \beta kj}$ are their side projections on the w_β axes, corresponding to the orientation ω; $\sigma_{\beta hkij}^+$ and $\sigma_{\beta hkij}^- \in \{0, 1\}$.

The constraints (4.7-1) and (4.7-2) have been introduced with the purpose of *tightening* the model (they are not taken account of in the following). It is worth noticing that, in some particular situations, the above *non-intersection* constraints ((4.5-1), (4.5-2) and (4.6)) should be properly complemented, in order to avoid solutions that could hardly be considered as appropriate in practice (see [1]). Nonetheless, these aspects will be omitted here.

The most straightforward formulation relevant to the *objective* function, to maximize the volume loaded, is the following:

$$max \sum_{i \in I} V_i \chi_i, \tag{4.8}$$

where V_i represents the volume of item i.

The formulation represented by expressions (4.1)–(4.8) (with possible variants regarding the constraints) is notoriously inefficient, even when restricted to single parallelepipeds only, and the situation tends to become even worse when tetris-like items are involved.

The following expression has thus been suggested [1] as a promising alternative to (4.8):

$$max \sum_{\substack{i \in I, \ h \in C_i}} \frac{V_{hi}}{\sum_{\alpha \in A} L_{\alpha hi}} \sum_{\substack{\beta \in B, \\ \omega \in \Omega}} L_{\omega\beta hi}\vartheta_{\omega i}, \tag{4.9}$$

where $L_{\alpha hi}$, $\alpha \in \{1,2,3\} = A$, are the sides of the generic component h of item i. It is assumed, without loss of generality, that $L_{1hi} \leq L_{2hi} \leq L_{3hi}$.

As easily seen, the functions (4.8) and (4.9) are equivalent for any *integer-feasible* solution. Indeed, the following implications hold:

$$\forall i \in I, \forall h \in C_i \quad \chi_i = 0 \iff \frac{\sum\limits_{\substack{\beta \in B, \\ \omega \in \Omega}} L_{\omega\beta hi}\vartheta_{\omega i}}{\sum\limits_{\alpha \in A} L_{\alpha hi}} = 0, \tag{4.10-1}$$

$$\forall i \in I, \forall h \in C_i \quad \chi_i = 1 \iff \frac{\sum\limits_{\substack{\beta \in B, \\ \omega \in \Omega}} L_{\omega\beta hi}\vartheta_{\omega i}}{\sum\limits_{\alpha \in A} L_{\alpha hi}} = 1. \tag{4.10-2}$$

Both derive from (4.1), the second, in particular, is true in virtue of the fact that, in any *integer-feasible* solution: $\forall i \in I / \chi_i = 1, \exists! \omega \in \Omega / \vartheta_{\omega i} = 1$.

Since *objective* functions (4.8) and (4.9) are equivalent, they give rise to the same optimal (or sub-optimal) integer solutions. Nonetheless, quite different behaviors occur when dealing with (partial or total) *LP-relaxations* of the MILP model (as usually utilized by the solvers), making the choice for the second one highly preferable. Some considerations follow, in support of this point.

First of all, it is worth recalling that non-trivial intrinsic difficulties make the MILP approach very intricate, per se [1]. This is the case, for instance, of the implicit *transitivity* conditions. Considering, indeed, the generic triplet of components h, h', h'' of items i, i', i'', respectively, these can be expressed as follows: *if, along the axis w_β, h precedes h' and h' precedes h'' then h precedes h'', along the same axis.* A major concern, moreover, is certainly represented by the *non-intersection* constraints (4.5-1) and (4.5-2), since they are of the *big-M* typology (well known for being, in general, very tough to cope with). Consequently, it is not surprising at all that a strong tendency to item overlapping prevails in the *LP-relaxed* solutions, making the task of finding an *integer-feasible* solution (albeit sub-optimal) demanding. As an immediate consideration, it should be noticed that the MILP model, related to (4.8), is characterized by a very weak correlation of the *non-intersection* constraints (4.5-1) and (4.5-2) with the χ_i variables appearing in the *objective* function (the association is attained only indirectly through (4.1) to (4.4) and (4.6)). On the contrary, (4.9) acts directly on the terms $L_{\omega\beta hi}\vartheta_{\omega i}$, appearing in (4.5-1) and (4.5-2), "minimizing" (in terms of a *surrogate objective* function), the overall overlapping between items. In order to see this point better, it is useful to introduce the variables $l_{\beta hi} = \sum_{\omega \in \Omega} L_{\omega \beta hi} \vartheta_{\omega i}$. For each component h of the generic item i, they represent indeed the lengths of the sides parallel to each w_β axis, respectively (and consequently $l_{\beta hi} \in [0, L_{3hi}]$, when an *LP-relaxation* is applied).

In order to go deeper into this matter, it is worth pointing out that a necessary condition for *integer-feasibility* is provided by the following (cf. 4.10-2):

$$\forall i \in I, \forall h \in C_i \quad \chi_i = 1 \Rightarrow \sum_{\substack{\beta \in B, \\ \omega \in \Omega}} L_{\omega \beta hi} \vartheta_{\omega i} = \sum_{\alpha \in A} L_{\alpha hi}. \qquad (4.11)$$

When an *LP-relaxation* is applied to the model associated with *objective* function (4.9), the inequalities below hold:

$$\forall i \in I, \forall h \in C_i \quad \sum_{\substack{\beta \in B, \\ \omega \in \Omega}} L_{\omega \beta hi} \vartheta_{\omega i} \leq \sum_{\alpha \in A} L_{\alpha hi}, \qquad (4.12)$$

with $\vartheta_{\omega i} \in [0, 1]$.

In order to show this, a single component h of item i is selected. As easily gathered, depending on the specific orientation $\vartheta_{\omega i}$ taken by item i, each variable $l_{\beta hi}$ can assume only one value out of the following: L_{1hi}, L_{2hi} and L_{3hi}. More precisely, the following logical conditions hold:

$$\forall \alpha \in A \quad (l_{1hi} = L_{\alpha hi}) \veebar (l_{2hi} = L_{\alpha hi}) \veebar (l_{3hi} = L_{\alpha hi}), \qquad (4.13-1)$$

$$\forall \beta \in B \quad (l_{\beta hi} = L_{1hi}) \veebar (l_{\beta hi} = L_{2hi}) \veebar (l_{\beta hi} = L_{3hi}), \qquad (4.13-2)$$

where "$\veebar$" represents the "*exclusive or*." As a straightforward consequence of what is specified above, for each α and β, there are eight cases in which $l_{\beta hi} = L_{\alpha hi}$, implying that the component side of length $L_{\alpha hi}$ is parallel to the reference axis w_β. The subsets $\Omega_{\alpha \beta hi} \subset \Omega$, with $\alpha \in A$ and $\beta \in B$, are hereafter introduced: they represent, for each α and β all the orientations $\omega \in \Omega$ such that $l_{\beta hi} = L_{\alpha hi}$. Evidently, the following conditions hold (with h and i fixed):

$$\forall \alpha \in A \quad \bigcup_{\beta \in B} \Omega_{\alpha \beta hi} = \Omega, \qquad (4.14-1)$$

$$\forall \alpha \in A, \forall \beta, \beta' \in B \quad \Omega_{\alpha \beta hi} \cap \Omega_{\alpha \beta' hi} = \emptyset. \qquad (4.14-2)$$

The equalities below are thus respected, in virtue of (4.1):

$$\sum_{\beta \in B} l_{\beta h i} = \sum_{\alpha \in A} \sum_{\substack{\omega \in \Omega_{\alpha \beta h i, \\ \beta \in B}}} L_{\alpha h i} \vartheta_{\omega i} = \sum_{\alpha \in A} L_{\alpha h i} \chi_i, \tag{4.15}$$

with $\vartheta_{\omega i}, \chi_i \in [0, 1]$. This proves the validity of inequalities (4.12).

The key point associated with *objective* function (4.9) may be summarized as follows. It induces, indeed, in any *LP-relaxation*, to attain the upper bounds corresponding to (4.12), and thus to satisfy the (necessary) *integer-feasibility* conditions (4.11). On the other hand, the overall item overlapping, controlled by (4.5-1) and (4.5-2) is (indirectly) "minimized." The adoption of *objective* function (4.9) has proved very efficient in practice [11].

4.3 An MINLP Model for the Feasibility Sub-problem

Non-linear formulations addressing the orthogonal placement of rectangles inside convex domains are available in the literature (e.g., [2, 12, 13]). The following section recalls an MINLP approach put forward in [1, 14], to which the reader is referred for a more in-depth discussion. The general packing problem, as stated in Sect. 4.1, is considered here in terms of *feasibility* only, i.e. it is expected that a number of preselected items can be loaded (otherwise the problem is infeasible).

For this purpose, all the variables χ_i, corresponding to the given set of items, are set to one, keeping the *orthogonality* and *domain* constraints (4.1), (4.2), (4.3) and (4.4) unaltered. Since no *objective* function is provided a priori, an ad hoc one is introduced. It consists of *penalty functions*, representing the *non-intersection* constraints (4.5-1), (4.5-2) and (4.6) that are eliminated from the model. The corresponding expression is shown below:

$$min \left\{ \sum_{\substack{\beta \in B, \\ i,j \in I/i<j, \\ h \in C_i, k \in C_j}} max \left\{ -\left(w_{\beta 0 h i} - w_{\beta 0 k j} \right)^2 + \left[\frac{1}{2} \sum_{\omega \in \Omega} \left(L_{\omega \beta h i} \vartheta_{\omega i} + L_{\omega \beta k j} \vartheta_{\omega j} \right) \right]^2 \right. \right.$$

$$\left. \left. -r_{\beta h k i j}, 0 \right\} + K_P \sum_{\substack{i, j \in I/i<j, \\ h \in C_i, k \in C_j,}} \prod_{\beta \in B} r_{\beta h k i j} \right\}. \tag{4.16}$$

Here $r_{\beta hkij} \in \left[0, D_\beta^2\right]$, whilst K_P is a positive coefficient (that represents an appropriate "weight"); the other terms have been defined in Sect. 4.2. The general problem, as stated in Sect. 4.1, has thus been reformulated in terms of an MINLP.

It is immediately seen that the *objective* function (4.16) is non-negative and that a zero-global-optimal solution of the above defined model exists if and only if the constraints (4.1), (4.2), (4.3), (4.4), (4.5-1), (4.5-2) and (4.6) (with all variables χ_i set to one) define a feasible region. This *objective* function, indeed, "minimizes" the intersection between items (indirectly) and any global optimum provides a solution to the *feasibility* sub-problem under discussion.

The MINLP model outlined in this section, even if theoretically suitable for solving the general problem stated in Sect. 4.1, when a given set of items is requested to be loaded, is, per se, very hard to solve. Search for sub-optimal solutions can however be profitably adopted to improve the initial or intermediate ones, obtained by heuristic procedures [1], where intersection between items is admitted. In such a case, the MINLP model is utilized to reduce the overall overlapping.

4.4   Grid-Based Position MILP Model

The *space-indexed* approach (e.g., [9, 10]) can be advantageously reconsidered to include operational scenarios that are quite frequent in practice. Relevant extensions, albeit still addressed to box-shaped items and domains, are aimed at allowing for additional conditions, such as *stability* and *load bearing* (cf. [15]). This section focuses instead on a *grid-based-position* MILP model, conceived as an alternative to the one discussed in Sect. 4.2, focusing on the orthogonal packing of tetris-like items, inside a convex region.

The given domain (of Sect. 4.1) is discretized, so that it is associated with a set of internal points whose coordinates are supposed to be integer. The main reference frame, still defined as in Sect. 4.1, thus becomes a unit-cube grid, whose node coordinates are indicated as $(n_1, n_2, n_3) \in D$. Tetris-like items are grouped on a typology basis. The set of all types τ is denoted by T.

The following assumptions relevant to each tetris-like item are made:

- *the local reference frame has a pre-fixed orientation (orthogonal with respect to the main one);*
- *the local reference frame origin can only be positioned on grid points; all component vertices have integer coordinates.*

Remark 4.1 It should be observed that the prefixed orientation assumption does not represent an actual limitation. Orthogonal rotations of the same object can, indeed, simply be considered by introducing a set of pre-oriented items (one for each possible orthogonal orientation).

For each type τ, the sub-set of grid points in which the local frame origin can be positioned (so that the corresponding item is entirely inside the domain D) is introduced. It is denoted hereinafter by D_τ.

The binary variables $\chi_{\tau n_1 n_2 n_3} \in \{0, 1\}$ are then defined, with the following meaning:

$\chi_{\tau n_1 n_2 n_3} = 1$ if one item of type τ is positioned with its local reference origin in the grid node of coordinates (n_1, n_2, n_3);

$\chi_{\tau n_1 n_2 n_3} = 0$ otherwise.

A possible modeling of the general problem (of Sect. 4.1) is shown next, considering the *orthogonality*, *domain*, and *non-intersection* conditions. The first are implicitly respected by the orientation of each item type that is imposed a priori. The second ones are stated by introducing, for each type τ, the grid point sub-sets D_τ. The *non-intersection* conditions, instead, need to be expressed through dedicated constraints.

The following inequalities prevent the positioning of more than one local reference frame in the same grid points:

$$\forall n_1, n_2, n_3 \in D \quad \sum_{\substack{\tau \in T/ \\ n_1, n_2, n_3 \in D_\tau}} \chi_{\tau n_1 n_2 n_3} \leq 1. \tag{4.17}$$

Furthermore, for each pair (τ, τ') of item types (including the case when $\tau' = \tau$) and each grid node $(n_1, n_2, n_3) \in D_\tau$, the set $F_{\tau/\tau n_1 n_2 n_3}$ is introduced. Except for point (n_1, n_2, n_3), it contains all the forbidden positions, for all item types, when a τ one is assumed to be placed in (n_1, n_2, n_3). Each set $F_{\tau/\tau n_1 n_2 n_3}$ is built as follows:

- *position virtually any item i of type τ (indicated as i_τ) in node $(n_1, n_2, n_3) \in D_\tau$;*
- *identify for any item $i_{\tau'}$ all the surrounding nodes $(n'_1, n'_2, n'_3) \in D_{\tau'}$ where overlapping between $i_{\tau'}$ and i_τ would occur (at least partially), should $i_{\tau'}$ be positioned in (n'_1, n'_2, n'_3).*

The inequalities below prevent the overlapping of items, on the basis of the forbidden positions:

$$\forall \tau, \tau' \in T, \forall n_1, n_2, n_3 \in D_\tau \tag{4.18}$$

$$\sum_{n'_1, n'_2, n'_3 \in F_{\tau'/\tau n_1 n_2 n_3}} \chi_{\tau'n'_1 n'_2 n'_3} \leq (1 - \chi_{\tau n_1 n_2 n_3}) \left| F_{\tau'/\tau n_1 n_2 n_3} \right|,$$

where $\left| F_{\tau'/\tau n_1 n_2 n_3} \right|$ indicate the cardinalities of the corresponding sets.

For each typology τ, a maximum number N_τ of items are available. These conditions are represented as follows:

$$\forall \tau \in T \quad \sum_{n_1,n_2,n_3 \in D_\tau} \chi_{\tau n_1 n_2 n_3} \leq N_\tau. \tag{4.19}$$

The *objective* function has the following form:

$$max \sum_{\substack{\tau \in T, \\ n_1,n_2,n_3 \in D_\tau}} V_\tau \chi_{\tau n_1 n_2 n_3}, \tag{4.20}$$

denoting by V_τ the volume associated with each item type τ.

It should be noticed that, whilst the discretized model discussed in this section is very simple, since it consists of three groups of constraints only, the generation of both sets D_τ and $F_{\tau' \tau n_1 n_2 n_3}$ is, instead, non-trivial. An ad hoc preprocessing phase has to be envisaged, in order to generate the model instances in practice. These quite tricky aspects are not discussed here.

As for the model discussed in Sect. 4.2, also in this case additional conditions, such as balancing, could quite easily be introduced. They are, however, not taken into account here. It should, moreover, be observed, that the *grid-based position* model, as formulated in this section is (at least) theoretically susceptible to extensions contemplating any irregularly shaped item type. In such cases, the above-mentioned pre-processing phase should be carried out appropriately.

4.5 An MILP Approach for the Tetris-Like Approximation of Irregular Items

The problem of approximating irregular objects, in terms of covering, by means of tetris-like items, can be regarded per se as an optimization problem. This section provides some topical insights, restricting the discussion to the two-dimensional case of convex polygons (the three-dimensional generalization is quite straightforward). More precisely, the issue under consideration can be stated as follows:

Given a convex polygon, cover it with a minimum-surface tetris-like item, consisting of N_R components (rectangles).

Evidently, the larger N_R is, the better approximation of the polygon is possible. Moreover, in the problem general statement formulated above, it could be implicit that the dimensions of each rectangle may vary with continuity within given ranges. The formulation provided hereinafter, however, is based on quite a simplified approach. It restricts the selection of the rectangles to a finite number of possibilities, resulting from a proper discretization carried out a priori.

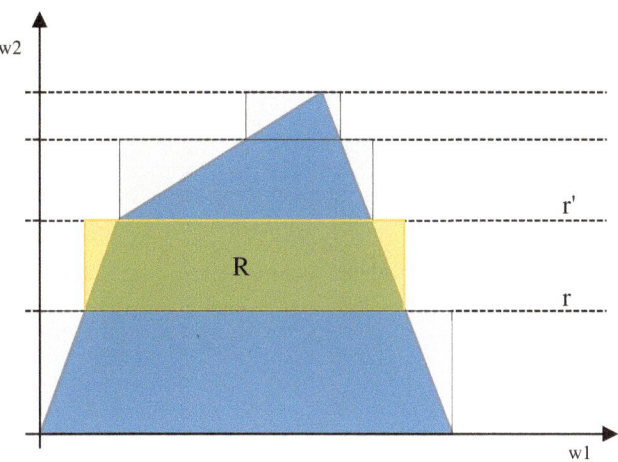

Fig. 4.2 Polygon covering by a (2D-)tetris-like item

Given a pre-oriented polygon, we shall consider it with respect to an orthogonal frame with origin O and axes w_β, $\beta \in \{1, 2\} = B$ (the same symbolism already utilized in the three-dimensional case is maintained, as no ambiguity occurs). The axis w_1 will represent the "horizontal," while w_2 the "vertical" one. The edges of the polygon are subsequently discretized, by drawing "horizontal" straight lines that identify a set of border points including all polygon vertices, see Fig. 4.2.

The sets of all such lines and points are indicated as H and Γ respectively, corresponding to generic indexes r and γ. For each pair of lines $(r, r') \in H$, all the enclosed border points determine the set $\Gamma_{rr'}$. The relevant coordinates are referred to as $W_{rr'\beta\gamma}$, with $\beta \in B$ and $\gamma \in \Gamma_{rr'}$.

For each $\Gamma_{rr'}$, the following lower and upper bounds are defined:

$$\forall \beta \in B \quad \underline{W}_{rr'\beta} = \min_{\gamma \in \Gamma_{rr'}} \left\{ W_{rr'\beta\gamma} \right\}, \quad \overline{W}_{rr'\beta} = \max_{\gamma \in \Gamma_{rr'}} \left\{ W_{rr'\beta\gamma} \right\}. \tag{4.21}$$

The rectangle $R_{rr'}$, corresponding to the straight lines r and r', delimited by the vertices listed here, is introduced:

$$V_{rr'LL} \left(\underline{W}_{rr'1}, \underline{W}_{rr'2} \right), \quad V_{rr'LU} \left(\underline{W}_{rr'1}, \overline{W}_{rr'2} \right), \tag{4.22}$$

$$V_{rr'UL} \left(\overline{W}_{rr'1}, \underline{W}_{rr'2} \right), \quad V_{rr'UU} \left(\overline{W}_{rr'1}, \overline{W}_{rr'2} \right).$$

Next, the binary variables $\chi_{rr'} \in \{0, 1\}$ are defined as:

$\chi_{rr'} = 1$ if rectangle $R_{rr'}$ is selected as a component of the covering tetris-like item;

$\chi_{rr'} = 0$ otherwise.

The (continuous) variables below are also introduced:

$$\forall \gamma \in \Gamma, \forall r, r' \in H \quad \chi_{\gamma r r'} \in [0, 1] \ .$$

They are assigned as per the following condition:

$\chi_{\gamma r r'} = 1$ if the border point $\gamma \in \Gamma$ is covered by the selected rectangle $R_{rr'}$;
$\chi_{\gamma r r'} = 0$ otherwise.

The following inequalities correlate the variables $\chi_{\gamma r r'}$ and $\chi_{rr'}$:

$$\forall r, r' \in H / r < r' \quad \sum_{\gamma \in \Gamma_{rr'}} \chi_{\gamma r r'} = |\Gamma_{rr'}| \, \chi_{rr'}, \tag{4.23}$$

where $|\Gamma_{rr'}|$ indicates the cardinality of $\Gamma_{rr'}$. These expressions highlight the obvious implication that if a rectangle $R_{rr'}$ is selected, then all the associated border points are covered by it (and vice versa). With this in mind, the inequalities below are introduced to guarantee that each border point is actually covered:

$$\forall \gamma \in \Gamma \quad \sum_{r,r' \in H / r < r'} \chi_{\gamma r r'} \geq 1 \ . \tag{4.24}$$

Since the number of selected rectangles has to be equal to N_R, the following equations hold:

$$\sum_{r,r' \in H / r < r'} \chi_{rr'} = N_R. \tag{4.25}$$

The *objective* function is stated below:

$$min \sum_{\substack{r,r' \in H / \\ r < r'}} S_{rr'} \chi_{rr'}, \tag{4.26}$$

where the terms $S_{rr'}$ represent the surfaces associated with each rectangle, respectively.

Remark 4.2 Ingenuity is needed to extend the approach proposed to non-convex polygons. As a first consideration, the rectangles $R_{rr'}$ should be split in the corresponding sub-rectangles actually covering parts of the polygon. This way, each term $S_{rr'}$ would be calculated (more precisely) as the sum of the sub-rectangles' surfaces. The situation is even more complicated when the non-convexities are related to the presence of internal "holes." All these aspects may well become the subject of a dedicated research.

4.6 Heuristics

An overall modeling-based heuristic methodology has been developed to tackle real-world scenarios, generally consisting of large-scale instances, characterized by tricky geometries dealt with by tetris-like approximations, in the presence of additional conditions such as balancing. In [1] a range of models and procedures were discussed in a general framework, providing the basis to build alternative solution strategies. A novel and promising approach, representing the objective of ongoing research, is, instead, discussed here (see [11] for experimental results). Prior to proceeding with the topical discussion, the basic concept of *abstract configuration* [1] is recalled, providing the following two definitions.

Constraints of the types

$$w_{\beta 0hi} - w_{\beta 0kj} \geq \frac{1}{2} \sum_{\omega \in \Omega} \left(L_{\omega\beta hi}\vartheta_{\omega i} + L_{\omega\beta kj}\vartheta_{\omega j} \right),$$

$$w_{\beta 0kj} - w_{\beta 0hi} \geq \frac{1}{2} \sum_{\omega \in \Omega} \left(L_{\omega\beta hi}\vartheta_{\omega i} + L_{\omega\beta kj}\vartheta_{\omega j} \right),$$

corresponding to either $\sigma^+_{\beta hkij} = 1$ or $\sigma^-_{\beta hkij} = 1$ in (4.5-1) and (4.5-2), respectively, are called relative position constraints.

Given a set of N items and the corresponding N_C pairs of components belonging to different items, an abstract configuration consists of N_C relative position constraints, exactly one for each pair, giving rise to a feasible solution in any unbounded domain.

A method to extract an *abstract configuration* from any approximate solution, with intersections between items, has been shown [1]: this subject is not discussed here, referring to the cited work. As previously, the whole process discussed in this section is essentially based on the following modules: *Initialization*, *Packing*, *Item-exchange*, and *Hole-filling*. In the versions investigated here, they are based on the MILP model presented in Sect. 4.2. In the following, the heuristic overall logic is outlined first and then the basic modules are considered.

4.6.1 Overall Logic

As in the heuristics looked into in the previous work, the search algorithm consists of a recursive procedure that, at each step, activates one of the above-mentioned modules. An *abstract configuration* is generated at each step tentatively improving the previous one; the best-so-far solution is retrieved when the current step does not meet its objective. The search process is terminated when a satisfactory, albeit non-optimal proven solution (in terms of loaded volume) is found. Since for real-world instances the computational task is quite demanding, at each step, only sub-optimal solutions are sought, interrupting the optimization on the basis of suitable stopping rules.

The *Initialization* phase is aimed at solving a *feasibility* sub-problem, with the scope of providing a good starting *abstract configuration*. An *LP-relaxation* of the general MILP model of Sect. 4.2 is adopted. All the *N* items available are considered, although some of them may be rejected subsequently, during the search process. This module seeks for a first approximate solution, enclosing all the items inside the domain and "minimizing" their total overlapping indirectly. An *abstract configuration* is directly provided by the solution obtained. The MINLP model of Sect. 4.3 may be adopted, if opportune, to further reduce (although without a guarantee for eliminating) the intersection between items. In this case, a procedure able to extract *abstract configurations* from approximate solutions with overlapping has to be available.

The *abstract configuration* derived from the *Initialization* step is imposed to the *Packing* module that offers, by means of the general MILP model of Sect. 4.2, a non-approximate (albeit usually still sub-optimal) solution, maximizing the loaded volume and rejecting items if necessary. Both *Item-exchange* and *Hole-Filling* phases are devoted to the improvement, if possible, of the *Packing* solution, providing (if successful) upgraded *abstract configurations*. Also for these steps the general MILP model of Sect. 4.2 is utilized, and non-approximate solutions are found.

The *Item-exchange* module is aimed at carrying out advantageous exchanges between non-loaded and loaded items. Two subsets of non-loaded and loaded items, respectively, are selected. The *relative positions* (corresponding to the current *abstract configuration*) relevant to both subsets are set free. A further optimization step, aimed at maximizing the loaded volume, is subsequently performed. If, in the thus obtained solution, the loaded volume has been increased, the current *abstract configuration* is upgraded correspondingly. Otherwise, the best-so-far solution is retrieved. Alternatively, *relative position* exchanges can be activated among a subset of non-loaded items only, in order to perturb the current *abstract configuration*.

The *Hole-filling* module has the scope of incrementing the loaded volume, by exploiting the empty spaces still available. For this purpose, a subset of unloaded items is selected. All *relative positions* (corresponding to the current *abstract configuration*), relevant to them are set free and a further optimization step performed (to maximize the loaded volume). Again, the current *abstract configuration* is upgraded only if an improvement has been obtained with the new solution.

The four modules discussed above can be activated repeatedly, following different strategies (e.g., the *Initialization* itself could, time after time, be executed also during the process, with the imposition of "partial" *abstract configurations*, restricted to subsets of items already loaded). In the following, the use of the general MILP model of Sect. 4.2, corresponding to each phase, is illustrated.

4.6.2   Use of the General MILP Model

The *Initialization* module, in the version considered here, focuses on the use of a specific *LP-relaxation* of the general MILP model of Sect. 4.2. As the relevant sub-problem is expressed in terms of *feasibility*, all variables χ_i ($\forall i \in I$) are set to 1. The $l_{\beta hi}$ variables, introduced in Sect. 4.2, are reconsidered instead. These are not defined any longer as $l_{\beta hi} = \sum_{\omega \in \Omega} L_{\omega \beta hi} \vartheta_{\omega i}$, but simply as continuous variables subject to the following bounds:

$$\forall \beta \in B, \forall i \in I, \forall h \in C_i \quad L_{1hi} \leq l_{\beta hi} \leq L_{3hi}. \tag{4.27}$$

Here, as previously specified, L_{1hi} and L_{3hi} represent the sides associated with h, of minimum and maximum length, respectively. The *non-intersection* conditions (4.5-1) and (4.5-2) and the *objective* function (4.9) are rewritten as follows:

$$\forall \beta \in B, \forall i, j \in I / i < j, \forall h \in C_i, \forall k \in C_j \tag{4.28-1}$$

$$w_{\beta 0 h i} - w_{\beta 0 k j} \geq \frac{1}{2} \sum_{\omega \in \Omega} \left(l_{\beta h i} + l_{\beta k j} \right) - D_\beta \left(1 - \sigma^+_{\beta h k i j} \right),$$

$$\forall \beta \in B, \forall i, j \in I / i < j, \forall h \in C_i, \forall k \in C_j \tag{4.28-2}$$

$$w_{\beta 0 k j} - w_{\beta 0 h i} \geq \frac{1}{2} \sum_{\omega \in \Omega} \left(l_{\beta h i} + l_{\beta k j} \right) - D_\beta \left(1 - \sigma^-_{\beta h k i j} \right),$$

$$max \sum_{i \in I, h \in C_i} \frac{V_{hi}}{\sum_{\alpha \in A} L_{\alpha h i}} \sum_{\beta \in B} l_{\beta h i}. \tag{4.29}$$

If the sub-problem related to the model above is infeasible, then all lower bounds L_{1hi} (in 4.27) are subsequently reduced until a feasible solution is obtained. The variables $\sigma^{+/-}_{\beta h k i j}$, for which in the obtained solution $\sigma^{+/-}_{\beta h k i j} = 1$, directly provide an *abstract configuration* for the subsequent steps of the heuristic procedure. They are referred to as $\tilde{\sigma}^{+/-}_{\beta h k i j}$.

The *Packing, Item-exchange,* and *Hole-filling* modules exploit, totally or partially, the currently available *abstract configuration*. The *non-intersection* inequalities (4.5-1) and (4.5-2) corresponding to the above-mentioned $\tilde{\sigma}^{+/-}_{\beta h k i j}$ variables are maintained in the model (in addition to (4.6)), whilst the others are eliminated together with all the redundant $\sigma^{+/-}_{\beta h k i j}$ variables (i.e., those that are not correlated to any $\tilde{\sigma}^{+/-}_{\beta hkij}$). The *non-intersection* constraints, relative to the (thus "imposed") *abstract configuration*, are hence rewritten, for the relevant indexes, in the following form:

$$w_{\beta 0hi} - w_{\beta 0kj} \geq \frac{1}{2} \sum_{\omega \in \Omega} \left(L_{\omega \beta hi} \vartheta_{\omega i} + L_{\omega \beta kj} \vartheta_{\omega j} \right) - D_\beta \left(1 - \sigma^+_{\beta hkij} \right), \qquad (4.30\text{-}1)$$

$$\vee$$

$$w_{\beta 0kj} - w_{\beta 0hi} \geq \frac{1}{2} \sum_{\omega \in \Omega} \left(L_{\omega \beta hi} \vartheta_{\omega i} + L_{\omega \beta kj} \vartheta_{\omega j} \right) - D_\beta \left(1 - \sigma^-_{\beta hkij} \right) \qquad (4.30\text{-}2)$$

$$\sigma^{+/-}_{\beta hkij} \geq \chi_i + \chi_j - 1, \qquad (4.31)$$

with $\sigma^{+/-}_{\beta hkij} \in [0, 1]$ (i.e. they are no longer considered as binary variables).

4.7 Conclusion

Non-standard packing problems that involve non-box-shaped items and domains, in the presence of additional constraints, are usually very tough to solve. This chapter, extending the author's previous work, discusses the issue of placing tetris-like items orthogonally into a convex domain. A Global Optimization point of view, focused on MILP/MINLP formulations, is looked into for the purpose of providing models that are suitable for treating additional loading restriction rules and global conditions such as balancing.

An efficient heuristic procedure, aimed at finding satisfactory solutions to real-world instances, is proposed. This approach will be the objective of future investigation, focused on the MILP/MINLP search strategies.

The issue of covering irregularly shaped objects with tetris-like items consisting of a given number of components of minimum total volume, itself, leads to a non-trivial optimization problem. Insights on its two-dimensional version, relevant to the optimal outer approximation of polygons, are provided. A further contribution appearing in this book is dedicated to the computational aspects relevant to the MILP model discussed in this chapter.

Acknowledgements The author wishes to thank Janos D. Pintér for discussing the manuscript in depth. His suggestions have significantly contributed to improve the original version of the work, making several parts easier to read. Special thanks are due to Jane Evans for her invaluable support in revising the whole text, as well as to Alessandro Castellazzo for his accurate review.

References

1. Fasano, G.: Solving Non-standard Packing Problems by Global Optimization and Heuristics. SpringerBriefs in Optimization, Springer Science + Business Media, New York (2014)
2. Cassioli, A., Locatelli, M.: A heuristic approach for packing identical rectangles in convex regions. Comput. Oper. Res. **38**(9), 1342–1350 (2011)
3. Chen, C.S., Lee, S.M., Shen, Q.S.: An analytical model for the container loading problem. Eur. J. Oper. Res. **80**, 68–76 (1995)
4. Padberg, M.W.: Packing small boxes into a big box. Office of Naval Research, N00014-327, New York University (1999)
5. Pisinger, D., Sigurd, M.: The two-dimensional bin packing problem with variable bin sizes and costs. Discret. Optim. **2**(2), 154–167 (2005)
6. Stoyan, Y.G., Chugay, A.M.: Packing cylinders and rectangular cuboids with distances between them into a given region. Eur. J. Oper. Res. **197**, 446–455 (2009)
7. Stoyan, Y.G., Zlotnik, M.V., Chugay, A.M.: Solving an optimization packing problem of circles and non-convex polygons with rotations into a multiply connected region. J. Oper. Res. Soc. **63**(3), 379–391 (2012)
8. Egeblad, J., Nielsen, B.K., Odgaard, A.: Fast neighborhood search for two-and three-dimensional nesting problems. Eur. J. Oper. Res. **183**(3), 1249–1266 (2007)
9. Beasley, J.E.: An exact two-dimensional non-guillotine cutting tree search procedure. Oper. Res. **33**(1), 49–64 (1985)
10. Hadjiconstantinou, E., Christofides, N.: An exact algorithm for general, orthogonal, two-dimensional knapsack problems. Eur. J. Oper. Res. **83**(1), 39–56 (1995)
11. Gliozzi, S., Castellazzo, A., Fasano, G.: Container loading problem MIP-based heuristics solved by CPLEX: an experimental analysis. In: Fasano, G., Pintér, J.D. (eds.) Optimized Packings and Their Applications. Springer Optimization and Its Applications. Springer Science + Business Media, New York (2015)
12. Birgin, E.G., Lobato, R.D.: Orthogonal packing of identical rectangles within isotropic convex regions. Comput. Ind. Eng. **59**(4), 595–602 (2010)
13. Birgin, E., Martinez, J., Nishihara, F.H., Ronconi, D.P.: Orthogonal packing of rectangular items within arbitrary convex regions by nonlinear optimization. Comput. Oper. Res. **33**(12), 3535–3548 (2006)
14. Castellazzo, A., Fasano, G., Pintér, J.D.: An MINLP Formulation for The Container Loading Problems: An Experimental Analysis. Working paper, Thales Alenia Space (2015)
15. Junqueira, L., Morabito, R., Yamashita, D.S., Yanasse, H.H.: Optimization models for the three-dimensional container loading problem with practical constraints. In: Fasano, G., Pintér, J.D. (eds.) Modeling and Optimization in Space Engineering, pp. 271–294. Springer Science +Business Media, New York (2013)

Chapter 5
CAST: A Successful Project in Support of the International Space Station Logistics

Giorgio Fasano, Claudia Lavopa, Davide Negri, and Maria Chiara Vola

Abstract The International Space Station (ISS) is one of the most challenging currently ongoing space programs. It has led to a number of very demanding logistic issues, in particular in relation to the on-orbit maintenance and resource resupply.

A fleet of launchers and vehicles is periodically made available by the most prominent space agencies in order to serve this scope. An overall traffic plan schedules the recurrent upload and download interventions. The relevant Cargo Manifest (delivered by NASA) establishes, for each carrier launch and re-entry, the shipment that is supposed to be transported from Earth to orbit and vice versa.

The European Space Agency (ESA) contributed annually to the ISS logistics from 2008 to 2014, by accomplishing five Automated Transfer Vehicle (ATV) missions. The ATV transportation system was conceived to support the recurrent upload phases from Earth to the ISS.

Within the relevant cargo accommodation context, in addition to tight balancing conditions, intricate three-dimensional packing issues arose. Furthermore, besides the remarkable complexity related, per se, to the loading aspects, very strict deadlines were usually imposed to accomplish the task. Last minute upgrades or even significant changes, moreover, often were expected to take place.

CAST (Cargo Accommodation Support Tool) is a dedicated optimization framework, funded by ESA and developed by Thales Alenia Space to carry out the whole analytical ATV cargo accommodation. This chapter describes the ATV loading problem first. The basic concept of CAST is further outlined, highlighting the advantages of the methodology adopted, both in terms of solution quality and time saving. Current extensions and possible future enhancements are investigated.

G. Fasano (✉) • C. Lavopa
Thales Alenia Space Italia S.p.A., Turin, Italy
e-mail: giorgio.fasano@thalesaleniaspace.com; claudia.lavopa@thalesaleniaspace.com

D. Negri
SSE – Sofiter System Engineering S.p.A. Consultant c/o Thales Alenia Space Italia S.p.A., Turin, Italy
e-mail: davide.negri@external.thalesaleniaspace.com

M.C. Vola
Altran Italia S.p.A. Consultant c/o Thales Alenia Space Italia, Turin, Italy
e-mail: mariachiara.vola@external.thalesaleniaspace.com

© Springer International Publishing Switzerland 2015
G. Fasano, J.D. Pintér (eds.), *Optimized Packings with Applications*, Springer Optimization and Its Applications 105, DOI 10.1007/978-3-319-18899-7_5

Keywords Space engineering • Space vehicle/module • Cargo accommodation • International Space Station (ISS) • Automated Transfer Vehicle (ATV) • Columbus Laboratory • Packing optimization • Static and dynamic balancing • Additional conditions • Mixed integer programming (MIP) • Heuristics

5.1 Introduction

This chapter focuses on the very challenging issue relevant to the (analytical) cargo accommodation of vehicles and modules, arising in space engineering. The operating context refers to that of the International Space Station (ISS) (see [1]) and, more precisely, of the Automated Transfer Vehicle (ATV) (see [2]), provided by the European Space Agency (ESA) [3] to support the ISS logistics. The CAST (Cargo Accommodation Support Tool) project, funded by ESA and achieved by Thales Alenia Space [4], was conceived with the specific objective of optimizing the ATV cargo accommodation, for each planned mission. The overall reference framework is described hereinafter, in order to introduce the operational scenario that motivated the dedicated effort. Section 5.2 describes the ATV cargo accommodation problem; Sect. 5.3 discusses the CAST tool in depth; Sect. 5.4 reports a real-world instance and the relevant results; Sect. 5.5 outlines current and prospective extensions.

The ISS (see Fig. 5.1) represents a paramount worldwide initiative involving the Brazil, Canada, Europe, Japan, Russia, and United States, through the respective

S128E009993

Fig. 5.1 The International Space Station (ISS, a pictorial view). Photo: NASA

participating space agencies: CSA, ESA, ASI, JAXA, Roscosmos, and NASA (see [3, 5–9]). At present, it is the largest spacecraft in orbit, consisting of a modular structure, whose first component was launched in 1998. The station encompasses pressurized/unpressurized modules, external trusses, solar arrays, and further components. The ISS is a manned platform that serves as a microgravity and space environment research laboratory in which experiments in astronomy, geology, meteorology, physics, material sciences, general/human biology, space medicine, and other disciplines are performed by the crew present on board (see [1]). It is further expected to provide essential testing of spacecraft systems and equipment required for the extremely challenging missions to the Moon and Mars in the near future.

The permanent human presence on board, as well as the demanding targets deriving from the requested experimentation, has led to the necessity of a continuous upload/download activity, implying recurrent transportation between the Earth and the ISS. A fleet of launchers and vehicles (i.e., Soyuz, Progress, ATV, H-II Transfer Vehicle, Dragon and Cygnus, see [1]) is made available to provide the ISS with the necessary logistic support. An overall traffic plan schedules the recurrent upload and download interventions. The Cargo Manifest (delivered by NASA), relevant to the current mission, establishes, in particular, for each carrier launch and re-entry, the shipment that is supposed to be transported from Earth to orbit and vice versa.

The ATV (see Fig. 5.2) concept placed itself in this very demanding operative context (see [2]). Developed by ESA, the ATV systems were designed to supply the ISS with propellant (to re-boost the station to the required orbit altitude), payloads (i.e., equipment devoted to the on-board experimentation), as well as material both for the crew and the ISS activity and maintenance. Five ATVs were successfully launched since March 2008: Jules Verne, Johannes Kepler, Edoardo Amaldi, Albert Einstein, and Georges Lemaître. In April 2012, the ESA announced that the ATV program would end after the fifth mission, in July 2014.

Fig. 5.2 The Automated Transfer Vehicle (ATV, a pictorial view). Photo: NASA

Each ATV weighed about 20.7 tons at launch and had an overall load capacity of 8 tons. The overall cargo was partitioned into the following categories:

- *pressurized* cargo, consisting of objects of different typologies (frequently also referred to as *dry* cargo, e.g. scientific equipment or resupply goods);
- *unpressurized* cargo, consisting of fluids;
- propellant for the re-boost maneuvers and station refueling.

With the ATV docked, the station crew used to enter the cargo section and remove the payloads. The ATV's liquid tanks were connected to the station and discharged. The station crew manually released air components directly into the ISS's atmosphere. For up to 6 months, the ATV remained attached to the ISS. The crew then steadily filled the cargo section with the station's waste. At intervals from 10 to 45 days, the ATV's thrusters were utilized for re-boosting. This vehicle was of an expendable type, i.e. at the end of each attachment period, the spacecraft was moved out of orbit to perform a controlled destructive re-entry high above the Pacific Ocean.

5.2   The ATV Cargo Accommodation Problem

The ATV transportation system was essentially aimed at supporting the ISS upload-ing phases, since it performed a destructive re-entry (the only download activity consisted of the trash destruction). Both fluids and cargo items were delivered on board the ISS, based on the current Cargo Manifest list. The amount of each fluid type could (generally) vary within an admissible range. Some items could, moreover, be rejected, on the basis of a given priority. The ultimate task was hence that of maximizing the overall load, in compliance with the accommodation rules (that could change, even quite significantly, from mission to mission). In addition to tight balancing conditions, deriving from the system control specifications, intri-cate three-dimensional packing issues arose, encompassing tricky accommodation scenarios, at item, bag, rack and system level. Furthermore, besides the remarkable complexity related per se to the loading aspects, very strict deadlines were usually imposed to accomplish each analytical cargo accommodation task relative to the whole mission preparation. These consisted, essentially, of three analysis-cycle periods (typically of 1–2 weeks each), per year, carried out on the basis of the current information available. Last minute upgrades or even significant changes, moreover, often occurred.

This section focuses on the ATV cargo accommodation problem (see [10]) that consisted of the loading of both the *unpressurized* (fluid) and *pressurized* (solid) material. The re-boost propellant amount simply represented an analysis input, since it was established a priori for each mission. The cargo accommodation task could, therefore, be summarized as follows:

Given the spacecraft overall mass distribution, including the re-boost propellant contribution, the unpressurized and pressurized cargo had to be accommodated, in order to fulfill the Cargo Manifest request as much as possible, in compliance with the balancing and operational requirements.

This section is aimed at describing the above outlined ATV cargo accommodation issue, at quite a detailed level, restricting the discussion to the upload phases only, since the loading aspects relevant to the destructive re-entry (limited to the waste load) were significantly less demanding. A description of the ATV cargo carrier features, concerning the cargo accommodation problem is provided first. The cargo typologies are considered afterwards and subsequently the accommodation rules to cope with.

5.2.1 The ATV Cargo Carrier

From the cargo accommodation standpoint, as defined above, the spacecraft consisted essentially of two components, see Fig. 5.3: the unpressurized module,

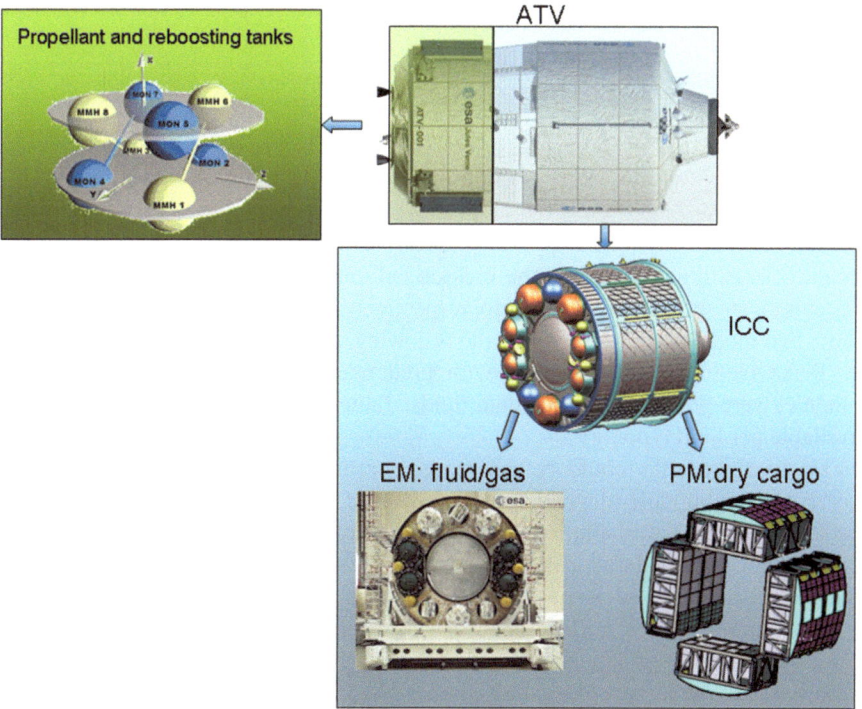

Fig. 5.3 ATV's external and pressurized modules

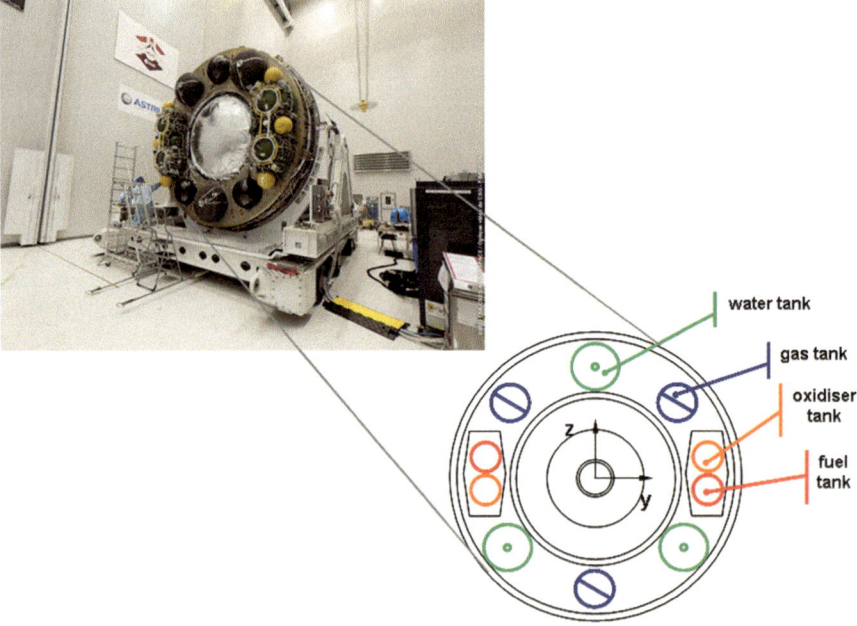

Fig. 5.4 Tanks inside the external module. Photo: ESA/CNES/Arianespace/Optique

Table 5.1 Fluid load capability

Fluid	Water	Gas	Fuel	Oxidizer
Tank availability	3 tanks (285 kg capacity each)	3 tanks (33.34 kg capacity each)	2 tanks (150 kg capacity each)	2 tanks (280 kg capacity each)

denoted as external (EM) and the pressurized one (PM). These were designed to accommodate the two kinds of cargo, respectively. In the following, the merging of these two modules is referred to, in the following, as the system.

Tanks (see Fig. 5.4), situated with predefined positions inside the external module, were predisposed to contain fluids. Table 5.1 reports the number of tanks available per fluid typology.

Up to eight racks could be put to use for the *pressurized* cargo, inside the corresponding module (depending on the specific mission, the utilized racks could be fewer than the available locations). They were inserted (when employed) into proper structural facilities, called rack locations, see Fig. 5.5. There were eight of these in all and they had prefixed (axially symmetrical) positions inside the module.

Each rack provided a plane anterior side, named rack front that was oriented towards the inside of the module, see Figs. 5.5 and 5.6. It could be equipped with four structural elements called adapter plates (of two possible different types,

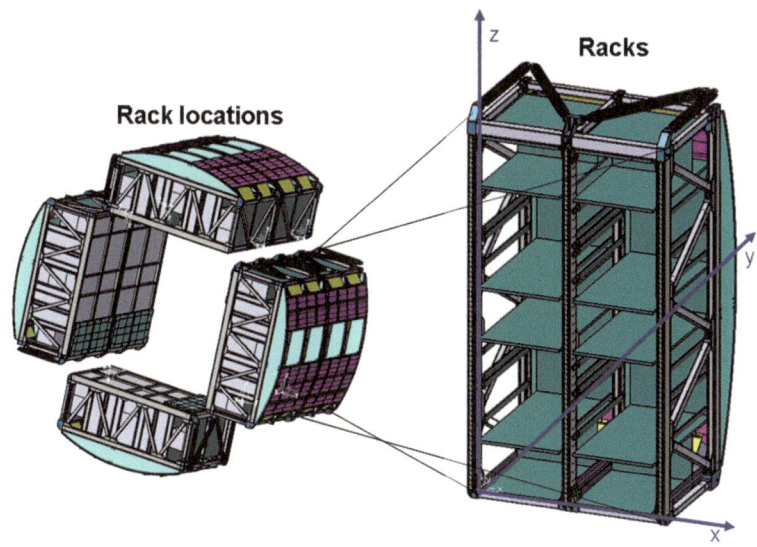

Fig. 5.5 Racks and rack locations

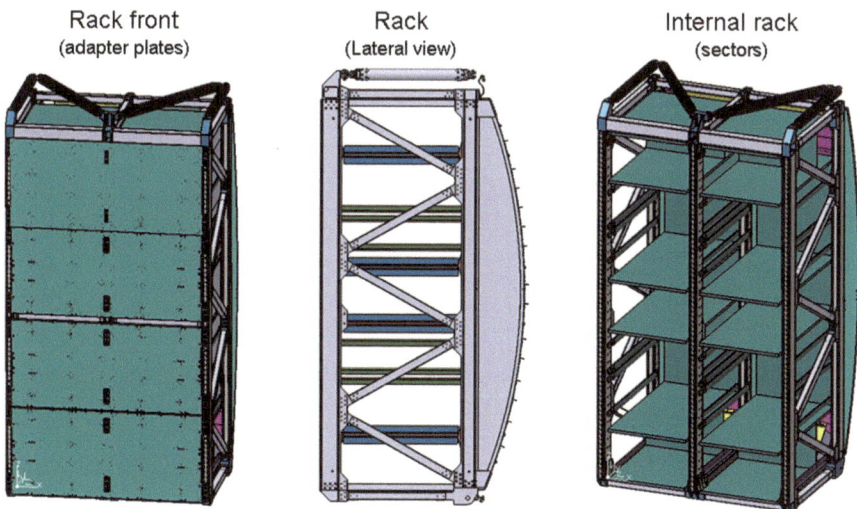

Fig. 5.6 Rack overall configuration

i.e. A and B), aimed at holding items externally. The posterior surface of the rack was curved, to match the cylindrical shape of the spacecraft. The internal rack volume was partitioned into sectors (of different kinds).

There were two types of racks, i.e. A and B, characterized by quite significant structural differences, implying, respectively, diverse cargo capacity and capability. In particular, the type B rack planes fitted the posterior surface in order to exploit the available volume as much as possible, see Fig. 5.7.

Rack type A   **Rack type B**

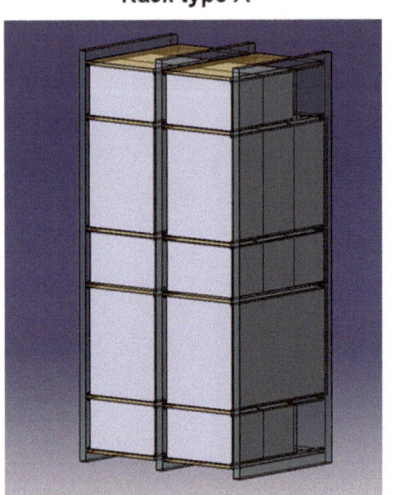

Fig. 5.7 Rack A and B volume exploitation

5.2.2   Cargo Typologies

Each Cargo Manifest could include the following fluids:

- water;
- air;
- oxygen;
- nitrogen;
- fuel;
- oxidizer

    (the fuel was different from the re-boost one and was utilized for payloads only).
    The two main classes of *pressurized* cargo were:

- cargo items;
- bags.

    The cargo items were partitioned into the following typologies:

- small items;
- large items;
- external large items;
- mid-deck lockers (special items, named MDLs);
- drawers.

Table 5.2 Tank mass capacities

Fluids	Tank capacity (kg)
Water	285
Air	33.34
Oxygen	33.34
Nitrogen	33.34
Fuel	150
Oxidizer	280

Small items were, initially, only box-shaped (and assumed to be of homogeneous density), as well as MDLs and drawers. In the last missions, however, non-box-shaped cases had to be taken into account. Large items, instead, were ordinarily characterized by complex shapes and non-homogeneous internal mass distribution.

5.2.3 Loading Rules

The *unpressurized* cargo was subject, for each fluid type, to given capacity limitations. Table 5.2 reports the relevant tank characteristics.

In addition, the following specific loading rules were imposed:

- if the same fluid was loaded in more than one tank, the difference of mass between each pair could not exceed a given amount (scattering rule);
- oxidizer/fuel amounts had to respect a given ratio (this requirement had been stated for stoichiometric reasons).

As far as the *pressurized* cargo was concerned, different classes of (soft) bags were available to contain small items, namely:

- standard;
- non-standard;
- internal;
- external.

As is understood, internal bags were supposed to be accommodated inside the racks, whilst the external ones on the rack fronts. For the first missions, only box-shaped internal bags were considered. Curved types were, however, introduced later, in order to exploit the rack internal volume as much as possible. External bags, instead, were only box-shaped.

The following types of internal standard/non-standard bags, classified on the basis of their shapes and dimensions, were available.

Box-shaped standard bags:

- half;
- single;
- double;
- internal triple.

Box-shaped non-standard bags:

- (CTB, Cargo Transfer Bag) type C;

Curve-shaped non-standard bags:

- (CTB) type A;
- (CTB) type B.

Box-shaped internal standard bags were modular. Consequently, two halves could be joined to replace a single, two singles a double, a single plus a double a triple, and so on (see Fig. 5.8).

The following types of external standard/non-standard bags were envisaged:

- external triple.

Box-shaped non-standard bags:

- M01;
- M02.

Internal/external bag mass capacities are reported in Table 5.3 and some typologies of bags are illustrated in Fig. 5.8.

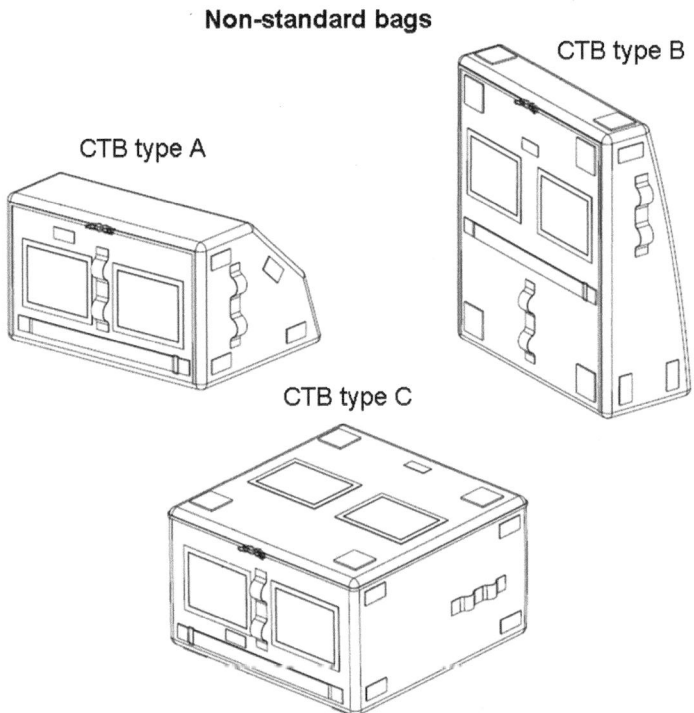

Fig. 5.8 Bag typologies

Standard bags

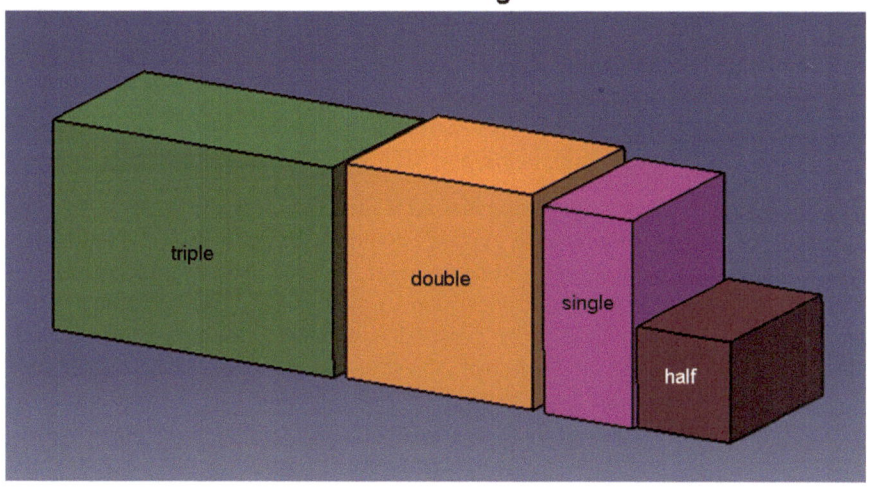

Fig. 5.8 (continued)

Table 5.3 Bag mass capacities

	Types	Mass capacity (kg)
Internal bags	CTB_type_A	16.27
	CTB type B	17.27
	CTB type C	24.15
	Half	13.62
	Single	27.24
	Double	54.48
	Triple	81.72
External bags	Triple	80
	M01	120
	M02	90.8

The major overall accommodation rules adopted are summarized as follows, in terms of item typology and possible locations:

- small items had to be accommodated into standard bags (of different types);
- large items either into racks or on the rack fronts;
- internal bags and drawers into rack sectors;
- external bags, external large items, and MDLs on the rack fronts.

Type A rack (utilized only for the first ATV missions) was able to accommodate all kinds of cargo items, while type B could only house bags. Compatibility/incompatibility conditions could be imposed, for each specific mission, dealing with the accommodation of the bags, due to operational needs. Frequently, for instance, some of them had to be grouped within the same sector or, at least, allocated together inside the same rack. On the contrary, others were not allowed to be accommodated in the same sector/rack. Some internal bags were requested to be positioned in proximity of the rack front, with no obstructing external bags or adapter plates. This was usually due to accessibility reasons.

Small items had to be positioned, orthogonally (see Fig. 5.9), taking into account the mass capacity of the bags utilized. Quite often, additional conditions, such as the following, were posed:

- item prefixed position/orientation;
- presence of separation planes;
- minimum gap between items;
- minimum gap between items and bag sides;
- *static* balancing.

Conditions on prefixed position/orientation could, for instance, be motivated by manageability reasons. The presence of separation planes (inside the bags), usually with non-prefixed positions, forced the item accommodation in different internal sectors. Using separation planes made their handling by astronauts easier (Fig. 5.9 provides an example of a bag with two separation planes).

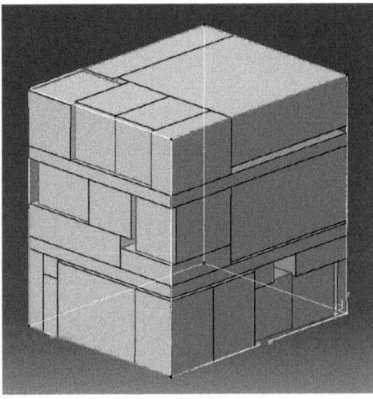 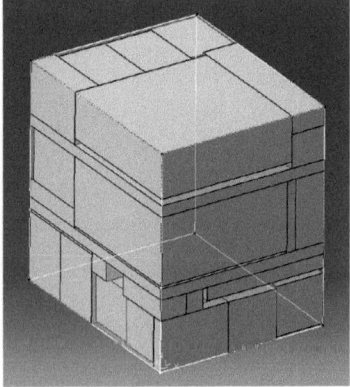

Fig. 5.9 Small items placed orthogonally in a bag (with two separation planes)

The *static* balancing restriction, at bag level, had two different statements, depending either on the internal or external connotation. In the case of internal bags, their center of mass was requested to stay within a box-shaped domain (of given dimensions), positioned in the center of the box. The center of mass of the external bags, on the other hand, had to stay within a box-shaped domain (of given dimensions), adjacent to the box side in contact with the adapter plate. This rule was posed in order to reduce, as much as possible, the bag unbalancing towards the rack-front outside. Depending on the incumbent Cargo Manifest to satisfy, a number of internal/external bags could already be pre-integrated.

A number of additional conditions, at system level, had to be taken into account:

- overall mass capacity;
- overall *static* balancing;
- overall *dynamic* balancing.

The total cargo (both *unpressurized* and *pressurized*), indeed, was not allowed to exceed a given threshold. The *static* balancing condition meant that the system center of mass coordinates had to stay within given ranges that depended on the total mass loaded. More precisely, the following constraints were introduced:

$$\forall \beta \in \{1, 2, 3\} \quad \underline{c}_\beta(m) \leq \frac{\sum_\iota m_\iota w_{\iota\beta}}{m} \leq \overline{c}_\beta(m). \tag{5.1}$$

Here, the indexes β indicate the system reference frame axes; m_ι are the involved (*unpressurized* and *pressurized* cargo) masses, assumed as points, associated with the locations $w_{\iota\beta}$ (with respect to the system reference frame); $m = \sum_\iota m_\iota$ is the total mass loaded; $\underline{c}_\beta(m)$ and $\overline{c}_\beta(m)$ are given lower and upper (piecewise linear) functions, delimiting, for each mass m, the center of mass ranges, see Fig. 5.10.

In expressions (5.1), the terms $\frac{\sum_\iota m_\iota w_{\iota\beta}}{m}$ represent the coordinates of the overall center of mass with respect to the system reference frame.

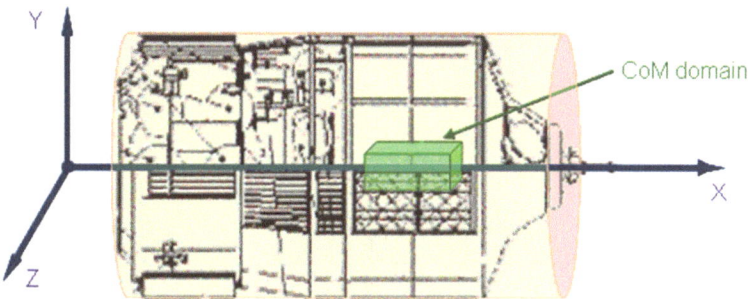

Fig. 5.10 Overall center of mass domain

The *dynamic* balancing conditions essentially compelled the system inertia matrix, defined with respect to the barycentric reference frame, to assume a quasi-diagonal form (providing the system with a mechanical behavior approximately equivalent to that of a homogeneous cylinder). To this purpose, lower and upper bounds, expressed as piecewise linear functions of the loaded mass, were posed (neglecting a priori, for the sake of simplicity, the inertia property relative to each mass m_l that was considered as concentrated in a single point). Figure 5.11

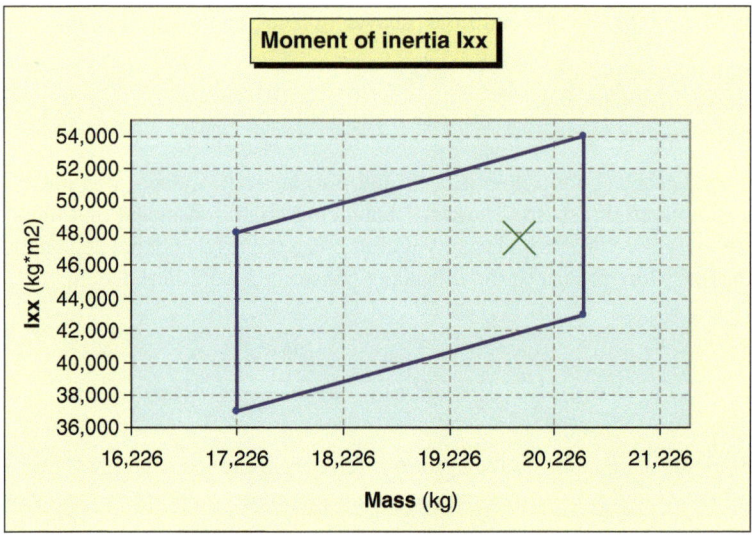

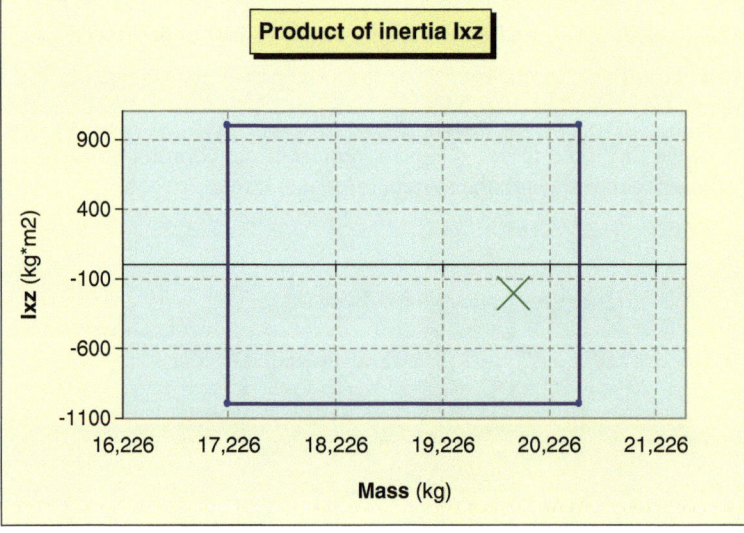

Fig. 5.11 *Dynamic* balancing bounds

provides illustrative examples (concerning the moment of inertia, with respect to the longitudinal axis X, and the XZ product). They are reported below (neglecting all secondary details of the actual situation):

$$\forall \beta, \beta' \in B/\beta < \beta' \quad \left| \sum_\iota m_\iota w^*_{\beta_\iota} w^*_{\beta_\iota} \right| \leq \bar{I}_{\beta\beta'}(m), \tag{5.2}$$

$$\forall \beta, \beta', \beta'' \in B/\beta < \beta', \beta, \beta' \neq \beta'' \quad \sum_\iota m_\iota \left(w^{*2}_{\beta_\iota} + w^{*2}_{\beta'_\iota} \right) \geq \underline{I}_{\beta''}(m), \tag{5.3-1}$$

$$\forall \beta, \beta', \beta'' \in B/\beta < \beta', \beta, \beta' \neq \beta'' \quad \sum_\iota m_\iota \left(w^{*2}_{\beta_\iota} + w^{*2}_{\beta'_\iota} \right) \leq \bar{I}_{\beta''}(m). \tag{5.3-2}$$

Here, in addition to the symbols defined hitherto, $w^*_{\beta_\iota}$ represent the mass location coordinates with respect to the barycentric reference frame, whilst $\bar{I}_{\beta\beta'}(m)$, $\underline{I}_{\beta''}(m)$ and $\bar{I}_{\beta''}(m)$ are (non-negative piecewise linear) functions of the total loaded mass m.

The following overall conditions had moreover to be considered at rack level (for each rack type):

- overall mass capacity;
- sector mass capacity (depending on the specific rack configuration adopted, e.g. with/without adapter plates);
- rack-front mass capacity;
- adapter-plate mass capacity (for each type);
- heavy-light bag adjacency incompatibility (heavy bags were not allowed to be adjacent to lightweight ones);
- *static* balancing.

The *static* balancing restriction, in this case, required that the rack center of mass was contained inside a given convex domain, see Fig. 5.12.

The loading of cargo items and bags, either inside the racks or on the rack fronts, was regulated by specific conditions (not reported here). The internal bags had to be accommodated into sectors, on the basis of predefined patterns (i.e., admissible sector configurations). Accessibility conditions could be posed, requiring specific placement ordering (with respect to the rack front) or fixed position/orientation for some bags. The rules for the external accommodation, posed for structural reasons, were very complicated. Figure 5.13 shows the maximum mass (expressed in kilograms) loadable externally, depending on the number/typology of bags and adapter plates utilized.

Fig. 5.12 Rack center of mass (convex) domain

A number of grouping conditions could moreover be present, mission by mission, for particular cargo items. For example, some could have been requested to be accommodated into the same rack sector. On the contrary, others had to be placed in different ones (these additional accommodation rules are not detailed here).

5.3 A Dedicated Cargo Accommodation Tool

The cargo accommodation issue described in Sect. 5.2 represented a very challenging non-standard three-dimensional packing problem, with difficult additional conditions to be compliant with. Complex geometries had to be considered, both in terms of objects and containers/locations, fluid masses were present, specific accommodation rules had to be taken account of, in addition to tight balancing requirements at different levels.

Attention has been paid quite recently, in the vast literature of packing (e.g., [11–13]), to non-standard problems, in addition to the usual issue of loading

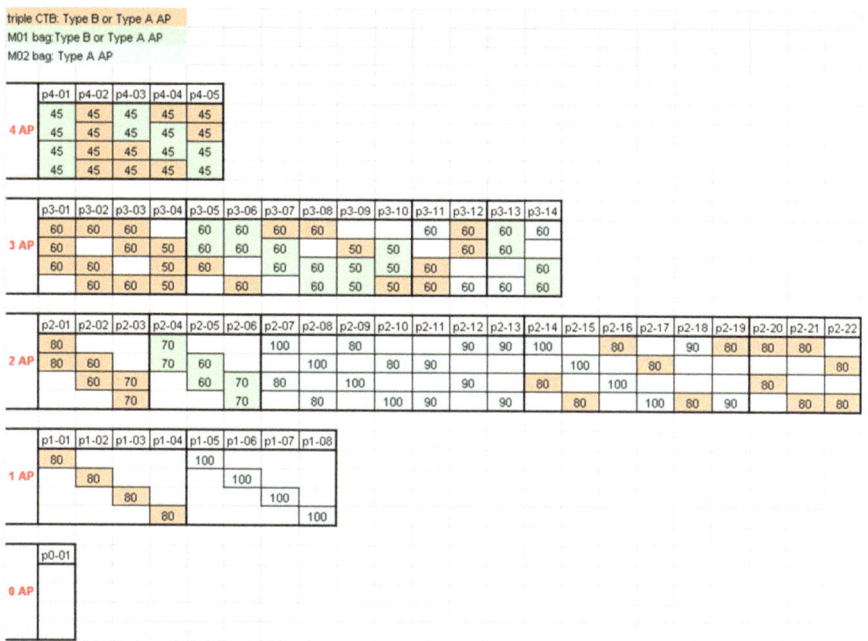

Fig. 5.13 Accommodation rules for external bags

(orthogonally) "small boxes" into "big boxes," with no extra conditions (e.g., [14]). Topical works tackle very complex issues involving intricate geometries and balancing conditions (e.g., [15–17]). Studies on cargo accommodation in space engineering, based on artificial intelligence [18] and multi-agent methods [19] are also available.

Dealing with the ATV issue, an overall heuristic approach, based on mathematical-programming (e.g., [20]), nonetheless, definitely seemed more suitable than merely non-deterministic methodologies, including meta-heuristics in general (e.g., [21, 22]). The ATV specific case, indeed, presented a very significant number of "strong" constraints (for which no relaxation was admitted) to cope with, in particular those derived from the tight balancing restrictions. As the given problem could unquestionably not be figured out tout court, taking into account all the relevant accommodation levels (i.e., bag, rack, system) contemporarily, a decomposition of the whole problem into sub-problems, was devised (see [23]).

This concept, in addition to a strong modeling-based approach, was therefore embraced by our specialist team (Giorgio Fasano, Claudia Lavopa, Davide Negri, and Maria Chiara Vola) that since 2000 has been developing the CAST system, supporting its ongoing upgrading and utilization from the first ATV mission (Jules Verne 2008) until the last (Georges Lemaître 2014, see [2]). This section is devoted to provide the reader with quite detailed insights on CAST's basic concept and major features.

CAST is characterized by an overall architecture, see Fig. 5.14, based on a mathematical library that represents the core of the entire optimization framework.

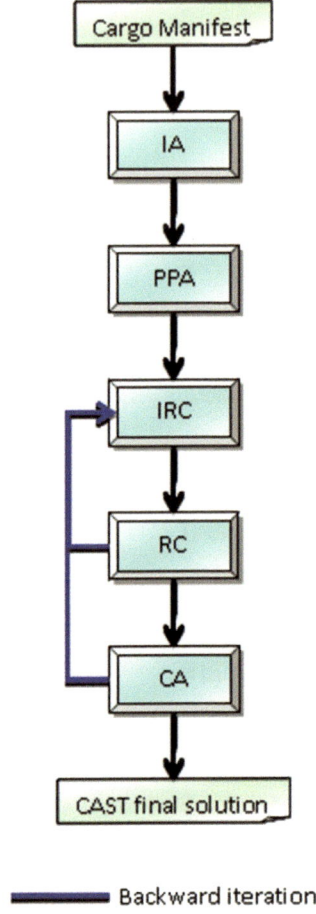

Fig. 5.14 CAST's overall architecture

Sub-problems are solved iteratively by adopting, step by step, the relevant mathematical library module, consisting of specific MIP (Mixed Integer Programming, e.g. [24]) models and heuristic algorithms. These are handled by an overall System Management module (CAST-SM) and a (3D) Graphical User Interface (CAST 3D-GUI).

Backward iterations are admitted, when the desired solution is not attained, performing a recursive process. In its latest version, CAST's mathematical library consists of the following modules:

- Item Accommodation (IA);
- Preprocessing Assessment (PPA);
- Item-Rack Correlation (IRC);
- Rack Configuration (RC);
- (Overall) Cargo Accommodation (CA).

The above-mentioned modules are outlined hereinafter, in a simplified manner, to provide the reader with a wide-ranging description of the basic concepts. Most of the aspects relevant to the mathematical model formulation are not taken into account in this chapter (the reader is referred to [23, 25] for a more in-depth discussion).

The IA module is employed to accommodate the small items inside bags, on the basis of the general packing rules and possible additional ones, such as *static* balancing, if any (see Sect. 5.2.3). When particular cases, involving items that can hardly be modeled as single boxes, arise an ad hoc tetris-like representation is adopted. More precisely, items that are not simply box-shaped (and cannot reasonably be assumed of homogeneous density) are substituted with clusters of (mutually orthogonal rectangular) parallelepipeds. The available volume is moreover assumed to consist of a convex domain, in general non box-shaped. The underlying packing problem may, therefore, be referred to as a non-standard container loading one, where items to accommodate are tetris-like shaped and the container of a convex type, in the presence of possible additional conditions. An MIP-based formulation of this problem has been provided (see [25] for a detailed discussion). The relevant general MILP (Mixed Integer Linear Programming, see [20, 24]) model consists of three blocks of linear constraints, respectively devoted to guarantee that items:

- are placed orthogonally (*orthogonality* conditions);
- are inside the given domain (*domain* conditions);
- do not overlap (*non-intersection* conditions).

The *orthogonality* constraints require that each item, if loaded, has to assume one of the possible orthogonal positions, with respect to the main reference frame, associated with the domain (being in general asymmetric, a tetris-like item is expected to have up to 24 admissible orientations). The *domain* constraints require that each vertex of each tetris-like item must stay within the convex container volume. The *non-intersection* constraints exclude the intersection of any two components, belonging to different items.

The following sets of binary variables are introduced:

- $\chi_i \in \{0, 1\}$ controlling the presence of item i, inside the container;
- $\vartheta_{\omega i} \in \{0, 1\}$ (in association with the *orthogonality* and *domain* constraints) providing item i with an orthogonal *orientation* ω, within the domain;
- $\sigma^+_{\beta hkij}, \sigma^-_{\beta hkij} \in \{0, 1\}$ (in association with the *non-intersection* constraints, with respect to each axis β of the main reference frame) preventing components (h and k) of different items (i and j) from overlapping.

The positions of items' local reference frames are represented by continuous variables. A number of additional conditions, such as *static* balancing, may be modeled, by means of linear constraints (see [25]).

The resulting MILP model is, in general, very difficult to solve. Therefore, heuristic procedures have been designed, in order to obtain efficient (albeit not proven to be optimal) solutions, in an acceptable time (see [25]).

The PPA module is aimed at providing a fast approximate solution to the whole problem, in order to attain a preliminary feasibility check. This module serves also the scope of identifying an upper bound, relevant to the overall loadable mass. An MILP model solves a multiply constrained, multiple continuous knapsack problem with additional conditions. The set of knapsacks is represented by both tanks and racks with given mass capacities. Additional conditions include:

- the specific rules for fluids;
- the compatibility between rack and rack location;
- the compatibility (both for *unpressurized* and *pressurized* cargo) between mass and container type;
- the compatibility between mass and rack-front type;
- the *static* and *dynamic* balancing requirements at system level.

The *static* balancing constraints (being nonlinear) are controlled by appropriate binary variables (utilized for linearization purposes). In those relevant to the *dynamic* balancing, the barycentric reference frame, as a further approximation, is supposed to be fixed in a "central" position. This corresponds to the center of mass domain (as delimited by the *static* balancing constraints). The compliance with the actual barycentric reference frame is verified "a posteriori" and the terms $\bar{I}_{\beta\beta'}(m)$, $\underline{I}_{\beta''}(m)$ and $\bar{I}_{\beta}''(m)$ re-adjusted appropriately, if necessary, executing an opportune number of iterations.

The MILP model considers, as quite a rough approximation, both the *unpressurized* and *pressurized* cargo in the same way as fluid mass. Continuous variables represent the amount of both *pressurized* and *unpressurized* mass, associated with the destination containers and locations (tanks, racks, and rack fronts). Binary variables control:

- the correlation of racks to rack-locations;
- the utilization of the rack fronts;
- the compliance with the specific rules for fluids.

The IRC module has the scope of obtaining an initial correlation between integrated bags (by means of the IA module) or pre-integrated bags and the rack locations. A number of tasks not discussed here are associated with this module. These are executed by means of an MILP model (whose utilization mode depends, time after time, on the specific task). Multiple constrained, multiple (non-continuous) knapsack problems with additional conditions are solved. The model is based on a more sophisticated formulation with respect to the PPA. Items (i.e., the *pressurized* cargo) are considered in terms of distinct (flexible) objects, characterized by their mass and volumes (their actual dimensions are, however, still neglected).

All conditions contemplated by the PPA model, in particular the *static* and *dynamic* balancing, are taken into account. In this case, nonetheless, in order to overcome the difficulties associated with the nonlinearity of the *static* balancing constraints, the overall mass loaded is considered as a constant (a first approximation of its value is derived by the PPA step itself). The error, expressing the difference between the constant mass estimation and the actual amount loaded by the IRC model is, time after time, minimized. A number of iterations, aimed at tuning the constant mass estimation are performed as necessary. The total mass loaded is, as a matter of fact, maximized only indirectly. Additionally, the presence of sectors inside the racks is considered, together with the relevant accommodation rules. *Pressurized* cargo items may be rejected, if necessary, on the basis of their priority. The overall logic utilized by the IRC module is quite complicated and is not discussed here.

The RC module has the task of determining the internal/external rack loading, in compliance with the given accommodation rules. Integrated/pre-integrated bags, drawers, large items, and MDLs are accommodated into the racks or on the rack fronts, on the basis of the designations provided by the previous IRC stage. All cargo items and bags involved have, in this phase, their actual shapes and dimensions. The *static* balancing restriction (at rack level) is respected. In the original version of CAST, the RC module consisted, essentially, of two major components, directed to solve the internal/external accommodation of types A and B racks, respectively. The rack A type accommodation task was carried out by means of an ad hoc heuristic procedure, aimed at solving the related non-standard container loading problem with additional conditions (the procedure, being now obsolete, shall not be discussed in this chapter). The component relevant to the internal/external accommodation rack of type B has been based on an MILP model that solves the related multiply constrained, multiple knapsack problems with additional conditions. The knapsacks, here, represent the rack (internal) sectors and the (external) adapter plates. The additional conditions (see Sect. 5.2.3), at this stage, consist of:

- the rules for the internal accommodation (e.g., bags into sectors);
- the mass capacity both at sector and adapter-plate levels;
- the rules for the external accommodation;
- the *static* balancing of the whole rack.

Since during the accommodation process several approximations are - step by step, directly or indirectly - introduced, the CA module has the objective of re-arranging all the partial so-far obtained solutions, in order to attain an ultimate result. The assignment of the already accommodated racks to rack locations is reconsidered, looking into a final accommodation, compliant with the given *static* and *dynamic* balancing conditions (at system level). At this step, errors, with respect to the mass loaded in each rack are admitted. They are minimized by the CA model objective function. If the outcome obtained is acceptable (in terms of error tolerance) the final solution is attained. Otherwise, backward/forward iterations are executed throughout the entire accommodation process, until a satisfactory result is obtained.

The major functions associated with the CAST-SM module are (see Fig. 5.15):

- importing and handling the ATV and Cargo Manifest data, relative to the current mission;
- executing, step by step, the predisposed mathematical models and algorithms;
- showing the ongoing and final accommodation analysis state;
- exporting the final accommodation solution to the database.

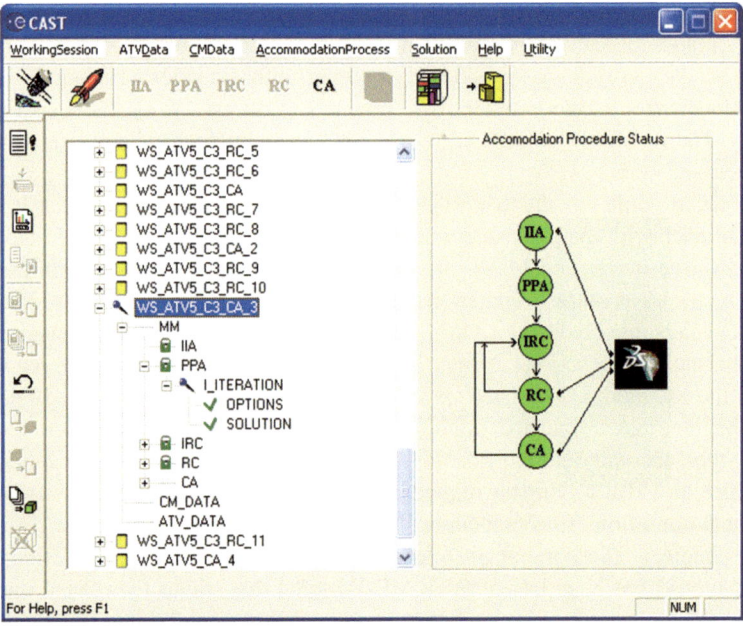

Fig. 5.15 Example of CAST-SM control panel

The CAST 3D-GUI provides the cargo engineer with graphical representations of the current and final outcomes, at bag and rack level, with the possibility of interacting during the entire process or making desirable changes in the final solutions. The main reason for this interactive option is that some human-perception-based evaluative criteria can hardly be contemplated by mathematical models.

5.4 Real-World Instances and Solutions

This section focuses on the real-world framework that characterized the whole history of the ATV missions, with respect to the cargo accommodation task. The cost-effectiveness of the approach followed, as well as its promptness to compare, when necessary, different operational scenarios and work out last-minute requests, resulted first and foremost in a relatively limited human commitment. Each cargo accommodation cycle was carried out, for all missions, by the CAST team in not more than a fortnight. This included all interactions with the current ATV program activity, necessary design re-adjustments, trade-offs between alternative solutions and Cargo Manifest changes. The high-quality outcomes attained for all the missions accomplished, moreover, were made possible by the advanced optimization methodology put into action.

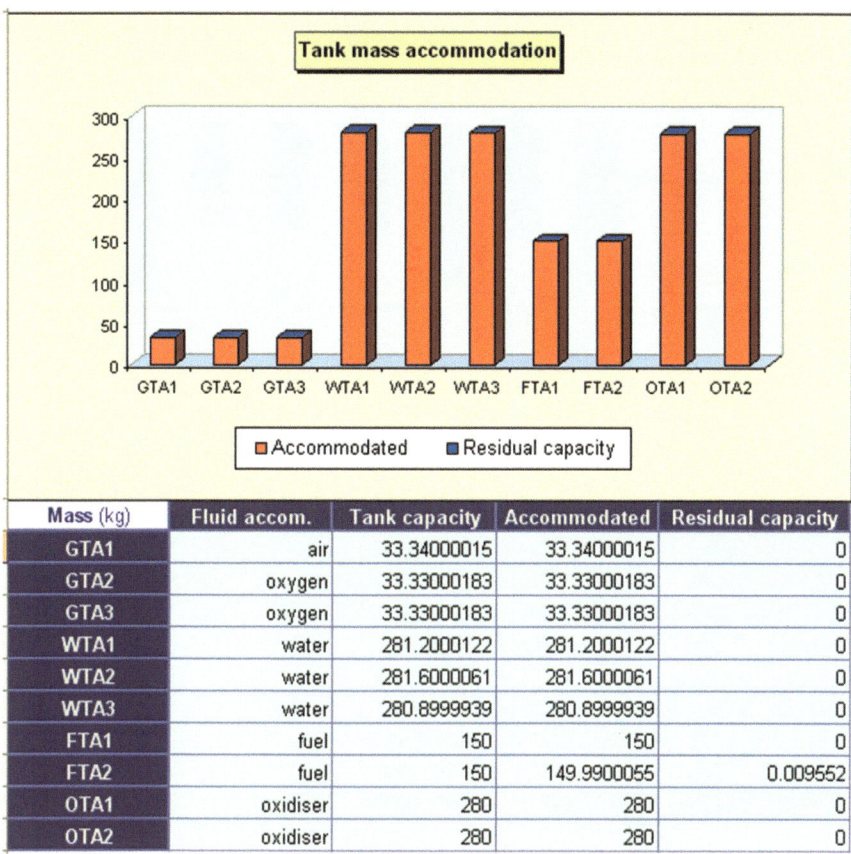

Mass (kg)	Fluid accom.	Tank capacity	Accommodated	Residual capacity
GTA1	air	33.34000015	33.34000015	0
GTA2	oxygen	33.33000183	33.33000183	0
GTA3	oxygen	33.33000183	33.33000183	0
WTA1	water	281.2000122	281.2000122	0
WTA2	water	281.6000061	281.6000061	0
WTA3	water	280.8999939	280.8999939	0
FTA1	fuel	150	150	0
FTA2	fuel	150	149.9900055	0.009552
OTA1	oxidiser	280	280	0
OTA2	oxidiser	280	280	0

Fig. 5.16 *Unpressurized* cargo mass distribution

CAST was originally designed to cope with extremely complex large-scale instances, involving up to 1,000 items, 8 racks and 80 rack sectors. Real-world scenarios were usually less demanding, albeit never trivial at all.

The case regarding the ATV5 mission that occurred in 2014 is reviewed in this section, in order to provide insights on a typical operative scenario. The final solution, corresponding to the third analysis cycle, is considered hereinafter.

All the *static/dynamic* balancing conditions were fully satisfied, both at system and rack level (the relative graphical representations are, however, not included here). Figures 5.16 and 5.17 report the overall mass/volume distribution, both for the *unpressurized* and *pressurized* cargo, respectively. It should be noticed that the mass capacity, at rack level was in general not fully exploited, since the volume capacity, represented the more restrictive condition.

Figure 5.18 illustrates a (type B) rack loaded both internally and externally, whilst the one shown in Fig. 5.19 (also of type B) contains internal cargo only.

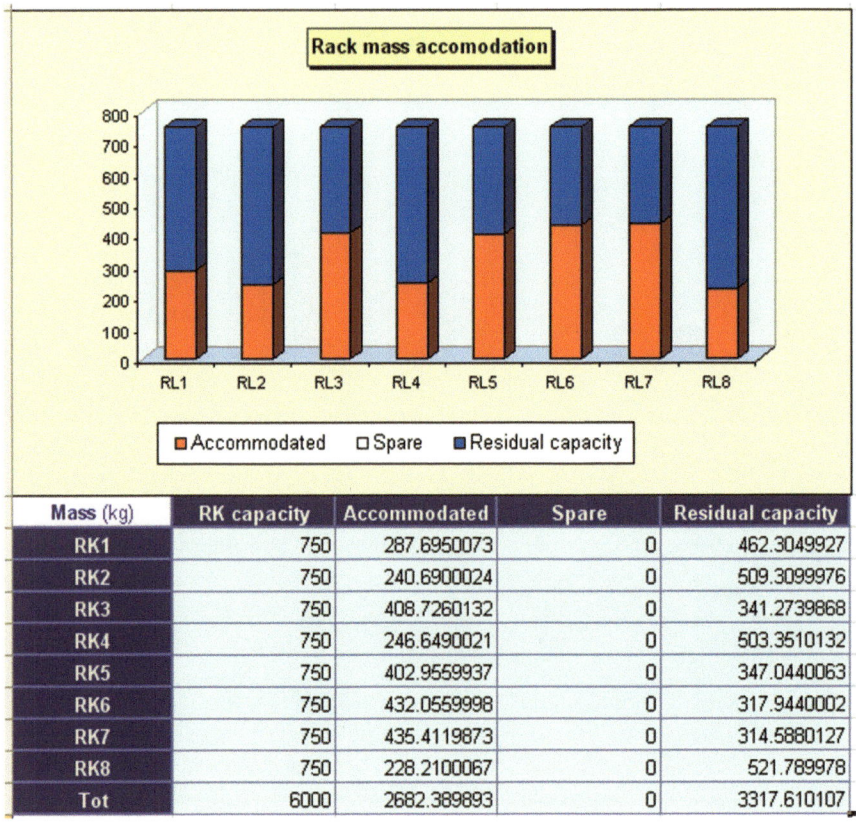

Mass (kg)	RK capacity	Accommodated	Spare	Residual capacity
RK1	750	287.6950073	0	462.3049927
RK2	750	240.6900024	0	509.3099976
RK3	750	408.7260132	0	341.2739868
RK4	750	246.6490021	0	503.3510132
RK5	750	402.9559937	0	347.0440063
RK6	750	432.0559998	0	317.9440002
RK7	750	435.4119873	0	314.5880127
RK8	750	228.2100067	0	521.789978
Tot	6000	2682.389893	0	3317.610107

Fig. 5.17 *Pressurized* cargo mass and volume distribution

5.5 Extensions

A number of further potential ATV-derived developments have recently been put forward. For instance, an appropriate adaptation of the spacecraft concept to serve as the service module of the NASA Orion spacecraft has been proposed (see [1]). Other suggestions have included the CArgo Return Version (CARV), conceived to provide ESA with the capability to transport payloads and cargo from the ISS to Earth, as well the realization of a Crew Transport Vehicle (CTV), in addition to ATV-like systems, in the context of the commercial orbital transportation services. Proper extensions/adaptations of the present version of CAST are hence foreseen. A specific application of the tool (see [26]) is, indeed, already in use, in support of the Columbus Laboratory logistic utilization (see [27]). Insights on this subject are given hereinafter.

The Columbus module, initially intended as an attached laboratory, with the exclusive scope of supporting the ISS experimental activity (fluid physics, new

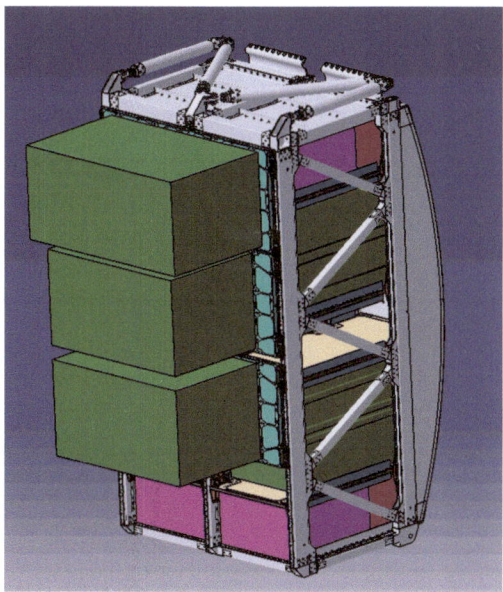

Fig. 5.18 Rack with internal and external cargo

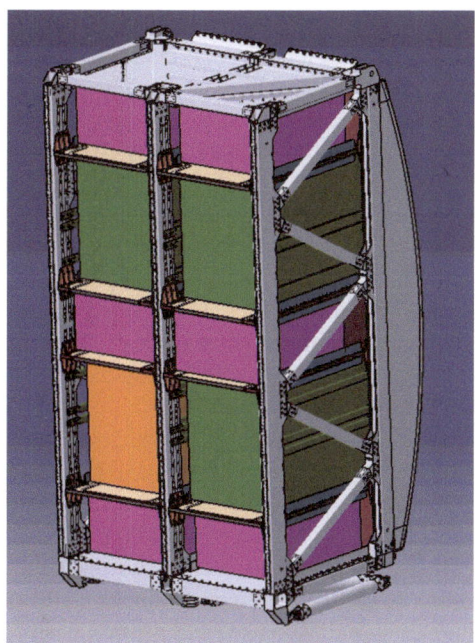

Fig. 5.19 Rack with internal cargo only

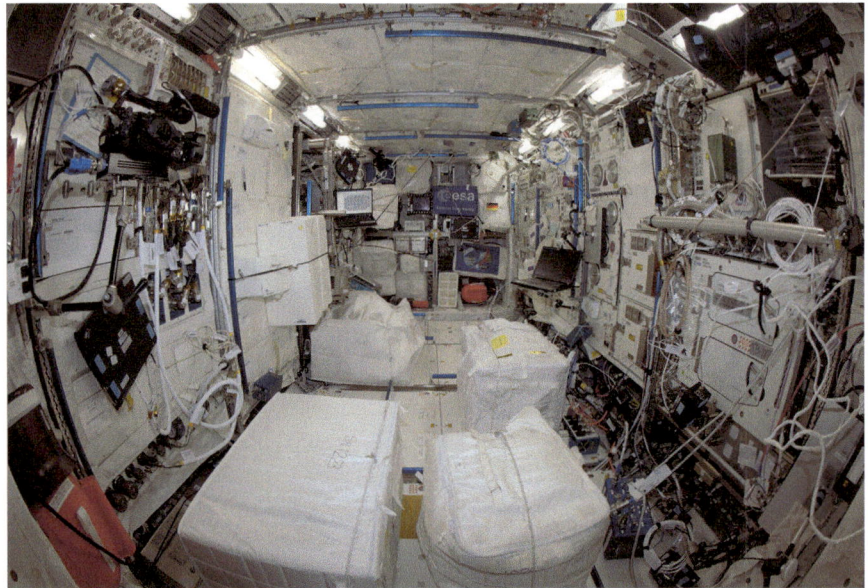

Fig. 5.20 The Columbus Laboratory. Photo: ESA

materials, life science and earth observation), is currently also utilized as a stowage facility. The related accommodation issue is deemed as very demanding, as is, the more general one, concerning the whole ISS.

As is well known, this overall task creates a very strong impact on habitability and crew productivity on board. As a consequence, when exploiting the volume left available by the downloaded cargo, safety, ergonomic needs and operational feasibility must be considered. A number of mandatory stowage constraints, applicable to the whole space station's framework have therefore been stated. It is clear that the stored material is not allowed, for instance, to inhibit emergency interventions, to interfere with the equipment designated for critical safety operations nor to reduce the usability of devices. All these kinds of stumbling blocks, directly or indirectly, give rise to non-trivial packing issues.

As in the ATV case, Columbus has racks available (see Fig. 5.20) that are partially exploitable for the above-mentioned stowage purposes. They are likewise provided with internal sectors of different types and corresponding mass capacities. There are two accommodation levels, i.e.: items into bags/sectors and bags into sectors. Items/bags are placed into bags/sectors, on the basis of their partial or total availability. In order to exploit the empty volumes inside the module as much as possible, also non-rack-based accommodations are considered. An interesting instance deals with the Columbus starboard end cone, see Fig. 5.21. Items may be placed inside, taking advantage of the volumes left empty. The presence of structural elements and the resulting clearance zones have to be taken into account. Figure 5.21

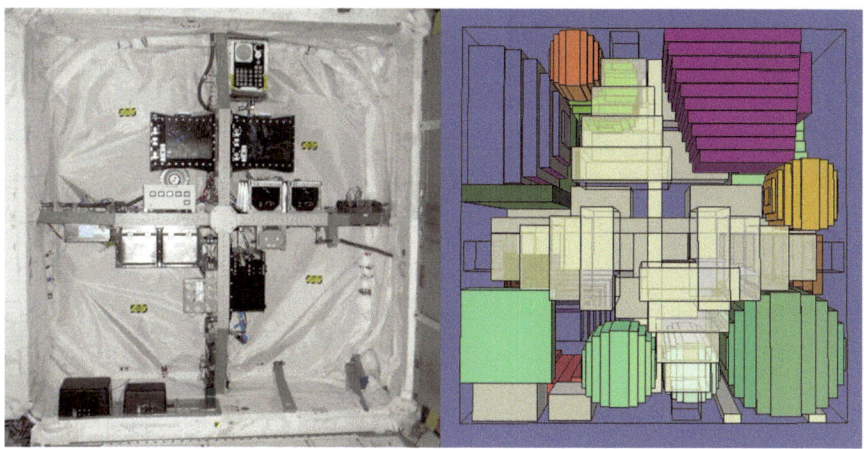

Fig. 5.21 Columbus's starboard end cone utilization

depicts the exploitable room on the left. On the right, items are placed in compliance with the forbidden spaces (represented in the figure by transparent boxes).

In the general Columbus stowage problem, a first optimization objective is evidently that of fulfilling, as much as possible, the current Cargo Manifest request, minimizing, also in this case, the number of the rejected items (on the basis of their priority).

Mandatory requirements, usually deriving from ergonomic and operational needs, quite often impose the grouping of some items inside the same sector/rack, or, on the contrary, define incompatibility conditions. A number of forbidden positions, moreover, can sometimes make the accommodation task quite problematic.

The concern of sparing the crew workload, as much as possible, induces, as a first attempt, to keep the items already uploaded in their acquired positions, within their assigned racks. If strictly necessary, solely in order to prevent the rejection of some items, they can, nonetheless, be re-allocated differently.

Once an optimal solution has been attained (minimizing the amount of cargo unloaded), a post-analysis is performed in order to facilitate the subsequent accommodation task (relevant to the forthcoming Cargo Manifest). Some items are re-allocated, when opportune, to reduce the number of sectors only partially utilized, as much as possible. Afterwards, if residual volumes are still available inside some of these, then the cargo engineer is asked to advise the Cargo Manifest team how these empty spaces could be suitably exploited in the next resupply steps. In such a case, a further optimization process is carried out to identify a number of hole-filling *virtual* items (see [25]).

A case study, aimed at re-allocating items inside the same rack, following the rationale outlined above, is shown by Fig. 5.22 that compares a handmade solution (left) with the optimized one (right). Figure 5.23 displays the concept of *virtual* items (shown externally).

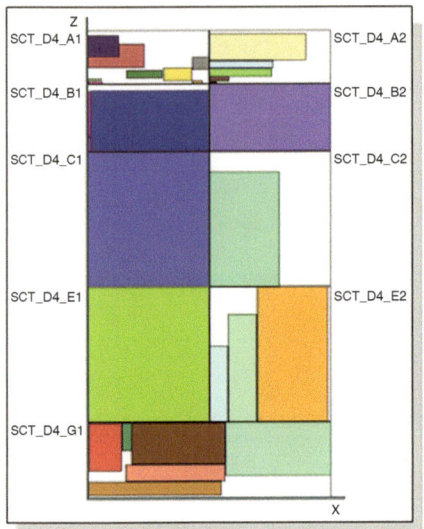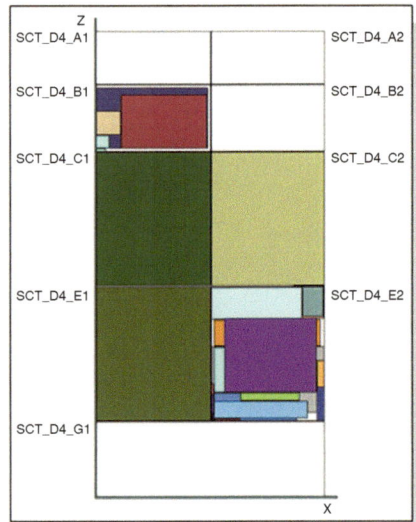

Fig. 5.22 A case study

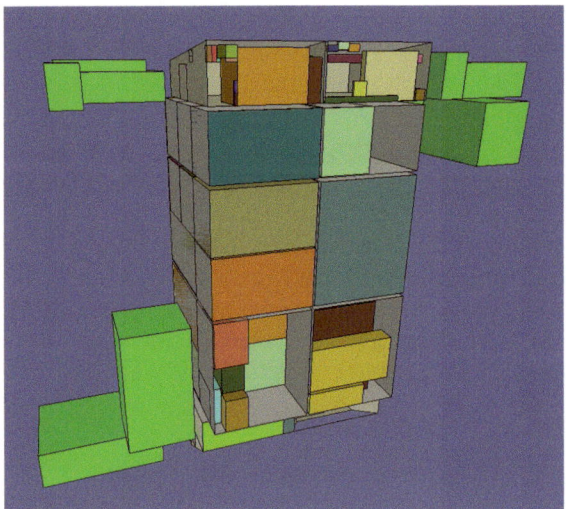

Fig. 5.23 Identification of *virtual* items

On the basis of the expertise gained so far, satisfactory accommodations are expected in less than 1 h in all, opposed to the 2/3 day elapse time needed by following a manual approach. This dramatically reduces the cargo engineer effort, while also significantly increasing the average filling coefficient of the utilized sectors (with a typical increment of more than 20 %).

The overall methodology referred to in this chapter is quite clearly subject to possible extensions to closely related issues of interest, in addition to the already mentioned ATV-like context. Relevant insights are briefly mentioned here, in order to suggest prospective development directions.

Inflatable systems will present an innovative line of development, posing specific cargo accommodation issues. Additionally, a leading role, also for the subject in question, will certainly be taken by the future manned space missions, such as the ones concerning the lunar bases, for which the optimization of the inhabited volumes will represent a major objective.

Further interesting applications could deal with the so-called payload accommodation issue, inside space modules, usually implying both scheduling and packing aspects. A payload consists of a set of facilities, with specific resource requirements (generally related to crew time, electrical power and water/air cooling). Payloads usually have to be accommodated into predefined positions, provided with different resource availability. This has to be done as efficiently as possible, in order to accomplish the experimental tasks requested. As a consequence, the payload assignment to the available locations gives rise to a non-trivial optimization problem. A major objective, in this context, consists of finding satisfactory time-dependent solutions, i.e. feasible accommodations, in compliance with the payload operational constraints and the overall availability of resources.

An important topic, seemingly barely correlated to the cargo accommodation one, concerns the on-orbit unloading of modules/vehicles. This problem, for instance, arose in the ATV mission scenario, during its attachment phase, before the destructive re-entry. Since (for emergency reasons) the vehicle had to be able to depart from the station at any moment, it was permanently requested to be compliant with all the given balancing requirements. Whilst the cargo accommodation analysis certainly represents a demanding task, the issues related to the on-orbit unloading are, altogether, not any easier. In this context, indeed, accurate operational paradigms are mandatory. These can be achieved by proper cargo-removal procedures, implying, if necessary, temporary exchanges/repositioning of items, so that, at each step, no constraint/bound violations can occur.

In addition to what has been briefly mentioned hitherto, some considerations can be made, focusing on aspects still related to the cargo accommodation overall task, albeit addressing different points of view. Indeed, when dealing with the early phases of the entire cargo accommodation process, or even during the design of the spacecraft itself (where sizing investigations of the system are needed), a significant level of uncertainty, concerning the prospective load characteristics, has to be expected. On the basis of the information, available time after time, a dedicated worst-case analysis could be carried out by readapting the overall optimization methodology presented here. To this purpose, proper (opposite-oriented) objective functions, aimed at looking for the worst solutions, e.g. accommodations attaining the maximum overall center of mass distance from the desired position, should be introduced.

5.6 Conclusions

This chapter has illustrated a recurrent packing task taking place in space engineering and logistics. It concerns the so-called cargo accommodation issue, well known for being, in general, very challenging. A significant number of operational conditions, in addition to *static* and *dynamic* balancing, are usually posed, in the presence of complex geometries relating both to the items that are to be loaded and the exploitable volumes.

The thought-provoking ATV case that occurred in support of the ISS logistics, encompassing all its five successful missions, has been considered thoroughly here, in order to show the complexity of this arduous real-world application. An ad hoc methodology has been thought up with the scope of looking into the relevant cargo accommodation problem satisfactorily, bearing in mind both solution quality and cost-effectiveness.

CAST is the in-house cargo accommodation support environment, designed and developed by the company to serve the scope adequately. Its overall features have been outlined in a dedicated section, enucleating the underlying modeling-based heuristic approach adopted. Although no longer employed for the ATV programs, this tool, as per a tailored version, at present addresses the non-less-demanding issue of the stowage on board the Columbus Laboratory (permanently attached to the ISS). Insights on this application have been provided as well, pointing out possible extensions to a number of forthcoming challenges in space.

The overall cargo accommodation approach followed to date has proved to be up to tackling the intricate scenarios arising in the related contexts appropriately. CAST is paving the way for a number of intriguing applications in space engineering and not only.

Acknowledgements Thanks are due to Janos D. Pintér for his in-depth review of the manuscript. We wish to thank Jane Evans for her accurate review of the text.

References

1. National Aeronautics and Space Administration (NASA): International Space Station (ISS). www.nasa.gov/mission_pages/station
2. European Space Agency (ESA): Automated Transfer Vehicle (ATV). www.esa.int/Our_Activities/Human_Spaceflight/ATV
3. European Space Agency (ESA). www.esa.int
4. Thales Alenia Space. www.thalesgroup.com/Markets/Space/Home
5. Canadian Space Agency (CSA). www.asc-csa.gc.ca/eng
6. Italian Space Agency (ASI). www.asc-csa.gc.ca/eng
7. Japan Aerospace Exploration Agency (JAXA). www.jaxa.jp
8. Russian Federal Space Agency (ROSCOSMOS). www.roscosmos.ru
9. National Aeronautics and Space Administration (NASA). www.nasa.gov
10. Fasano, G., Gastaldi, C., Piras, A., Saia, D.: An optimization framework to tackle challenging cargo accommodation tasks in space engineering. Adv. Aircr. Spacecr. Sci. **1**(2), 197–218

(2014)
11. Bortfeldt, A., Wäscher, G.: Container Loading Problems – A State-of-the-Art Review. No 120007, FEMM Working Papers, Otto-von-Guericke University Magdeburg, Faculty of Economics and Management (2012)
12. Cagan, J., Shimada, K., Yin, S.: A survey of computational approaches to three-dimensional layout problems. Comput.-Added Des. **34**, 597–611 (2002)
13. Coffman Jr., E.G., Csirik, J., Galambos, G., Martello, S., Vigo, D.: Bin packing approximation algorithms: survey and classification. In: Pardalos, P.M., Du, D.Z., Graham, R.L. (eds.) Handbook of Combinatorial Optimization, pp. 455–531. Springer, New York (2013)
14. Pisinger, D.: Heuristics for the container loading problem. Eur. J. Oper. Res. **141**(2), 382–392 (2002)
15. Egeblad, J.: Placement of two- and three-dimensional irregular shapes for inertia moment and balance. In: Morabito, R., Arenales, M.N., Yanasse, H.H. (eds.) International Transactions in Operational Research; Special issue on Cutting Packing and Related Problems, vol. 16, no. 6, pp. 789–807. International Federation of Operational Research Societies (2009)
16. Stoyan, Y.G., Romanova, T.: Mathematical models of placement optimisation: two-and three-dimensional problems and applications. In: Fasano, G., Pintér, J.D. (eds.) Modeling and Optimization in Space Engineering, pp. 363–388. Springer Science + Business Media, New York (2013)
17. Fadel, G.M., Wiecek, M.M. In: Fasano, G., Pintér, J.D. (eds.) Packing Optimization of Free-Form Objects in Engineering Design. Springer Science + Business Media, New York (2015)
18. Daughtrey, R.S.: A Simulated Annealing Approach to 3-D Packing with Multiple Constraints. Cosmic Program MFS28700, Boeing Huntsville AI Center Huntsville, Alabama (1991)
19. Takadama, K., Tokunaga, F., Shimohara, K.: Capabilities of a multiagent-based cargo layout system for H-II transfer vehicle. In: 16th IFAC Symposium on Automatic Control in Aerospace (ACA'04), St. Petersburg, June 14–18 (2004)
20. INFORMS Computing Society: Mathematical Programming Glossary. www.glossary.computing.society.informs.org
21. Glover, F., Kochenberger, G.A.: Handbook of metaheuristics 57. International Series in Operations Research & Management Science, Springer, New York (2003)
22. Ibaraki, T., Imahori, S., Yagiura, M.: Hybrid metaheuristics for packing problems. In: Blum, C., Aguilera, M.J., Roli, A., Sampels, M. (eds.) Hybrid Metaheuristics: An Emerging Approach to Optimization. Studies in Computational Intelligence (SCI), vol. 114, pp. 185–219. Springer, Berlin (2008)
23. Fasano, G.: A multi-level MIP-based heuristic approach for the cargo accommodation of a space vehicle. In: 6th ESICUP Meeting, Valencia, March 25–29 (2009)
24. Jünger, M., Liebling, T.M., Naddef, D., Nemhauser, G.L., Pulleyblank, W.R., Reinelt, G., Rinaldi, G., Wolsey, L.A. (eds.): 50 Years of Integer Programming 1958–2008: From the Early Years to the State-of-the-Art. Springer, Berlin (2010)
25. Fasano, G.: Solving Non-Standard Packing Problems by Global Optimization and Heuristics. SpringerBriefs in Optimization, Springer Science + Business Media, New York (2014)
26. Fasano, G., Saia, D., Piras, A.: Columbus stowage optimization by CAST (Cargo Accommodation Support Tool). Acta Astronaut. **67**(3–4), 489–495 (2010)
27. European Space Agency (ESA): Columbus Laboratory. www.esa.int/Our_Activities/Human_Spaceflight/Columbus

Chapter 6
Cutting and Packing Problems with Placement Constraints

Andreas Fischer and Guntram Scheithauer

Abstract In real-life problems of cutting and packing very often placement constraints are present. For instance, defective regions of the raw material (wooden boards, steel plates, etc.) shall not become part of the desired products. More generally, due to different quality demands, some products may contain parts of lower quality which are not allowed for other goods. Within this work we consider one- and two-dimensional rectangular cutting and packing problems where items of given types have to be cut from (or packed on) raw material such that an objective function attains its maximum. In the one-dimensional (1D) case, we assume for each item type that allocation intervals (regions of the raw material) are given so that any item of the same type must be completely contained in one of the corresponding allocation intervals. In addition, we deal with problems where the lengths of the 1D items of a given type may vary within known tolerances. In the two-dimensional (2D) case, where rectangular items of different types have to be cut from a large rectangle, we investigate guillotine cutting under the condition that defective rectangular regions are not allowed to be part of the manufactured products (even not partially). For these scenarios we present solution strategies which rely on the branch and bound principle or on dynamic programming. Based on properties of the corresponding objective functions we discuss possibilities to reduce the computational complexity. This includes the definition of appropriate sets of potential allocation (cut) points which have to be inspected to obtain an optimal solution. By dominance considerations the set of allocation points is kept small. In particular, the computational complexity becomes independent of the unit of measure of the input data. Possible generalizations will be discussed as well.

Keywords Cutting and packing • Placement constraints • Quality demands • Defective regions • Set of allocation points

A. Fischer (✉) • G. Scheithauer
Institute of Numerical Mathematics, Technische Universität Dresden, 01062 Dresden, Germany
e-mail: Andreas.Fischer@tu-dresden.de; Guntram.Scheithauer@tu-dresden.de

© Springer International Publishing Switzerland 2015
G. Fasano, J.D. Pintér (eds.), *Optimized Packings with Applications*, Springer
Optimization and Its Applications 105, DOI 10.1007/978-3-319-18899-7_6

6.1   Introduction

In real-life problems of cutting and packing placement constraints are present very often. For instance, defective regions of the raw material (wooden boards, steel plates, etc.) shall not become part of desired products. More generally, due to different quality demands, some products may contain parts of lower quality which are not allowed for other goods. For packing problems forbidden regions may exist where no objects must be placed. Due to the strong relationship between cutting and packing problems this paper mostly concentrates on cutting problems.

6.1.1   Aims and Scope

We consider one- and two-dimensional rectangular cutting problems where items of given types have to be cut from raw material such that an objective function attains its maximum. Such problems are also called 1D or 2D rectangular knapsack problems. Two scenarios will be discussed in detail. In the first, some rectangular parts of the raw material are not allowed to be used at all. In the second more general scenario, different quality demands are considered.

In the 1D case, we assume for each item type that allocation intervals (regions of the raw material) are given so that any item of the same type must be completely contained in one of the corresponding allocation intervals. In the 2D case, where rectangular items of different types have to be cut from a large rectangle, we investigate guillotine cutting under the condition that defective rectangular regions are not allowed to be part of the manufactured products (even not partially). Different qualities of the raw material are described by allocation areas.

Furthermore, we also deal with problems where the lengths (or width) of the (one- or two-dimensional) items of a given type may vary within known tolerances.

For these scenarios we present solution strategies which rely on the branch and bound (B&B) principle or on dynamic programming (DP). Based on properties of the corresponding objective functions we discuss possibilities to reduce the computational complexity. This includes the definition of appropriate sets of potential allocation (cut) points which have to be inspected to obtain an optimal solution. By dominance considerations the set of allocation points is kept small. In particular, the computational complexity becomes independent of the unit of measure of the input data.

Some generalizations will be discussed as well. Finally, we hope that techniques from this area can be used and extended to new fields, for example for the placement of chips and other electronic parts on boards.

6.1.2 Related Work

Cutting and packing (C&P) problems with defective or forbidden regions were studied in the past. In the earlier survey paper by Sweeney and Paternoster [15] some work related to this topic is referenced, whereas in the recent typology of C&P [17] the topic is only briefly addressed. Therefore, we give a short overview on articles that are relevant for our work.

Hahn [6] presented a recursive DP-based procedure to solve a 2D cutting problem with defects. She suggested a three-stage guillotine cutting scenario with vertical cuts in the first stage.

Herz [7] presented a recursive B&B-based procedure for the 2D rectangular knapsack problem (without defects) to obtain *canonical* patterns by introducing *discretization points*, see the definition of allocation points in Sect. 6.2. Dowsland [4] used certain discretization points to analyze the structure of optimal (and nearly optimal) solutions and the objective function for the manufacturer's pallet loading problem, a special 2D knapsack problem where only one type of pieces (rotatable by 90°) has to be packed.

Beasley [2] presented a 0/1 model and a tree-search procedure for 2D non-guillotine rectangle packing including the occurrence of forbidden regions. Upper bounds are computed from a Lagrangian relaxation problem which are improved by the help of a subgradient ascent method. We will not use the approach in [2] since it requires a very large number of 0/1-variables.

In Terno et al. [16] a principle used by Nicholson [8] was applied to 2D rectangle cutting and packing problems leading to the concept of reduced sets of allocation (or cut) points (cf. Sect. 6.2). The book (Scheithauer [12]) presents a renewed description of this concept.

For the three-stage guillotine cutting of defective boards a recursive procedure was developed by Scheithauer and Terno [13]. In particular, appropriately reduced sets of allocation points were applied.

Algorithmic approaches for 1D cutting problems with different quality demands were addressed by Sweeney and Haessler [14]. Such problems which are modeled by allocation intervals and pieces of variable length were also investigated in Scheithauer [11]. The latter paper generalizes real-world problems in hardwood cutting. Similar problems are considered in Rönnqvist [9] and Rönnqvist and Åstrand [10], where a discretization of the board is used.

6.1.3 General Notation and Assumptions

We assume throughout the paper that all input data are positive integers. The set of positive integers is denoted by $\mathbb{Z}_>$. In the 1D case, the length of the raw material is given by L. The pieces $i \in I := \{1, \ldots, m\}$ which shall be cut have the lengths ℓ_i.

Moreover, profit coefficients γ_i for $i \in I$ are known. Additionally, in the 2D case, the raw material has width W, whereas the rectangular pieces are of widths w_i. It is always assumed that

$$\max\{\ell_i \mid i \in I\} \leq L \quad \text{and} \quad \max\{w_i \mid i \in I\} \leq W$$

holds. For later use we define

$$\ell_{min} := \min\{\ell_i \mid i \in I\}, \quad w_{min} := \min\{w_i \mid i \in I\}.$$

and

$$\ell := (\ell_1, \ldots, \ell_m)^\top, \quad w := (w_1, \ldots, w_m)^\top, \quad \gamma := (\gamma_1, \ldots, \gamma_m)^\top.$$

Moreover, in case that a maximum of indexed numbers is taken over an empty index set the maximum is set to 0. In order to describe (parts of) objects we use

$$[a, b] \times [c, d] := \{(x, y) \in \mathbb{R}^2 \mid a \leq x \leq b, \, c \leq y \leq d\}$$

and, for short, $b \times d := [0, b] \times [0, d]$, where $a, b, c, d \in \mathbb{Z}_+$ with $a \leq b$ and $c \leq d$. By $\mathbb{Z}_+$ the set of all non-negative integers is denoted. If not stated otherwise, we allow that several copies of a piece can be obtained from the raw material. For the sake of simplicity, we do not consider a (positive) kerf nor least distances between two allocation points within a pattern. Furthermore, in the 2D case we do not allow rotation of pieces for the same reason.

6.2 Reduced Set of Potential Allocation Points

The (standard) *Knapsack Problem* (KP) is a basic problem also within the field of cutting and packing. This problem consists of finding a vector $x^* \in \mathbb{Z}_+^m$ such that x^* satisfies the capacity constraint $a^\top x \leq b$ and the objective $c^\top x$ attains its maximum for $x = x^*$. For short, we write

$$\text{KP}(c, a, b): \quad \sum_{i \in I} c_i x_i \to \max \quad \text{subject to} \quad \sum_{i \in I} a_i x_i \leq b, \, x_i \in \mathbb{Z}_+ \text{ for } i \in I, \tag{6.1}$$

where $a = (a_1, \ldots, a_m)^\top \in \mathbb{Z}_>^m$, $c = (c_1, \ldots, c_m)^\top \in \mathbb{Z}_>^m$, and $b \in \mathbb{Z}_>$ are given.

If we consider $a^\top x \leq y$ with a parameter $y \in \mathbb{R}$, then we obtain the following optimal value function $f : \mathbb{R} \to \mathbb{Z}_+ \cup \{-\infty\}$ related to KP(c, a, b):

$$f(y) := \max \left\{ \sum_{i \in I} c_i x_i \mid \sum_{i \in I} a_i x_i \leq y, \, x_i \in \mathbb{Z}_+ \text{ for } i \in I \right\} \quad \text{for all } y \in \mathbb{R}, \tag{6.2}$$

where $f(y) := -\infty$ for any $y < 0$. It is well known that f is piecewise constant and non-decreasing. Its jump discontinuities are non-negative integers. The set of all these jump discontinuities of f depends on c and a and is a subset of

$$S(a) := \left\{ r = \sum_{i \in I} a_i x_i \mid x_i \in \mathbb{Z}_+, \, i \in I \right\}. \tag{6.3}$$

The fact that the optimal value function of $KP(c, a, b)$ changes (increases) only at discrete points can be used in B&B or DP approaches to reduce the computational complexity of solving the knapsack problem.

If the knapsack problem (6.1) is used to model a cutting problem we call $S(a)$ *set of potential allocation points*. Replacing c by γ, a by ℓ, and b by L we see that $KP(\gamma, \ell, L)$ models a 1D cutting problem. In this case, x_i denotes the number how often piece i is cut. The set $S(\ell)$ contains infinitely many elements whereas the points (coordinates) for cutting the raw material are bounded by L. Therefore, we introduce the finite set

$$S(\ell, L) := \{ r \in S(\ell) \mid r \leq L \}, \tag{6.4}$$

Of course, $S(\ell, L)$ contains all those jump discontinuities of the optimal value function arising from $KP(\gamma, \ell, L)$ which are not larger than L. Depending on γ, additional points may belong to $S(\ell, L)$ as well. Thus, the question arises whether one can describe the set of jump discontinuities exactly. This is possible in the important case when $\gamma = \ell$. Then, the knapsack problem $KP(\ell, \ell, L)$ has the optimal value function f given by

$$f(y) = \max \left\{ \sum_{i \in I} \ell_i x_i \mid \sum_{i \in I} \ell_i x_i \leq y, \, x_i \in \mathbb{Z}_+, \, i \in I \right\} \quad \text{for all } y \in \mathbb{R}$$

and $S(\ell, L)$ is exactly the set of those jump discontinuities of f which are not larger than L.

Let x^* denote a solution of $KP(\gamma, \ell, L)$ with $\ell^\top x^* < L$, i.e., there is some waste of raw material. Then, to cut the items according to x^*, infinitely many possibilities exist to choose a pattern, i.e. the coordinates of the items of the solution. If the items are placed as left as possible on the raw material, the number of such patterns is finite, all the waste lies right of the items, and all coordinates of the cut positions belong to $S(\ell)$. Such patterns are often called *normalized* or *left-justified*. Herz [7] used the terms *discretization point* for $r \in S(\ell)$ and *canonical* for left-justified patterns.

For a set $T \subset \mathbb{Z}_+$ and $y \in \mathbb{R}_+$ let $p_T(y)$ and $s_T(y)$ denote the *predecessor of y with respect to T* and the *successor of y with respect to T*, respectively, i.e.,

$$p_T(y) := \max\{ r \in T \mid r \leq y \} \quad \text{and} \quad s_T(y) := \min\{ r \in T \mid r \geq y \}$$

for all $y \in [\min\{r \in T\}, \max\{r \in T\}]$. In terms of the 1D cutting problem, $p_{S(\ell)}(y)$ denotes the largest allocation point less than or equal to y, and $s_{S(\ell)}(y)$ denotes the least length of raw material needed to obtain a length of y. With other words, $p_{S(\ell)}(y)$ is the maximum usable length when the raw material has length y. Obviously, we have

$$y - \ell_{min} < p_{S(\ell)}(y) \le y \quad \text{for all } y \ge \ell_{min}$$

and

$$p_{S(\ell)}(y) + p_{S(\ell)}(L - y) \le p_{S(\ell)}(L) \quad \text{for all } y \in [0, L].$$

The knapsack problem $KP(c, a, b)$ can be solved by means of the following backward dynamic programming (BDP) algorithm. If set T used in this algorithm contains at least all jump discontinuities of the optimal value function of $KP(c, a, b)$, then Algorithm BDP provides a function $g : T \to \mathbb{Z}_+$ by which a solution of $KP(c, a, b)$ can be easily determined. For example, $T := S(a, b)$ would do the job. Later on, it will turn out that Algorithm BDP can even successfully be used for solving knapsack problems if T contains only a certain subset of jump discontinuities.

Algorithm BDP

Input: c, a, b, T; Output: g

(1) Set $g(0) := 0$, $y := 0$.

(2) **While** $y < p_T(b)$ **do**

(3) $y := s_T(y + 1)$,

(4) $g(y) := \max_{i \in I}\{c_i + g(p_T(y - a_i)) \mid y \ge a_i\}$.

Theorem 1. *Let T contain at least all jump discontinuities of the optimal value function f of $KP(c, a, b)$. Then, if Algorithm BDP is used for determining $g : T \to \mathbb{Z}_+$, it holds*

$$f(y) = g(p_T(y)) \quad \text{for all } y \in [0, b].$$

This well-known result can be also obtained for the following forward dynamic programming (FDP) algorithm.

Algorithm FDP

Input: c, a, b, T; Output: g

(1) Set $g(0) := 0$, $y := 0$.

(2) **While** $y \le p_T(b - \min\{a_i \mid i \in I\})$ **do**

(3)   **For all** $i \in I$ with $y + a_i \le p_T(b)$ **do**

(4) $g(s_I(y + a_i)) := \max\{g(s_T(y + a_i)), c_i + g(y)\}$,

(5) $\bar{y} := y$,

(6)   **Repeat** $y := s_T(y + 1)$ **until** $g(\bar{y}) < g(y)$.

Note that $T = S(a, b)$ implies $s_T(y + a_i) = y + a_i \in T$ for all $i \in I$ and all $y \in T$ with $y + a_i \leq b$. Thus, $s_T(y + a_i)$ can be replaced by $y + a_i$ in Step (4) of Algorithm FDP. Moreover, because of the repeat-loop in Step (6), some updates in Steps (3) and (4) can probably be saved compared to Step (4) of Algorithm BDP.

The worst-case complexity of both algorithms, BDP and FDP, is $O(b + m|T|)$ since y is increased at most b times, and at most m comparisons are done in the max-terms for each element of T. Thus, both are pseudo-polynomial algorithms.

In order to determine a reduced set of allocation points that is sufficient to obtain an optimal solution of $KP(c, a, b)$ by Algorithm BDP or FDP we will apply some dominance condition (cf. [12, 16]). For that purpose we let $b > \max_{i \in I} a_i$ be satisfied. In view of the *separability* of the optimal value function f we have

$$f(b) = \max_{0 < y \leq b/2} \{f(y) + f(b - y)\}.$$

Since f is piecewise constant with jump discontinuities in $S(a)$ it further follows that

$$\begin{aligned} f(b) &= \max_{0 < y \leq b/2} \{f(p_{S(a)}(y)) + f(p_{S(a)}(b - y))\} \\ &= \max_{r \in S(a), 0 < r \leq b/2} \{f(r) + f(p_{S(a)}(b - r))\}. \end{aligned} \tag{6.5}$$

By $p_{S(a)}(r) \leq p_{S(a)}(b - p_{S(a)}(b - r))$, we obtain

$$f(b) = \max_{r \in S(a), 0 < r \leq b/2} \{f(p_{S(a)}(b - p_{S(a)}(b - r))) + f(p_{S(a)}(b - r))\}.$$

This formula motivates the definition of the *reduced set of potential allocation points* by

$$S^{red}(a, b) := \{p_{S(a)}(b - r) \mid r \in S(a, b)\}.$$

Consequently, we have

$$f(b) = \max_{r \in T, 0 < r \leq b/2} \{f(r) + f(p_T(b - r))\} \quad \text{with} \quad T := S^{red}(a, b). \tag{6.6}$$

Theorem 2. *Let f denote the optimal value function of $KP(c, a, b)$. If Algorithm BDP (or Algorithm FDP) with $T = S^{red}(a, b)$ is used to determine $g : T \to \mathbb{Z}_+$, then*

$$f(r) = g(r) \quad \text{for all } r \in T \qquad \text{and} \qquad f(y) \geq g(p_T(y)) \quad \text{for all } y \in [0, b]$$

holds.

Due to $S^{red}(a, b) \subseteq S(a, b)$, the recursion (6.6) might be less expensive than the one in (6.5). Moreover, dependent on the instance, significant savings are possible if Algorithms BDP or FDP are applied for the solution of $KP(c, a, b)$ with

Table 6.1 Potential allocation points for Example 1

y	0	1	2	3	4	5	6	7	8	9	10	11	12	13	14	15
$S(a,b)$	★				★			★	★	★		★	★	★	★	★
$S^{red}(a,b)$	★				★			★	★			★				★
$f(y)$	0				5			10		12		15		17	20	

$T = S^{red}(a, b)$. Note that using $S^{red}(a, b)$ instead of $S(a, b)$ is an application of the Nicholson-principle [8]. Since $p_{S(a)}(y) \geq p_{S^{red}(a,b)}(y)$ and $s_{S(a)}(y) \leq s_{S^{red}(a,b)}(y)$ for all $y \in [0, b]$, the application of Algorithm FDP with $T = S^{red}(a, b)$ does, in general, not any longer provide left-justified patterns.

Example 1. Let us consider the instance of the knapsack problem $KP(c, a, b)$ with $c := (12, 10, 5)^\top$, $a := (9, 7, 4)^\top$, and $b := 15$. In Table 6.1, the elements of the sets $S(a, b)$ and $S^{red}(a, b)$ are marked by ★. Additionally, the optimal value function f is tabulated at their jump discontinuities.　　　　　　　　　　　　　　□

In Example 1, we have $|S^{red}(a, b)| < |S(a, b)| < b$. Moreover, we see that in this example the set of jump discontinuities is not a subset of $S^{red}(a, b)$. In general, the cardinality of $S(a, b)$ and $S^{red}(a, b)$ strongly depends on the input data. Nevertheless, there is a high potential to save memory and computation time by using $S^{red}(a, b)$ instead of $S(a, b)$. Moreover, the cardinality does not change if the unit of measure is changed, for instance from cm to mm.

Investigations how to compute $S(a, b)$ efficiently can be found in [3]. The computational amount for determining $S(a, b)$ and $S^{red}(a, b)$ is bounded from above by $O(mb)$. More precisely, it is bounded by $O(b + m|S(a, b)|)$ due to the application of Algorithm FDP for $KP(a, a, b)$.

6.3　The 1D Cutting Problem with Fix-Lengths

In the 1D case the presence of defective parts which are not allowed for any piece leads to independent smaller problem instances. Therefore, we consider only the scenario with different quality demands.

6.3.1　Problem Description

The following 1D cutting problem is considered: Pieces of various lengths ℓ_i, $i \in I$, and different quality demands $q \in Q$ have to be cut from a non-homogeneous raw material of length L in such a way that all allocation conditions (i.e., quality demands) are met and the total value of obtained pieces is maximal.

Such problems arise, for instance, in timber cutting. In that case, the length of the pieces is, in general, assumed to be variable within a given range but in this section we restrict the pieces to have fix-lengths, which is also of high interest. The more general case of pieces with variable lengths will be considered in the next section.

In order to formulate the cutting problem precisely, we consider several quality types $q \in Q$. To each quality type $q \in Q$ there is at least one piece $i \in I$ which is of this type. Conversely, each piece $i \in I$ is assigned to a quality type $q(i) \in Q$. Let $I_q \subset I$ denote the set of all pieces of quality type q, i.e., $I_q = \{i \in I \mid q(i) = q\}$. Hence, we have $\cup_{q \in Q} I_q = I$ and $I_q \cap I_p = \emptyset$ for $p, q \in Q$ with $q \neq p$.

Intervals of the raw material where exactly one quality demand is fulfilled will be called *allocation intervals* $A_k \subset [0, L]$ with $k \in K := \{1, \ldots, |K|\}$. These intervals are considered as given. The quality demand satisfied in A_k is denoted by $\widetilde{q}(k) \in Q$. Any allocation interval A_k can be described by

$$A_k := [b_k, e_k] \subseteq [0, L] \quad \text{with} \quad e_k - b_k \geq \min\{\ell_i \mid i \in I_{\widetilde{q}(k)}\}.$$

It is possible that different allocation intervals are of the same quality type. Two intervals A_j and A_k ($j \neq k$) may overlap, even if they fulfill the same quality demand. In the latter case, we assume $A_j \nsubseteq A_k$. Without loss of generality it can be assumed that

$$b_1 \leq b_2 \leq \cdots \leq b_{|K|} \quad \text{and} \quad e_k \leq e_{k+1} \text{ if } b_k = b_{k+1}.$$

For any piece $i \in I$ we require without loss of generality that there is a

$$k \in K_{q(i)} := \{k \in K \mid \widetilde{q}(k) = q(i)\}$$

so that $\ell_i \leq e_k - b_k$. This means, any piece $i \in I$ with *allocation point* $y_i \in [b_k, e_k - \ell_i]$ will be completely contained in an allocation interval with quality type $q(i)$.

For example, in hardwood cutting a quality demand could be that no more than one sound knot per reference length is allowed. Then, the occurrence of two sound knots within the reference length causes two partially overlapping allocation intervals.

The value of a piece $i \in I$ is again denoted by γ_i. In general, pieces of the same quality type may be obtained several times from the raw material, either from the same allocation interval (if it is sufficiently large) or from different allocation intervals of this quality type.

The cutting problems we consider consist of determining a (cutting) pattern that has a maximal total value of the obtained pieces.

A pattern π can be described by a (finite) sequence of triples (i_t, y_t, k_t), $t = 1, \ldots, t_\pi$ with $y_t + \ell_{i_t} \leq y_{t+1}$ for $t = 1, \ldots, t_\pi - 1$ where i_t denotes the index of the t-th placed piece, y_t is the allocation point of piece i_t, and k_t gives the corresponding allocation interval. Hence,

$$0 \le y_1, \; y_1 + \ell_{i_1} \le y_2, \cdots, y_{t_\pi} + \ell_{i_{t_\pi}} \le L,$$

$$b_{k_t} \le y_t, \quad y_t + \ell_{i_t} \le e_{k_t}, \quad q(i_t) = \widetilde{q}(k_t) \qquad \text{for } t = 1, \ldots, t_\pi.$$

Obviously, all allocation points can be considered to be integral.

6.3.2 Modeling

In order to formulate an integer optimization model it is assumed in this subsection that each piece is packed at most once. This can be done without loss of generality defining piece i several times with different indexes. We describe the allocation of a piece $i \in I$ by means of a 0/1-variable z_i as follows:

$$z_i := \begin{cases} 1 & \text{if piece } i \in I \text{ is allocated (should be cut)}, \\ 0 & \text{otherwise.} \end{cases}$$

If the allocation point of piece i is at y_i, the piece covers the interval $T_i(y_i) := [y_i, y_i + \ell_i]$ but only if it has been packed, i.e., if $z_i = 1$. Hence, the allocation problem can be modeled as follows where $\text{int} A$ denotes the interior of set A:

$$\sum_{i \in I} \gamma_i \cdot z_i \to \max \tag{6.7}$$

subject to

$$z_i \in \{0, 1\}, \; y_i \in \mathbb{Z}_+, \quad \text{for all } i \in I, \tag{6.8}$$

$$\text{int } T_i(y_i) \cap \text{int } T_j(y_j) = \emptyset \qquad \text{for all } i, j \in I \text{ with } i < j \text{ and } z_i + z_j = 2, \tag{6.9}$$

$$\begin{array}{c} \text{for each } i \in I \text{ with } z_i = 1 \text{ there are } q \in Q \text{ and } k \in K_q \\ \text{with } i \in I_q \text{ and } T_i(y_i) \subseteq A_k. \end{array} \tag{6.10}$$

Condition (6.9) ensures that the packed pieces do not overlap each other. Condition (6.10) guarantees that the allocation is done within an appropriate allocation interval.

To transform the previous model into an integer linear program (ILP) we define 0/1-variables u_{ij}, $i, j \in I$, $i < j$, and v_{ik}, $i \in I$, $k \in K_{q(i)}$ as follows:

$$u_{ij} := \begin{cases} 0 & \text{if piece } i \text{ is packed left to piece } j, \text{ i.e., } y_i + \ell_i \le y_j, \\ 1 & \text{if piece } i \text{ is packed right to piece } j, \text{ i.e., } y_j + \ell_j \le y_i, \end{cases}$$

and

$$v_{ik} := \begin{cases} 1 & \text{if piece } i \in I \text{ is packed within } A_k, \\ 0 & \text{otherwise.} \end{cases}$$

Conditions (6.9) and (6.10) can now be replaced by

$$y_i + \ell_i \leq y_j + L(2 - z_i - z_j + u_{ij}) \quad \text{for all } i, j \in I \text{ with } i < j,$$
$$y_j + \ell_j \leq y_i + L(3 - z_i - z_j - u_{ij}) \quad \text{for all } i, j \in I \text{ with } i < j, \tag{6.11}$$

$$y_i \geq b_k + L(z_i + v_{ik} - 2) \quad \text{for all } i \in I, \, k \in K_{q(i)},$$
$$y_i + \ell_i \leq e_k + L(2 - z_i - v_{ik}) \quad \text{for all } i \in I, \, k \in K_{q(i)}, \tag{6.12}$$

$$y_i \leq L z_i \quad \text{for all } i \in I, \tag{6.13}$$

$$\sum_{k \in K_{q(i)}} v_{ik} = z_i \quad \text{for all } i \in I, \tag{6.14}$$

$$u_{ij} \in \{0, 1\} \quad \text{for all } i, j \in I \text{ with } i < j,$$
$$v_{ik} \in \{0, 1\} \quad \text{for all } i \in I, k \in K_{q(i)}. \tag{6.15}$$

Conditions (6.11) are redundant if piece i or j is not allocated. Otherwise, if $z_i + z_j = 2$, because of $u_{ij} \in \{0, 1\}$ one of the two conditions in (6.11) is non-trivial. If item i is not packed, or if i is not packed within A_k, then restrictions (6.12) are redundant. If an item is not used, then condition (6.13) ensures that the allocation point of this item is set 0. By (6.14) it is required that a corresponding allocation interval exists if item i is packed.

The number of binary and integer variables can become very large in general. To solve the ILP (6.7), (6.8), (6.11)–(6.15) within a real time application scenario, we will describe B&B or DP approaches. B&B with depth first search (LIFO) has the advantage that good feasible solutions are found quickly so that time termination criteria can be applied. For a DP algorithm the computational amount (run time needed to solve an instance) can be well estimated because of its pseudo-polynomiality.

6.3.3   Sets of Potential Allocation Points

For the allocation problem (6.7)–(6.10) the optimal value function $v : [0, L] \to \mathbb{Z}_+$ is defined by

$$v(y) := \max_{y, z} \left\{ \sum_{i \in I} \gamma_i z_i \mid (6.8)\text{–}(6.10) \text{ hold and } y_i + \ell_i \leq y \text{ for all } i \in I \text{ with } z_i = 1 \right\}$$

for all $y \in [0, L]$. The function v is non-decreasing since the feasible region enlarges if y increases. The optimal value function is piecewise constant since only a finite number of different sequences of packed pieces exists. Moreover, v is continuous from the right.

In case that

$$(0, L) \setminus \bigcup_{k \in K} \operatorname{int} A_k \neq \emptyset$$

the packing problem can be separated into some smaller problems which can be solved independently from each other. The optimal value of the original problem is the sum of the optimal values of the smaller problems. In the following it is always assumed that the packing problem is not separable, i.e.,

$$\bigcup_{k \in K} \operatorname{int} A_k = (0, L). \tag{6.16}$$

Hence, $b_1 = 0$ and $\max\{e_k : k \in K\} = L$.

Let $S^*(\gamma, \ell)$ denote the set of jump discontinuities of v for an instance with input data $L \in \mathbb{Z}_>$, $\ell \in \mathbb{Z}_>^m$, $\gamma \in \mathbb{Z}_>^m$, and $A_k = [b_k, e_k]$ for $k \in K$. Our aim is to find a superset of $S^*(\gamma, \ell)$ which is independent of γ and as small as possible.

Theorem 3. *For any $\ell \in \mathbb{Z}_>^m$ and any $\gamma \in \mathbb{Z}_>^m$, the inclusion*

$$S^*(\gamma, \ell) \subseteq S_{ap}(\ell) := \bigcup_{k \in K} (b_k \oplus S(\ell)) \cap [0, L].$$

is fulfilled where $b_k \oplus S(\ell) := \{y \mid y = b_k + r, \, r \in S(\ell)\}$.

Note that set $S_{ap}(\ell)$ does not only depend on ℓ but also on the given allocation intervals. For simplicity we do not show this dependence in the notation of $S_{ap}(\ell)$ and of other sets that will be defined later.

Proof. To each jump discontinuity of the optimal value function belongs a left-justified pattern. Any left-justified pattern has allocation points only at the beginning of an allocation interval, i.e. at b_k for some $k \in K$, or at points $b_k + r$ with $r \in S(\ell)$. All these points define set $S_{ap}(\ell)$. □

Corollary 1. *For all $r \in [0, L-1] \cap \mathbb{Z}$ and all $y \in (r, s_{S_{ap}(\ell)}(r+1))$ we have $v(y) = v(r) = v(p_{S_{ap}(\ell)}(y))$.*

In general, we can even obtain a set $\widehat{S}_{ap}(\ell)$ that is smaller than $S_{ap}(\ell)$ but still allows to obtain an optimal solution of problem (6.7)–(6.10). To this end, a more sophisticated procedure is used. Its basic principle is the construction of all possible combinations (patterns) in dependence of the quality demands and the corresponding items. For example, if we have an allocation interval A_k with $b_k = 0$ and an item $i \in \overline{I}_{\overline{q}(k)}$ with $\ell_i > e_k$, then item i cannot be placed with allocation point 0. Therefore, it might happen that $\ell_i \in S(\ell)$ is not a jump discontinuity of v. A similar situation arises if, for item i, no allocation interval A_k with $q(i) = \widetilde{q}(k)$ and $b_k = 0$ exists. Then, item i cannot be placed with allocation point 0. Therefore, in general, the use of $S(\ell)$ in the definition of $S_{ap}(\ell)$ leads to a proper superset of the jump discontinuities.

The description of the procedure to obtain a reduced set $\widehat{S}_{ap}(\ell)$ which still contains all jump discontinuities of v requires some notation. For $y \in \{b_k : k \in K\}$, let

$$K_b(y) := \{k \in K \mid b_k = y\}, \quad Q_b(y) := \{q \in Q \mid q = \widetilde{q}(k),\ k \in K_b(y)\},$$

and, for $y \in [0, L)$, let

$$Q(y) := \{q \in Q \mid \exists k \in K_q \text{ with } b_k \leq y,\ y + \min\{\ell_i \mid i \in I_q\} \leq e_k\},$$
$$k(y, q) := \max\{k \in K_q \mid b_k \leq y\} \quad \text{for all } q \in Q(y),$$
$$K(y) := \{k \in K \mid k = k(y, q),\ q \in Q(y)\}$$

be defined. The set $Q(y)$ represents all quality types $q \in Q$ for which a sufficiently large allocation interval A_k with $\widetilde{q}(k) = q$ exists such that a piece $i \in I_q$ with allocation point y can be obtained. If for $q \in Q(y)$ several allocation intervals contain points y and $y + \min\{\ell_i \mid i \in I_q\}$, then we take that with largest b_k and collect them in $K(y)$.

Then the procedure to construct the set $\widehat{S}_{ap}(\ell)$ starts at $\widehat{y} := 0$. Then, we begin to construct $\widehat{y} \oplus S(\ell)$ by successively adding the lengths of those items which can be placed because of an existing allocation interval. At each point b_k, $k \in K$, where an allocation interval begins, the construction of $b_k \oplus S(\ell)$ restricted to feasible placements has to be started.

Initialization The first jump discontinuities can arise when a leftmost piece is allocated at point $\widehat{y} := 0$:

$$\widehat{S} := \{\ell_i \mid i \in I_q,\ \ell_i \leq e_{k(0,q)},\ q \in Q_b(0)\} \cup \{0\}.$$

Since $Q_b(0)$ represents all qualities having an allocation interval beginning at 0, all pieces of I_q, $q \in Q_b(0)$, can be placed at 0 which fit within the corresponding allocation interval, i.e., which are not longer than $e_{k(0,q)}$.

General Step Let

$$\widehat{y}_b := \min\{L;\ b_k \mid b_k > \widehat{y},\ k \in K\}, \qquad \widehat{y}_s := \min\{y \in \widehat{S} \mid y > \widehat{y}\}.$$

Here, $\widehat{y}_b$ denotes the coordinate of the next allocation interval which allows the placement of further items, whereas for $\widehat{y}_s$ there is already a feasible pattern which can possibly be extended.

If $\widehat{y}_b < \widehat{y}_s$, then $\widehat{y} := \widehat{y}_b$ and, because of the new allocation interval, all corresponding pieces are placed:

$$\widehat{S} := \widehat{S} \cup \{\widehat{y}\} \cup \{\widehat{y} + \ell_i \mid i \in I_q,\ \widehat{y} + \ell_i \leq e_{k(\widehat{y},q)},\ q \in Q_b(\widehat{y})\}.$$

Otherwise, if $\widehat{y}_b \geq \widehat{y}_s$, then $\widehat{y} := \widehat{y}_s$ and all pieces belonging to $Q(\widehat{y})$ and of suitable length are placed:

$$\widehat{S} := \widehat{S} \cup \{\widehat{y} + \ell_i \mid i \in I_q, \widehat{y} + \ell_i \leq e_{k(\widehat{y},q)}, q \in Q(\widehat{y})\}.$$

The algorithm terminates if no further piece can be placed, i.e., if

$$\widehat{y} > L - \min\{\ell_i \mid i \in I_L\}, \quad \text{where } I_L := \bigcup_{k \in K: e_k = L} I_{q(k)}.$$

Then, the set $\widehat{S}_{ap}(\ell)$ is given by the lastly obtained $\widehat{S}$.

The time for determining $\widehat{S}_{ap}(\ell)$ is bounded by $O((|K| + m)|\widehat{S}_{ap}(\ell)|)$ since $\widehat{y}$ is increased at most $|\widehat{S}_{ap}(\ell)|$ times and, for each such $\widehat{y}$, the identification of $Q_b(\widehat{y})$ or $Q(\widehat{y})$ costs at most $O(|K|)$ and not more than m pieces are considered.

According to the previous procedure the next result follows.

Theorem 4. *For any $\ell \in \mathbb{Z}_>^m$ and any $\gamma \in \mathbb{Z}_>^m$, the inclusions*

$$S^*(\gamma, \ell) \subseteq \widehat{S}_{ap}(\ell) \subseteq S_{ap}(\ell).$$

hold.

Thus, in analogy to Theorem 1, it is sufficient to use $\widehat{S}_{ap}(\ell)$ for a recursion based on DP for solving the 1D cutting problem with fix-lengths. To this end, let $T := \widehat{S}_{ap}(\ell)$ be defined.

Algorithm FDP-FL
Input: γ, ℓ, L, T; Output: g
(1) Set $g(0) := 0$, $y := 0$.
(2) **While** $y \leq p_T(L - \min\{\ell_i : i \in I_L\})$ **do**
(3)   **For all** $k \in K(y)$ and all $i \in I_{q(k)}$ with $y + \ell_i \leq e_k$ **do**
(4) $g(s_T(y + \ell_i)) := \max\{g(s_T(y + \ell_i)), \gamma_i + g(y)\}$,
(5) $\bar{y} := y$,
(6)   **Repeat** $y := s_T(y + 1)$ **until** $g(\bar{y}) < g(y)$ or $y \in \{b_k \mid k \in K\}$.

The worst-case complexity of Algorithm FDP-FL is $O((|K| + m)|T|)$ since y is increased at most $|T|$ times and, for each such y, the identification of $K(y)$ needs $O(|K|)$ time and at most m pieces are considered.

Theorem 5. *If Algorithm FDP-FL is used with $T := \widehat{S}_{ap}(\ell)$ to determine $g : T \to \mathbb{Z}_+$, then it holds*

$$v(y) = g(p_T(y)) \quad \text{for all } y \in [0, L],$$

where v is the optimal value function of problem (6.7)–(6.10).

Moreover, the set $\widehat{S}_{ap}(\ell)$ can be advantageously applied within B&B approaches for solving the 1D cutting problem with fix-lengths (6.7)–(6.10).

6.3.4 Applying the Nicholson Principle

In the following we apply the Nicholson principle [8, 16] to obtain a further reduction of the sets $S_{ap}(\ell)$ and $\widehat{S}_{ap}(\ell)$ of potential allocation points. Let

$$S_{ap}^{\leftarrow}(\ell) := \bigcup_{k \in K} (e_k \ominus S(\ell)) \cap [0, L] \quad \text{where } e_k \ominus S(\ell) := \{y \mid y = e_k - r, \, r \in S(\ell)\}.$$

denote the set of potential allocation points maximal in the following sense: for any $y \in S_{ap}^{\leftarrow}(\ell)$ there is a combination of piece lengths whose first (leftmost) piece, say i, has allocation point y and which is not feasible for allocation points for i larger than y. Then, a first reduced set of allocation points is obtained by the Nicholson principle as follows:

$$S_{ap}^{red}(\ell) := \{p_T(y) \mid y \in S_{ap}^{\leftarrow}(\ell)\} \quad \text{with} \quad T := S_{ap}(\ell).$$

Theorem 6. *If Algorithm FDP-FL is used with* $T := S_{ap}^{red}(\ell)$ *to determine* $g : T \to \mathbb{Z}_+$, *then it holds*

$$v(y) = g(y) \quad \text{for all } y \in T,$$

where v *is the optimal value function of problem* (6.7)–(6.10).

Proof. Similar to the optimal value function v defined in Sect. 6.3.3 for the allocation problem (6.7)–(6.10), but now looking from L to 0, we can define another optimal value function $\tilde{v} : [0, L] \to \mathbb{Z}_+$ by

$$\tilde{v}(y) := \max_{y,z} \left\{ \sum_{i \in I} \gamma_i z_i \mid (6.8)\text{–}(6.10) \text{ hold and } y_i \geq y \text{ for all } i \in I \text{ with } z_i = 1 \right\}.$$

The function $\tilde{v}$ is non-increasing since the feasible region shrinks if l increases. Since $v(L)$ and $\tilde{v}(0)$ are the optimal values of the same problem, obviously we have $v(L) = \tilde{v}(0)$. Let the sequence of triples (i_t, y_t, k_t), $t = 1, \ldots, t_\pi$ with $y_t + \ell_{i_t} \leq y_{t+1}$ for all t represent any normalized optimal pattern π of problem (6.7)–(6.10). If π does not consist of a single piece with length L, then there exists $y^* \in (0, L) \cap S_{ap}(\ell)$ with

$$v(L) = \tilde{v}(0) = v(y^*) + \tilde{v}(y^*) = \max\{v(y) + \tilde{v}(y) \mid y \in [0, L]\}.$$

As y^* any element in $\{y_t, y_t + \ell_{i_t} \mid t = 1, \ldots, t_\pi\} \cap (0, L)$ can be taken. Therefore, we have

$$\max\{v(y) + \tilde{v}(y) \mid y \in [0, L]\} = \max\{v(y) + \tilde{v}(y) \mid y \in S_{ap}(\ell)\}$$

but we have to prove

$$\max\{v(y) + \tilde{v}(y) \mid y \in S_{ap}(\ell)\} = \max\{v(y) + \tilde{v}(y) \mid y \in S_{ap}^{red}(\ell)\}.$$

To see this, we assume there exist $r \in S_{ap}^{red}(\ell)$ and $y \in S_{ap}(\ell) \setminus S_{ap}^{red}(\ell)$ with $r < y < s_T(r+1) =: r'$ and $v(y) + \tilde{v}(y) > \max\{v(y) + \tilde{v}(y) \mid y \in S_{ap}^{red}(\ell)\}$.

Assuming $v(y) = v(r)$ then $v(r) + \tilde{v}(r) \geq v(y) + \tilde{v}(y)$ since $\tilde{v}$ is non-increasing. Hence, we have $v(r) < v(y)$.

Assuming $\tilde{v}(y) = \tilde{v}(r')$ then $v(r') + \tilde{v}(r') \geq v(y) + \tilde{v}(y)$ since v is non-decreasing.

It remains the case that $\tilde{v}(y) > \tilde{v}(r')$. Then there is $y' \in S_{ap}^{\leftarrow}(\ell)$ with $y' \geq y$ and $\tilde{v}(y) = \tilde{v}(y')$. Since $y \in S_{ap}(\ell)$ and $y' \in S_{ap}^{\leftarrow}(\ell)$ with $y' \geq y$ we have a contradiction to $y \notin S_{ap}^{red}(\ell)$. $\qquad\square$

Corollary 2. *Among all optimal solutions of problem (6.7)–(6.10) there is a cutting pattern whose allocation points are all in $S_{ap}^{red}(\ell)$.*

In order to further reduce the set $S_{ap}^{red}(\ell)$ we will apply the Nicholson principle again by using $\widehat{S}_{ap}(\ell)$ instead of $S_{ap}(\ell)$. In analogy to the construction of $\widehat{S}_{ap}(\ell)$ in the previous subsection, a set $\widehat{S}_{ap}^{\leftarrow}(\ell)$ of rightmost allocation points can be constructed. Only those items are regarded which can be placed because of the existence of a corresponding allocation interval. For any $y \in \widehat{S}_{ap}^{\leftarrow}(\ell)$, there is a feasible pattern π, i.e., a sequence of triples (i_t, y_t, k_t), $k = 1, \ldots, t_\pi$ with $y_t + \ell_{i_t} \leq y_{t+1}$ for all t, whose first (leftmost) piece i_1 has allocation point $y_1 = y$. This pattern becomes infeasible for all y' with $y' > y$ if $y_1 := y'$ and any choice of the allocation points $y_2, \ldots, y_{t_\pi}$ with $y_t + \ell_{i_t} \leq y_{t+1}$ for all t. The construction requires some notation. For $y \in \{e_k \mid k \in K\}$, let

$$K_e(y) := \{k \in K \mid e_k = y\}, \quad Q_e(y) := \{q \in Q \mid q = \widetilde{q}(k), k \in K_e(y)\},$$

and, for $y \in (0, L]$, let

$$
\begin{aligned}
\overline{Q}(y) &:= \{q \in Q : \exists k \in K_q \text{ with } e_k \geq y, \, y - \min\{\ell_i \mid i \in I_q\} \geq b_k\}, \\
\overline{k}(y, q) &:= \min\{k \in K_q \mid e_k \geq y\} \quad \text{for all } q \in \overline{Q}(y), \\
\overline{K}(y) &:= \{k \in K \mid k = \overline{k}(y, q), q \in \overline{Q}(y)\}.
\end{aligned}
$$

be defined.

Initialization Rightmost allocation points are obtained if a piece is allocated at a point in

$$\widehat{S}^{\leftarrow} := \{L - \ell_i \mid i \in I_q, \, \ell_i \leq L - b_{\overline{k}(L,q)}, \, q \in Q_e(L)\} \cup \{L\}.$$

Since $Q_e(L)$ represents all qualities having an allocation interval ending at L, all pieces of I_q, $q \in Q_e(L)$, can be placed at $L - \ell_i$ which fit within the corresponding allocation interval, i.e. which are not longer than $L - b_{\overline{k}(L,q)}$. Let $\widehat{y} := L$.

General Step Let

$$\widehat{y}_e := \max\{0, e_k \mid e_k < \widehat{y}, k \in K\}, \quad \widehat{y}_s := \max\{y \in \widehat{S}^{\leftarrow} \mid y < \widehat{y}\}.$$

Here, $\widehat{y}_e$ denotes the coordinate of the next allocation interval which allows the placement of further items, whereas for $\widehat{y}_s$ there is already a feasible pattern in the interval $[\widehat{y}_s, L]$ which can possibly be extended. If $\widehat{y}_e > \widehat{y}_s$, then $\widehat{y} := \widehat{y}_e$ and, because of the new allocation interval, all corresponding pieces are placed:

$$\widehat{S}^{\leftarrow} := \widehat{S}^{\leftarrow} \cup \{\widehat{y} - \ell_i \mid i \in I_q, \widehat{y} - \ell_i \geq b_{\overline{k}(y,q)}, q \in Q_e(\widehat{y})\}.$$

Otherwise, if $\widehat{y}_e \leq \widehat{y}_s$, then $\widehat{y} := \widehat{y}_s$ and all pieces belonging to $\overline{Q}(\widehat{y}_s)$ and suitable length are placed:

$$\widehat{S}^{\leftarrow} := \widehat{S}^{\leftarrow} \cup \{\widehat{y} - \ell_i \mid i \in I_q, \widehat{y} - \ell_i \geq b_{\overline{k}(y,q)}, q \in \overline{Q}(\widehat{y})\}.$$

The algorithm terminates if no further piece can be placed, i.e., if

$$\widehat{y} < \min\{\ell_i \mid i \in I_0\}, \quad \text{where } I_0 := \bigcup_{k \in K: b_k = 0} I_{q(k)}.$$

Then, the set $\widehat{S}_{ap}^{\leftarrow}(\ell)$ is given by the lastly obtained $\widehat{S}^{\leftarrow}$.

The time to determine $\widehat{S}_{ap}^{\leftarrow}(\ell)$ is similar to that needed for $\widehat{S}_{ap}(\ell)$ and is bounded from above by $O(mL)$.

Now, we are able to define the announced reduced set of allocation points by

$$\widehat{S}_{ap}^{red}(\ell) := \{p_T(y) \mid y \in \widehat{S}_{ap}^{\leftarrow}(\ell)\} \quad \text{with} \quad T := \widehat{S}_{ap}(\ell) \quad \text{(cf. Sect. 6.3.3)}.$$

Now, because of construction, we have $\widehat{S}_{ap}^{red}(\ell) \subseteq \widehat{S}_{ap}(\ell)$ and moreover

Theorem 7. *If Algorithm FDP-FL with* $T := \widehat{S}_{ap}^{red}(\ell)$ *is used to determine* $g : T \to \mathbb{Z}_+$, *then it holds*

$$v(y) = g(y) \quad \text{for all} \ y \in T,$$

where v *is the optimal value function of problem* (6.7)–(6.10).

The theorem can be proved in analogy to Theorem 6. The time needed for computing all g-values according to Theorem 7 is bounded by $O(m|\widehat{S}_{ap}^{red}(\ell)|)$.

Corollary 3. *Among all optimal solutions of problem* (6.7)–(6.10) *there is a cutting pattern whose allocation points are all in* $\widehat{S}_{ap}^{red}(\ell)$.

Similar to Theorem 2, the use of $S_{ap}^{red}(\ell)$ or $\widehat{S}_{ap}^{red}(\ell)$ does not guarantee to obtain all values $v(y)$ of the optimal value function v for $y \in (0, L)$, nevertheless $v(L)$ and a corresponding optimal pattern can be determined.

Example 2. Let the unit of measurement be millimeter. An arbitrarily long wooden board of width $W = 300$ has to be cut into strips with widths of 40, 50, or 60. The cutting kerf is 2.5. Due to different quality demands, strips of width 50 can only be obtained within the interval [50, 200], and strips of width 60 only within [150, 225]. Multiplying all data by 2, adding 5 to the item widths and to the overall width to regard the kerf, and dividing all widths by 5 leads to a 1D cutting problem with the following input data: $L = 121$, item lengths $\ell_1 = 17$, $\ell_2 = 21$, $\ell_3 = 25$, and allocation intervals $A_1 = [0, 121]$, $A_2 = [20, 80]$, $A_3 = [60, 90]$. Then, we obtain $|S(\ell, L)| = 41$, $|S^{red}(\ell, L)| = 23$,

$$|S_{ap}(\ell)| = 67, \quad |S_{ap}^{\leftarrow}(\ell)| = 73, \quad |S_{ap}^{red}(\ell)| = 35,$$

$$|\widehat{S}_{ap}(\ell)| = 29, \quad |\widehat{S}_{ap}^{\leftarrow}(\ell)| = 36, \quad |\widehat{S}_{ap}^{red}(\ell)| = 12,$$

and

$$\widehat{S}_{ap}^{red}(\ell) = \{0, 17, 20, 34, 41, 51, 58, 62, 68, 87, 104, 121\}.$$

□

The computation of any of the introduced sets of potential allocation points takes a pseudo-polynomial amount of time. Due to its smaller cardinality, the application of $\widehat{S}_{ap}^{red}(\ell)$ can save computational effort in DP and B&B approaches if compared to the use of other sets of allocation points. If instances have to be solved which only differ in the profit coefficients γ, the construction of $\widehat{S}_{ap}^{red}(\ell)$ has to be done only once.

6.4   The 1D Cutting Problem with Variable Lengths

In some cutting tasks the lengths of desired items should not be fixed in advance. Instead, they can vary within known tolerances. For example, this is useful for producing finger joined lumber. There, items of various lengths (but with the same profile) are put together to obtain stripes of desired lengths.

6.4.1   Problem Formulation

Now, in contrast to the previous section, the lengths of the items to be cut are not fixed. Rather, it can take any value within a given range. More precisely, the length of piece i ($i \in I$) is again denoted by ℓ_i. However, ℓ_i is now a variable with

$$\ell_i \in [\underline{l}_i, \bar{l}_i] \quad (i \in I),$$

where $\underline{l}_i$ and $\bar{l}_i$ are given positive integers with $0 < \underline{l}_i \leq \bar{l}_i$, $i \in I$. Items with fix-lengths can also be considered by simply setting $\underline{l}_i = \bar{l}_i$.

The value of item i with length l is denoted by $\tilde{\gamma}_i(l)$. The function $\tilde{\gamma}$ is required to be affine, non-decreasing, and non-negative. For the sake of simplicity, we assume $\tilde{\gamma}_i(l) = \gamma_i \cdot l$ with some given $\gamma_i > 0$ for all $i \in I$.

Problems of this kind occur, for instance, related to hard wood cutting. There, pieces of various lengths (but with the same cross section) are put together using the finger-joining technology to get profiles of arbitrary length (see, e.g., [1]).

6.4.2 Modeling

In order to formulate a mixed-integer optimization model with 0/1-variables it is assumed in this subsection that each piece is allocated at most once (as in Sect. 6.3.2). The allocation of piece i is described by a 0/1-variable z_i defined as follows:

$$z_i = \begin{cases} 1 & \text{if piece } i \in I \text{ is allocated (should be cut)}, \\ 0 & \text{otherwise.} \end{cases}$$

The allocation point of piece i is again denoted by y_i. Then piece i with length ℓ_i covers the interval $T_i(y_i, \ell_i) := [y_i, y_i + \ell_i]$ if it has been placed, i.e., if $z_i = 1$. Hence, the cutting (allocation) problem can be modeled as follows:

$$\sum_{i \in I} \gamma_i \cdot \ell_i \cdot z_i \to \max \tag{6.17}$$

subject to

$$z_i \in \{0, 1\}, \ \ell_i, y_i \in \mathbb{R}_+ \quad i \in I, \tag{6.18}$$

$$\underline{l}_i z_i \leq \ell_i \leq \bar{l}_i z_i \quad i \in I, \tag{6.19}$$

$$\text{int } T_i(y_i, \ell_i) \cap \text{int } T_j(y_j, \ell_j) = \emptyset \quad \text{for all } i, j \in I \text{ with } i \neq j, \tag{6.20}$$

$$\begin{array}{c} \text{for each } i \in I \text{ with } z_i = 1 \text{ there are } q \in Q \text{ and } k \in K_q \\ \text{with } i \in I_q \text{ and } T_i(y_i, \ell_i) \subseteq A_k. \end{array} \tag{6.21}$$

Condition (6.20) ensures that the packed pieces do not overlap each other and condition (6.21) guarantees that the packing of a piece is done within an allocation interval of related quality.

Note that the optimization model (6.17)–(6.21) has a nonlinear objective function. Similar to Sect. 6.3.2, the restrictions (6.20) and (6.21) can be linearized using the same 0/1-variables u_{ij}, $i, j \in I$, $i < j$, and v_{ik}, $i \in I$, $k \in K_{q(i)}$.

6.4.3 Optimal Value Function

For the optimization problem (6.17)–(6.21) the optimal value function $v : [0, L] \to \mathbb{R}_+$ is defined by

$$v(y) := \max_{z,y,\ell} \left\{ \sum_{i \in I} \gamma_i \ell_i z_i \mid (6.18)\text{–}(6.21) \text{ hold and } y_i + \ell_i \leq y \text{ for all } i \text{ with } z_i = 1 \right\}.$$

The function v is continuous from the right and non-decreasing since the feasible region enlarges if y increases. Moreover, v is piecewise affine since only a finite number of different sequences of allocated pieces exists and the functions $\tilde{\gamma}_i$ providing the profit of pieces $i \in I$ were assumed to be linear. By the same reason, the domain of v can be partitioned into intervals where v is either constant or linearly increasing with slope in $\{\gamma_1, \ldots, \gamma_m\}$.

Example 3. Let the following instance of a cutting problem be given:

$$I := \{1, 2, 3\}, \quad Q := \{1, 2\},$$
$$\underline{l}_1 := \underline{l}_2 := 30, \bar{l}_1 := \bar{l}_2 := 50, \quad \underline{l}_3 := 20, \bar{l}_3 := 100,$$
$$\gamma_1 := \gamma_2 := 8, \gamma_3 := 5,$$
$$A_1 := [0, 60], \quad A_2 := [70, 100], A_3 := [0, 100],$$
$$I_1 := \{1, 2\}, \quad I_2 := \{3\},$$
$$K_1 := \{1, 2\}, \quad K_2 := \{3\}.$$

Figure 6.1 shows the optimal value function v for Example 3. The leftmost gap of v

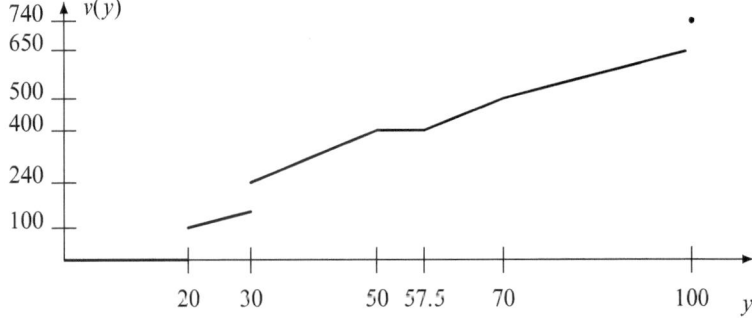

Fig. 6.1 Optimal value function v for Example 3

at $y = 20$ arises since item 3 with length $\ell_3 = \underline{l}_3 = 20$ and allocation point $y_3 = 0$ is placed. This leads to $v(20) = \gamma_3\ell_3 = 100$. For $y \in [20, 30)$, placing item 3 with length $\ell_3 = y$ and $y_3 = 0$ provides $v(y) = \gamma_3\ell_3 = 5y$. If $y = 30$, then item 1 (instead of item 3) is placed with $\ell_1 = 30$ and $y_1 = 0$. For $y \in [30, 50)$, placing item 1 with length $\ell_1 = y$ and $y_1 = 0$ yields $v(y) = \gamma_1\ell_1 = 8y$. If $y \in [50, 57, 5)$ then placing item 1 with $y_1 \in [0, y - 50]$ is optimal so that v remains constant in this interval. For $y \in [57.5, 70)$, both item 1 and item 3 are placed with $y_1 = 0$, $\ell_1 = y - 20$, $y_3 = y - 20$, $\ell_3 = 20$. This yields $v(y) = \gamma_1\ell_1 + \gamma_3\ell_3 = 8(y - 20) + 100$. For $y \in [70, 100)$, the length of item 1 becomes maximal, namely $\ell_1 = \bar{l}_1 = 50$ with $y_1 = 0$ and the length of item 3 is $\ell_3 = y - 50$ with $y_3 = 50$. Therefore, $v(y) = 400 + 5(y - 50)$. Finally, the rightmost gap occurs at $y = 100$ because of the optimal pattern with $y_1 = 0$, $y_2 = 70$, $y_3 = 50$, $\ell_1 = 50$, $\ell_2 = 30$, $\ell_3 = 20$ and $v(100) = 740$. □

Note that the allocation pattern with the optimal value $v(L)$ might be not unique. For example, the optimal lengths of two items of the same quality need not be unique but their sum is the same for all optimal patterns with the same sequence of items.

Since y_i and ℓ_i are non-negative **real numbers,** infinitely many points become potential allocation points. However, the subsequent theorem shows that a finite subset of allocation points suffices to define an optimal (allocation) pattern.

In case of variable lengths, a pattern π is a finite sequence of quadruples $(i_t, y_t, \ell_t, k_t)_{t=1}^{t_\pi}$, where i_t denotes the index of the t-th placed piece, y_t is the allocation point of piece i_t, ℓ_t is its length, and k_t gives the corresponding allocation interval.

Theorem 8. *Among all optimal patterns for problem* (6.17)–(6.21) *there is a pattern π with*

$$y_t \in S_{ap}(\underline{l}, \bar{l}) \quad \text{for all } t = 1, \dots, t_\pi$$

where $\mathscr{A} := \{b_k, e_k \mid k \in K\}$ and

$$S_{ap}(\underline{l}, \bar{l}) := \big((\mathscr{A} \oplus S(\underline{l}) \oplus S(\bar{l})) \cup (\mathscr{A} \ominus S(\underline{l}) \ominus S(\bar{l}))\big) \cap [0, L].$$

Note that the time required for determining $S_{ap}(\underline{l}, \bar{l})$ is bounded by $O(mL)$.

Proof. We consider the allocation of two items of not necessarily different qualities, say items 1 and 2, with corresponding allocation intervals $A_1 = [b_1, e_1]$ and $A_2 = [b_2, e_2]$, respectively. We show that optimal patterns exist having allocation points y_1 and y_2 belonging to $S_{ap}(\underline{l}, \bar{l})$. More general cases can be proved inductively. Without loss of generality we can assume that the allocation intervals overlap as in Fig. 6.2. Otherwise, the allocation problem can be separated into two smaller problems which can be dealt with independently. Moreover, to keep the case by case analysis short we only consider cases where additionally

$$\underline{l}_1 + \underline{l}_2 \leq e_2 - b_1 \leq \bar{l}_1 + \bar{l}_2 \quad \text{and} \quad 2\underline{l}_i > \bar{l}_i, \ e_i - b_i < \bar{l}_i + \underline{l}_i \text{ for } i = 1, 2$$

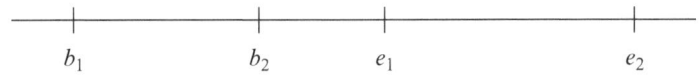

Fig. 6.2 Overlapping allocation intervals

holds. These conditions ensure that both items can be placed but not more than 2. Cases which do no fulfill these conditions can be handled analogously. Since an optimal pattern depends on the profit coefficients γ_1 and γ_2 several cases have to be considered:

Case 1: Let $\gamma_1 = \gamma_2$. As much as possible of the available length should be used, i.e., one of the two solutions

$$y_1 = b_1, \ell_1 = \min\{\bar{l}_1, e_1 - b_1\}, y_2 = b_1 + \ell_1, \ell_2 = \min\{\bar{l}_2, e_2 - y_2\},$$

or

$$\ell_2 = \min\{\bar{l}_2, e_2 - b_2\}, y_2 = e_2 - \ell_2, y_1 = b_1, \ell_1 = \min\{\bar{l}_1, y_2 - b_1\}$$

is optimal.

Case 2: Let $\gamma_1 > \gamma_2$. Then ℓ_1 should be as large as possible.

Subcase 2a: Let $e_1 - b_1 \geq \bar{l}_1$. If $e_2 \geq b_1 + \bar{l}_1 + \underline{l}_2$, then

$$y_1 = b_1, \quad \ell_1 = \bar{l}_1, \quad y_2 = \max\{b_2, b_1 + \bar{l}_1\}, \quad \ell_2 = e_2 - y_2.$$

describes an optimal pattern.
If $e_2 < b_1 + \bar{l}_1 + \underline{l}_2$, an optimal pattern is given by

$$\begin{array}{ll} y_1 = b_1, \ \ell_1 = e_2 - \underline{l}_2 - b_1, y_2 = e_2 - \underline{l}_2, \ \ell_2 = \underline{l}_2 & \text{if } \underline{l}_2\gamma_2 \geq (\bar{l}_1 - \ell_1)\gamma_1, \\ y_1 = b_1, \ \ell_1 = \bar{l}_1, & \text{item 2 is not allocated} & \text{if } \underline{l}_2\gamma_2 < (\bar{l}_1 - \ell_1)\gamma_1. \end{array}$$

In any case we have $y_1, y_2 \in S_{ap}(\underline{l}, \bar{l})$.

Subcase 2b: Let $e_1 - b_1 < \bar{l}_1$. If $e_2 - e_1 \geq \underline{l}_2$ then

$$y_1 = b_1, \quad \ell_1 = e_1 - b_1, \quad y_2 = e_1, \quad \ell_2 = \min\{\bar{l}_2, e_2 - e_1\}$$

provides an optimal pattern.
If $e_2 - e_1 < \underline{l}_2$, the pattern given by

$$\begin{array}{ll} y_1 = b_1, \ \ell_1 = e_2 - \underline{l}_2 - b_1, y_2 = e_2 - \underline{l}_2, \ \ell_2 = \underline{l}_2 & \text{if } \underline{l}_2\gamma_2 \geq (e_1 - b_1 - \ell_1)\gamma_1, \\ y_1 = b_1, \ \ell_1 = e_1 - b_1, & \text{item 2 is not allocated} & \text{if } \underline{l}_2\gamma_2 < (e_1 - b_1 - \ell_1)\gamma_1, \end{array}$$

is optimal as well. Again, we always have $y_1, y_2 \subset S_{ap}(\underline{l}, \bar{l})$.

Case 3: Let $\gamma_1 < \gamma_2$. Then ℓ_2 should be as large as possible. The corresponding subcases can be dealt with like in Case 2.   □

Similar to the case of only fix-lengths, the packing problem can be separated into some smaller problems if condition (6.16) is violated which can be solved independently from each other. So we assume again that (6.16) is fulfilled.

6.4.4 Packing a Single Piece

In order to compute the optimal value $v(L)$, the allocation of a single piece is considered in the following and will become a basic element in the DP and B&B solution approaches presented below.

Any feasible pattern within an interval $[0, l]$ yields a lower bound for $v(l)$. The current best lower bound for $v(l)$, obtained in a solution process, is denoted as $h(l)$.

The function h has analogous properties as v. Hence, there is a description of h by a finite sequence $(\underline{y}_j)$ of coordinates that at least contains all jump discontinuities and all kinks of h. Any two neighboring points $\underline{y}_j$, $\underline{y}_{j+1}$ define a so-called *basic* (or *reference*) *interval* $B_j := [\underline{y}_j, \underline{y}_{j+1})$. The end point $\underline{y}_{j+1}$ belongs to the basic interval B_{j+1} since h is continuous from the right.

For any interval B_j, the function h can be described by

$$h(y) = \alpha_j + \beta_j(y - \underline{y}_j) \quad \text{for all } y \in B_j.$$

To solve the cutting problem (6.17)–(6.21), a procedure is used which consists of successively placing single pieces. In principle, if piece i with length $\ell_i \in [\underline{l}_i, \bar{l}_i]$ and allocation point $y \in B_j$ is added to the current pattern (which determines h in B_j) then, depending on γ_i and β_j, an improved pattern might be obtained for the interval $[\underline{y}_j + \ell_i, \underline{y}_{j+1} + \bar{l}_i)$. Note that in general, due to the variability of $\ell_i \in [\underline{l}_i, \bar{l}_i]$ and $y \in B_j$, infinitely many patterns exist. In contrast to this, the number of different sequences of items is finite and can be reduced by means of upper bounds and dominance tests. Details can be found in [11].

Because of the variability just mentioned it is not possible to consider single allocation points, rather it is necessary to handle intervals of allocation points. To this end, the basic intervals B_j, defined by the current h-function, can be used.

In the following we provide a construction procedure in which a current solution (pattern) for $[0, \underline{y}_{j+1})$ is extended by placing a single piece i with allocation point in B_j, if possible. The placing of piece i is considered simultaneously for all allocation points $y \in B_j$ and all suitable lengths $\ell_i \in [\underline{l}_i, \bar{l}_i]$. For that, two cases have to be distinguished (see Sects. 6.4.4.1 and 6.4.4.2).

6.4.4.1 Packing a Piece i with $\gamma_i \geq \beta_j$

Let $i \in I_q$ for some $q \in Q$, and let $\gamma_i \geq \beta_j$. In order to find a pattern with value as large as possible, piece i has to be placed with an allocation point in $B_j = [\underline{y}_j, \underline{y}_{j+1})$ as small as possible because of $\gamma_i \geq \beta_j$. Hence, allocation points are both $\underline{y}_j$ (if there is $k \in K_q$ with $\underline{y}_j \in A_k$ and $e_k - \underline{y}_j \geq \underline{l}_i$) and the points b_k for all $k \in K_q$ with $\underline{y}_j < b_k < \underline{y}_{j+1}$. Without loss of generality, let $\xi_1, \dots, \xi_\kappa$ denote those allocation points with $\underline{y}_j \leq \xi_1 < \cdots < \xi_\kappa < \underline{y}_{j+1} =: \xi_{\kappa+1}$. Let $k(\xi_p) \in K_q$ denote the corresponding allocation interval, i.e., $b_{k(\xi_p)} = \xi_p$, for all $p = 1, \dots, \kappa$. The length l_i of piece i which should be placed with allocation point ξ_p, is bounded by the remaining length of the corresponding allocation interval and its maximal length, i.e., by

$$\min\{e_{k(\xi_p)} - \xi_p, \bar{l}_i\}, \quad \text{for } p = 1, \dots, \kappa.$$

For $p = 1, \dots, \kappa$, the following formula has to be applied to update the function h to possibly get an improved pattern with rightmost piece i:

$$h(y) := \begin{cases} \max\{h(y), h(\xi_p) + \gamma_i(y - \xi_p)\}, & \text{if } y \in [\xi_p + \underline{l}_i, \min\{e_{k(\xi_p)}, \xi_p + \bar{l}_i\}], \\ h(y) & \text{otherwise.} \end{cases}$$

$$(6.22)$$

Moreover, the placing of piece i with maximum length and variable allocation point has to be considered as well and leads to a further update of h for $p = 1, \dots, \kappa$:

$$h(y) := \begin{cases} \max\{h(y), h(y - \bar{l}_i) + \gamma_i \bar{l}_i\}, & \text{if } y \in [\xi_p + \bar{l}_i, \min\{e_k, \xi_{p+1} + \bar{l}_i\}], \\ h(y) & \text{otherwise.} \end{cases} \quad (6.23)$$

Formulas (6.22) and (6.23) are based on the following proposition.

Proposition 1. *Let $i \in I_q$ with $\gamma_i \geq \beta_j$. For any pattern $T_i(y, l) := [y, y + l] \subseteq A_k$ with $q = \widetilde{q}(k)$, which is caused by the allocation of piece i with allocation point $y \in B_j$ and length $l \in [\underline{l}_i, \bar{l}_i]$, there is a pattern $T_i(y^*, l^*)$ with $y^* = \xi_p$ for some $p \in \{1, \dots, \kappa\}$ or $l^* = \bar{l}_i$ so that $T_i(y^*, l^*)$ dominates $T_i(y, l)$ in respect to $\gamma_i l$.*

The proposition is a consequence of Theorem 8.

6.4.4.2 Packing a Piece i with $\gamma_i < \beta_j$

Let $i \in I_q$ with $\gamma_i < \beta_j$. In difference to above, the placement of piece i has to be done with an allocation point as large as possible and length as short as possible. For $p = 1, \dots, \kappa$, the following formula for updating $h(y)$ has to be applied.

$$h(y) := \begin{cases} \max\{h(y), h(y - \underline{l}_i) + \gamma_i \underline{l}\}, & \text{if } y \in [\xi_p + \underline{l}, \min\{e_{k(\xi_p)}, \xi_{p+1} + \underline{l}_i\}], \\ h(y) & \text{otherwise.} \end{cases}$$

$$(6.24)$$

Here, $y - \underline{l}_i$ is the related (varying) allocation point. Because of $\gamma_i < \beta_j$, the packing of piece i with a length $\ell_i > \underline{l}_i$ is only necessary for the allocation point $\xi_{\kappa+1} := \underline{y}_{j+1}$ if there exists $k \in K_q$ with $e_k - \underline{y}_{j+1} > \underline{l}_i$. But $\underline{y}_{j+1}$ belongs to the next basic interval, B_{j+1}, and will be considered there since $h(\underline{y}_{j+1}) \geq \lim_{y \uparrow \underline{y}_{j+1}} h(y)$.

Proposition 2. *Let $i \in I_q$ with $\gamma_i < \beta_j$. For any pattern $T_i(y, l) \subseteq A_k$ with $q = \widetilde{q}(k)$, which is caused by the allocation of piece i with allocation point $y \in B_j$ and length $l \in [\underline{l}_i, \bar{l}_i]$, there is a pattern $T_i(y^*, l^*)$ with $y^* = \underline{y}_{j+1}$ or $l^* = \underline{l}_i$ which dominates $T_i(y, l)$ in respect to $\gamma_i l$.*

Formulas (6.22)–(6.24) cause, in general, a change of the basic intervals $B_{j'}$ for $j' \geq j$. If the update of h by these formulas led to a new function h with (partially) increased function values, then the new h need not be monotonously increasing. Therefore, we have to further update this new h by

$$h(y) := \max\{h(y') \mid \underline{y}_j \leq y' \leq y\} \quad \text{for all } y \in [\underline{y}_j, L] \tag{6.25}$$

so that it becomes monotone again.

6.4.5 Solution Approaches

In the subsection we provide two solution approaches for the problem with pieces of variable length where we apply the update rules discussed in Sects. 6.4.4.1 and 6.4.4.2.

6.4.5.1 Branch and Bound Algorithm

The B&B algorithm presented below is based on the LIFO strategy. Appropriate upper bounds, denoted as $\tilde{u}(\cdot)$, are given in [11] where $\tilde{u}(y) \geq v(L) - v(y)$ for all $y \in [0, L]$. Without loss of generality, we assume $\gamma_1 \geq \gamma_2 \geq \cdots \geq \gamma_m$. Branching will be made with respect to

- the basic intervals and
- the pieces which can be allocated next according to a basic interval.

The index μ denotes the branching depth in the algorithm. For any basic interval B_j, the label v_j denotes whether the basic interval B_j is already investigated for further branching (then $v_j = 0$), or if it has still to be considered (then $v_j = 1$). Furthermore, $\tilde{B}_\mu = [\tilde{y}_\mu, \hat{y}_\mu)$ denotes the current basic interval. Note, the initial basic interval $\tilde{B}_0 = [0, L]$ is reduced in Step (4) when the first pieces are placed, and the same can happen for other basic intervals during the algorithm. The branching strategy presented here uses the LIFO principle (depth first search), but modifications are obviously possible.

B&B algorithm for 1D cutting problems with variable lengths

(1) *Initialization*
Set $B_0 := [0,L]$, $\alpha_0 := 0$, $\beta_0 := 0$, $\nu_0 = 1$, $\mu := 0$.

(2) *Selection of the next basic interval for branching*
Select the basic interval B_j with largest $\underline{y}_j$-coordinate with $\nu_j = 1$. If such an interval B_j does not exist then go to (5). Otherwise, set $\widetilde{B}_\mu := B_j$ and $\nu_j := 0$. If $\max_{y \in \widetilde{B}_\mu} \{h(y) + \widetilde{u}(y)\} \le h(L)$, then go to (2). If $L \in \widetilde{B}_\mu$ then go to (4).

(3) *Selection of the next piece for branching*
Select the next piece $i_\mu \in I$ (not already considered for branching) which can be allocated with an allocation point in B_μ. If such a piece does not exist then set $\mu := \mu - 1$ and go to (2).
If $\max_{y \in \widetilde{B}_{\mu-1}} \{h(y) + \gamma_i \underline{l}_i + \widetilde{u}(y + \underline{l}_i)\} \le h(L)$ then go to (3).

(4) *Allocating a piece*
If $L \in \widetilde{B}_\mu$ then allocate (one after another) all feasible pieces with allocation point $\widetilde{y}_\mu$ with all feasible lengths: a new partition of $\widetilde{B}_\mu = [\widetilde{y}_\mu, L]$ results; set $\nu_j := 1$ for all B_j with $\underline{y}_j \ge \widetilde{y}_\mu$. Otherwise, if $L \notin \widetilde{B}_\mu$ then allocate piece i_μ for all feasible lengths and all allocation points within $\widetilde{B}_\mu$ and update h in accordance with (6.22) – (6.25). If h is increased in some interval $B_{j'}$ during the update process then set $\nu_{j'} := 1$. If h has not been increased anywhere then go to (3), otherwise set $\mu := \mu + 1$ and go to (2).

(5) *Identify an optimal pattern.*

6.4.5.2   FDP Algorithm

For the solution of the considered cutting problem also a FDP method can be used. The general principle of the FDP is similar to the FDP algorithm for the knapsack problem in Sect. 6.2. It computes optimal values $v(y)$ and corresponding patterns for each $y \le L$, where y is successively increased. Thereby, the values $v(y)$ are obtained by updating the function h in such a way that, starting from a known optimal solution, feasible pieces are placed with all suitable lengths l to possibly get a better solution.

As in the B&B algorithm, intervals of allocation points are considered. In the B&B algorithm, after investigating the basic interval B_j other intervals $B_{j'}$ with $\underline{y}_{j'+1} \le \underline{y}_j$ have to be considered in general (due to backtracking), and therefore it can happen that B_j (or a subset of it) has to be considered anew if h_j has been (partially) increased. Using the FDP approach, the basic interval B_j is considered exactly once, namely if $h(y) = v(y)$ for all $y \in B_j$ holds. This can also be guaranteed in a B&B algorithm when an appropriate branching strategy (based on breadth first search) is used.

Note that during the allocation of a piece with allocation points in B_j further tests with upper bounds can be used as in the B&B algorithm. If piece i is feasible and $\underline{y}_j + \underline{l}_i < \underline{y}_{j+1}$ holds, then B_j can be split (i.e., $\underline{y}_{j+1}$ can be reduced) in a suitable way. Moreover, during the update process only partitions of $[0, L - \min\{\underline{l}_i \mid i \in I_L\}]$

FDP algorithm for 1D cutting problems with variable lengths

(1) *Initialization*

Set $B_0 := [0, L]$, $\alpha_0 := 0$, $\beta_0 := 0$. Allocate (one after another) all feasible pieces at $\xi = 0$ with all feasible lengths. A first partition of $[0, L]$, i.e., a first finite sequence $(\underline{y}_j)$, is obtained.

If $\underline{y}_1 > \min\{\min\{\underline{l}_i \mid i \in I\}, \min\{b_k \mid b_k > 0, k \in K\}\} =: \widetilde{y}$ then split B_0 into $[0, \widetilde{y})$ and $[\widetilde{y}, \underline{y}_1)$, and renumber. Set $j := 0$.

(2) *Selection of the next basic interval*

Set $j := j + 1$. If $\underline{y}_j + \min\{\underline{l}_i \mid i \in I_L\} > L$ then goto (4).

If $\max_{y \in B_j}\{h(y) + \widetilde{u}(y)\} \leq h(L)$ then go to (2).

(3) *Allocating pieces with allocation points in B_j:*

Allocate (one after another) all feasible pieces i at every allocation point $\xi \in B_j$ and with all feasible lengths according to (6.22) – (6.25).

A new partition of $[0, L]$, i.e., a new finite sequence $(\underline{y}_j)$, is obtained but changes of the $\underline{y}_{j'}$ are only possible for $j' \geq j + 1$. Go to (2).

(4) *Identify an optimal pattern.*

---

have to be considered since there is no feasible allocation point within the interval $(L - \min\{\underline{l}_i \mid i \in I_L\}, L)$.

In the worst case, $O(m|S_{ap}(\underline{l}, \overline{l})|)$ updates according to (6.22)–(6.25) have to be done in Step (3) of the algorithm. A single update requires at most $O(\overline{l}_i)$ time.

An advantage of the FDP approach can be the relatively constant and well assessable expense to solve an instance (pseudo-polynomiality of the algorithm). In general, this is not the case for the B&B method, where some examples can require much more computation time as in average. However, in general, good (near optimal) solutions are found quickly by a B&B algorithm with LIFO strategy so that a termination after a predefined time span is reasonable for on-line scenarios.

6.5 The 2D Cutting Problem with Quality Demands

In this section we consider 2D cutting problems. Rectangular pieces have to be cut from a larger rectangle of non-homogeneous raw material such that the yield is maximal. Thereby some rectangular parts of the raw material are not allowed to be used for some pieces because of bad quality. We investigate two cases: firstly, so-called defective regions, or simply defects, cannot be used to obtain desired pieces, and secondly, different quality demands are considered. We analyze the case of fixed dimensions of the pieces and give appropriate sets of allocation points, usable in DP and B&B approaches. We also discuss the case when one of the size parameters can vary.

6.5.1 Forbidden Regions

In this subsection we consider the case that some parts of the raw material cannot be used at all to obtain a desired piece. The next subsection is devoted to the discussion of different quality demands.

6.5.1.1 Problem Formulation

Let a rectangle of raw material (wood, metal, glass, etc.) of length L and width W be given. Moreover, the pieces $i \in I := \{1, \ldots, m\}$ to be cut are of length ℓ_i, width w_i and have profit coefficient γ_i. Only guillotine cuts are allowed to obtain desired pieces. The part of the raw material used for each piece has to be defect-free. The aim is to maximize to total yield of the cutting pattern. We denote by (x_i, y_i) the *allocation point* of piece i, i.e., if piece i is placed with allocation point (x_i, y_i) then it covers the rectangular region $[x_i, x_i + \ell_i] \times [y_i, y_i + w_i]$. Hence, a 2D pattern can be described by a set of triples (i_t, x_{i_t}, y_{i_t}), $t = 1, \ldots, \kappa$, where $i_t \in I$ denotes the t-th placed piece and (x_{i_t}, y_{i_t}) the corresponding allocation point.

Since only guillotine cuts can be applied we can assume, without loss of generality, that all defective parts of the raw material are described by rectangles or a finite union of rectangles. Let D_k, $k \in \overline{K} := \{1, \ldots, |\overline{K}|\}$, denote the defective parts with

$$D_k := D_k(a_k, b_k, c_k, d_k) := \{(x, y) \mid a_k \leq x \leq c_k, b_k \leq y \leq d_k\} \subset [0, L] \times [0, W],$$

and define

$$D := \bigcup_{k \in \overline{K}} D_k.$$

6.5.1.2 Sets of Allocation Points: No Defects

If $\overline{K} = \emptyset$, the well-known recurrence formula of Gilmore and Gomory [5] can be applied to obtain an optimal pattern with no restriction on the number of stages. Let $u(L', W')$, with $L' \in [0, L]$ and $W' \in [0, W]$, denote the optimal value for rectangle $L' \times W' = [0, L'] \times [0, W']$. Then, as a consequence of Theorem 1 we have

Theorem 9. *Let $T_\ell := S(\ell, L)$ and $T_w := S(w, W)$. Then,*

$$u(L', W') = \overline{u}(p_{S(\ell)}(L'), p_{S(w)}(W')) \quad \text{for all } L' \in [0, L], \ W' \in [0, W], \quad (6.26)$$

where $\overline{u}(L', W')$ is obtained by the following recursion:

$$\overline{u}(L', W') := \max\{\gamma(L', W'),\ g(L', W'),\ h(L', W')\} \quad \textit{for all } L' \in T_\ell,\ W' \in T_w$$

$$(6.27)$$

with

$$\gamma(L', W') := \max\{\gamma_i \mid i \in I,\ \ell_i \le L',\ w_i \le W'\},$$

$$g(L', W') := \max\{\overline{u}(r, W') + \overline{u}(p_{T_\ell}(L' - r), W') \mid 0 < r \le L'/2,\ r \in T_\ell\},$$

$$h(L', W') := \max\{\overline{u}(L', s) + \overline{u}(L', p_{T_w}(W' - s)) \mid 0 < s \le W'/2,\ s \in T_w\}.$$

Note that $u(L, W)$ can also be computed using the reduced sets of allocation points $T_\ell := S^{red}(\ell, L)$ and $T_w := S^{red}(w, W)$. In this case, the "=" in (6.26) has to be replaced by "≥" but "=" holds in particular for all $(L', W') \in S^{red}(\ell, L) \times S^{red}(w, W)$.

The time needed for computing $\overline{u}(L', W')$ for all $(L', W') \in T_\ell \times T_w$ according to formula (6.27) is bounded by $O(|T_\ell|\,|T_w|(m + |T_\ell| + |T_w|))$. A reduction to $O(|T_\ell|\,|T_w|(|T_\ell| + |T_w|))$ can be achieved by an appropriate initialization which avoids the consideration of $\gamma(L', W')$ for each (L', W'), see [12].

6.5.1.3 Sets of Allocation Points: With Defects

If $\overline{K} \ne \emptyset$, the sets of potential allocation points increase since every defective part D_k causes new potential allocation points, e.g., the right end c_k of D_k probably allows the allocation of pieces with x-coordinate c_k, and the left border a_k of D_k can cause that a piece i has x-allocation coordinate $a_k - \ell_i$. This is in difference to Theorem 9 since now regions are not allowed for allocation.

To simplify the description we define an artificial defect with coordinates $a_0 = L$, $b_0 = W$, $c_0 = d_0 = 0$ and set $\overline{K}_0 := \overline{K} \cup \{0\}$. Let $S_L^*(\gamma, \ell)$ and $S_W^*(\gamma, w)$ denote the sets of jump discontinuities in L- and W-direction of the optimal value function $v : [0, L] \times [0, W] \to \mathbb{Z}_+$ for the problem with defects.

Theorem 10. *For any $\ell \in \mathbb{Z}_>^m$, $w \in \mathbb{Z}_>^m$ and $\gamma \in \mathbb{Z}_>^m$, we have*

$$S_L^*(\gamma, \ell) \subseteq S_L^{ap}(\ell) := \bigcup_{k \in \overline{K}_0} (c_k \oplus S(\ell)) \cap [0, L],$$

$$S_W^*(\gamma, w) \subseteq S_W^{ap}(w) := \bigcup_{k \in \overline{K}_0} (d_k \oplus S(w)) \cap [0, W].$$

Moreover, with $T_\ell := S_L^{ap}(\ell)$ and $T_w := S_W^{ap}(w)$,

$$v(L', W') = v(p_{T_\ell}(L'), p_{T_w}(W')) \quad \textit{for all } (L', W') \in [0, L] \times [0, W].$$

holds.

The proof is similar to that of Theorem 3 but now, two dimensions have to be considered. The result of Theorem 10 can be strengthened with respect to the computational complexity by applying the Nicholson principle similar to Sect. 6.3.4. Let

$$S_L^{\leftarrow}(\ell) := \bigcup_{k \in K_0}(a_k \ominus S(\ell)) \cap [0, L], \quad S_W^{\leftarrow}(w) := \bigcup_{k \in K_0}(b_k \ominus S(w)) \cap [0, W]$$

denote the sets of potential allocation points maximal in the sense that, e.g., for any $x \in S_L^{\leftarrow}(\ell)$ there is a combination of piece lengths whose first (left-most) piece, say i, has allocation point x and which is not feasible for allocation points for i larger than x (and similar in W-direction). We define the reduced sets of allocation points by

$$S_L^{red}(\ell) := \{p_{S_L^{ap}(\ell)}(x) \mid x \in S_L^{\leftarrow}(\ell)\} \subseteq S_L^{ap}(\ell),$$

$$S_W^{red}(w) := \{p_{S_W^{ap}(w)}(y) \mid y \in S_W^{\leftarrow}(w)\} \subseteq S_W^{ap}(w).$$

Similar to the 1D case, in general $S_L^{red}(\ell)$ and $S_W^{red}(w)$ are not supersets of $S_L^*(\gamma, \ell)$ and $S_W^*(\gamma, w)$, respectively, but contain sufficiently many points to compute the optimal value $v(L, W)$.

In order to obtain $v(L, W)$ some modifications in comparison with the recursion (6.27) have to be done. The essential difference to the case $\overline{K} = \emptyset$ is that now the yield of a rectangular region $R := [L', L''] \times [W', W'']$ of raw material depends on its position because of the varying quality. That means, the yield function used in a DP recursion is now defined by

$$\gamma(R) := \begin{cases} u(L'' - L', W'' - W'), & \text{if } R \cap \text{int } D = \emptyset, \\ 0, & \text{otherwise.} \end{cases}$$

Thus, a DP recursion to compute v which uses the sets of allocation points $S_L^{red}(\ell)$ and $S_W^{red}(w)$ is given by the following procedure.

For all $R := [L', L''] \times [W', W'']$ with $L', L'' \in S_L^{red}(\ell)$ and $W', W'' \in S_W^{red}(w)$ set

$$\overline{v}(R) := \begin{cases} 0, & \text{if } L'' - L' < \ell_{min} \text{ or } W'' - W' < w_{min}, \\ \max\{\gamma(R), g(R), h(R)\} & \text{otherwise} \end{cases} \quad (6.28)$$

with

$$g(R) := \max\{\overline{v}(L', r, W', W'') + \overline{v}(r, L'', W', W'') \mid r \in S_L^{red}(\ell), L' < r < L''\},$$

$$h(R) := \max\{\overline{v}(L', L'', W', s) + \overline{v}(L', L'', s, W'')) \mid s \in S_W^{red}(w), W' < s < W''\}.$$

Theorem 11. *Let $T_\ell := S_L^{red}(\ell)$ and $T_w := S_W^{red}(w)$. Then,*

$$v(L', W') = \overline{v}(0, p_{S_L^{ap}}(L'), 0, p_{S_W^{ap}}) \quad \text{for all } (L', W') \in T_\ell \times T_w$$

holds, where $\overline{v}$ is defined by the recursion (6.28).

The proof is similar to that of Theorem 6 where the L- and W-directions have to be taken into account.

The computation of $\overline{v}(0, L', 0, W')$ for all $(L', W') \in T_\ell \times T_w$ according to (6.28) requires at most $O(|T_\ell|^2|T_w|^2(|K| + |T_\ell| + |T_w|))$ time, since, due to the dependence on the defective regions, $O(|T_\ell|^2|T_w|^2)$ optimal values $\overline{v}(R)$ have to be computed with a DP approach. Obviously, this estimation is rather rough. For instance, if R is defect-free and $u(R)$ is known, $v(R)$ can be determined in constant time.

Since, in general, a large number of $v(R)$-values is needed, the application of a B&B approach becomes more favorable. As upper bound for $v(R)$ we can simply use $u(L'' - L', W'' - W')$ as defined in (6.26) but tighter bounds which regard the existence of defects should be preferred.

The number of small rectangles $R = [L', L''] \times [W', W'']$ used to define subproblems in a B&B algorithm can be further reduced since, e.g., $[L', L''] \cap S_L^{red}(\ell)$ can contain allocation points which are not meaningful for dissecting $[L', L'']$. In principle, appropriate reduced sets of allocation points can be defined for each R similar to those for $L \times W$. Therefore, in order to keep the number of subproblems in a B&B approach small, for each R the reduced set of allocation points should be computed as follows. Let $K_a(R), \ldots, K_d(R)$ denote those defects which are relevant for allocating pieces into the rectangle R:

$$K_a(R) := \{k \in K \mid L' + \ell_{min} \leq a_k < L'', (W', W'') \cap [b_k, d_k] \neq \emptyset\},$$
$$K_b(R) := \{k \in K \mid W' + w_{min} \leq b_k < W'', (L', L'') \cap [b_k, c_k] \neq \emptyset\},$$
$$K_c(R) := \{k \in K \mid L' < c_k \leq L'' - \ell_{min}, (W', W'') \cap [b_k, d_k] \neq \emptyset\},$$
$$K_d(R) := \{k \in K \mid W' < d_k \leq W'' - w_{min}, (L', L'') \cap [b_k, c_k] \neq \emptyset\}.$$

The corresponding sets of allocation points are

$$\tilde{S}_L(R) := ((L' \oplus S(\ell)) \cup \{\bigcup(c_k \oplus S(\ell)) \mid k \in K_c(R)\}) \cap [L', L''],$$
$$\tilde{S}_W(R) := ((W' \oplus S(w)) \cup \{\bigcup(d_k \oplus S(w)) \mid k \in K_d(R)\}) \cap [W', W''],$$
$$\tilde{S}_L^\leftarrow(R) := ((L'' \ominus S(\ell)) \cup \{\bigcup(a_k \ominus S(\ell)) \mid k \in K_a(R)\}) \cap [L', L''],$$
$$\tilde{S}_W^\leftarrow(R) := ((W'' \ominus S(w)) \cup \{\bigcup(b_k \ominus S(w)) \mid k \in K_b(R)\}) \cap [W', W''].$$

Applying the Nicholson principle, we define reduced sets of allocation points for a single rectangular region R by

$$\tilde{S}_L^{red}(R) := \{p_{\tilde{S}_L(R)}(x) \mid x \in S_L^\leftarrow(R)\}, \quad \tilde{S}_W^{red}(R) := \{p_{\tilde{S}_W(R)}(x) \mid x \in S_W^\leftarrow(R)\},$$

and we have
|
$$\tilde{S}_L^{red}(R) \subseteq S_L^{red}(\ell) \cap [L', L''], \quad \tilde{S}_W^{red}(R) \subseteq S_W^{red}(w) \cap [W', W''].$$

By the above construction we obtain

Theorem 12. *Let us consider the 2D cutting problem for the rectangle* $R = [L', L''] \times [W', W'']$ *with defective parts. Then, among all optimal patterns of this problem, there is an optimal pattern having only allocation points with coordinates in* $\tilde{S}_L^{red}(R)$ *and* $\tilde{S}_W^{red}(R)$.

6.5.2 Allocation Areas

Here we investigate the more general case that the raw material consists of areas of different qualities. Obviously, the case with forbidden regions, as discussed in the previous subsection, can be seen as a special case, in which for all items the same parts of the raw material can be used.

6.5.2.1 Problem Formulation

The following 2D cutting problem is considered. Rectangular pieces i of various dimensions $\ell_i \times w_i$, $i \in I$, and different quality demands $q(i) \in Q$ have to be cut from a non-homogeneous raw material of size $L \times W$ in such a way that all allocation conditions (i.e., quality demands) are met and the total value of obtained pieces is maximal. It is allowed that pieces can be cut several times.

As in Sect. 6.4, the set Q denotes the set of all different quality demands. Moreover, let $I_q \subset I$ denote the set of all pieces with quality demand q, i.e., $I_q := \{i \in I \mid q(i) = q\}$. We assume $\cup_{q \in Q} I_q = I$ and $I_q \cap I_p = \emptyset$ for $q \neq p$, $q, p \in Q$.

Parts of the raw material, where a quality demand is fulfilled, are represented by an *allocation area* A_k, $k \in K := \{1, \ldots, |K|\}$. We assume that exactly one quality $q = \tilde{q}(k) \in Q$ is assigned to each $k \in K$, that the allocation areas are given in the form

$$A_k = \{(x, y) \mid a_k \leq x \leq c_k, b_k \leq y \leq d_k\} \subseteq [0, L] \times [0, W],$$

and that, for any $k \in K$, there is $i \in I_{\tilde{q}(k)}$ with

$$\ell_i \leq c_k - a_k, \quad w_i \leq d_k - b_k.$$

We allow that allocation areas can occur several times but then for different qualities, and that they can overlap although if they belong to the same quality demand. But we assume $A_k \not\subseteq A_j$ for all $k, j \in K_q$, $k \neq j$ and all $q \in Q$, where $K_q := \{k \in K \mid q = \tilde{q}(k)\}$.

For example, if in hardwood cutting a quality demand requires that no black knot is allowed, then the occurrence of a black knot causes up to four partially overlapping allocation areas.

6.5.2.2   Sets of Potential Allocation Points

For the 2D allocation (cutting or packing) problem let $v : [0, L] \times [0, W] \to \mathbb{Z}_+$, denote the optimal value function. This function is non-decreasing. Moreover, it is piecewise constant since only a finite number of different sequences of allocated pieces exists.

Clearly, if $\cup_{k \in K}(a_k, c_k) \neq (0, L)$ or $\cup_{k \in K}(b_k, d_k) \neq (0, W)$, then the problem can be split into subproblems or the size of the raw material can be reduced.

Let $S_L^*(\gamma, \ell)$ and $S_W^*(\gamma, w)$ denote the coordinates of the jump discontinuities in L- and W-direction of the optimal value function v. Our aim is to find supersets of $S_L^*(\gamma, \ell)$ and $S_W^*(\gamma, w)$ which are independent on γ_i, $i \in I$.

Theorem 13. *For any* $\ell \in \mathbb{Z}_>^m$, $w \in \mathbb{Z}_>^m$ *and* $\gamma \in \mathbb{Z}_>^m$, *we have*

$$S_L^*(\gamma, \ell) \subseteq S_L^{ap}(\ell) := \bigcup_{k \in K}(a_k \oplus S(\ell)) \cap [0, L],$$

$$S_W^*(\gamma, w) \subseteq S_W^{ap}(w) := \bigcup_{k \in K}(b_k \oplus S(w)) \cap [0, W].$$

Moreover, with $T_\ell := S_L^{ap}(\ell)$ *and* $T_w := S_W^{ap}(w)$, *for all* $(L', W') \in [0, L] \times [0, W]$,

$$v(L', W') = v(p_{T_\ell}(L'), p_{T_w}(W'))$$

holds.

The proof is similar to that of Theorem 3. The result of Theorem 13 can be strengthened with respect to the computational complexity by applying the Nicholson principle similar to Sect. 6.3.4. Let

$$S_L^\leftarrow(\ell) := \bigcup_{k \in K_0}(c_k \ominus S(\ell)) \cap [0, L], \quad S_W^\leftarrow(w) := \bigcup_{k \in K_0}(d_k \ominus S(\ell)) \cap [0, L].$$

denote the set of potential allocation points maximal in the sense that, for any $x \in S_L^\leftarrow(\ell)$, there is a combination of piece lengths whose first (left-most) piece, say i, has allocation point x and which is not feasible for allocation points for i larger than x (and similar in W-direction). Applying the Nicholson principle, we define reduced sets of allocation points

$$S_L^{red}(\ell) := \{p_{S_L^{ap}(\ell)}(x) \mid x \in S_L^\leftarrow(\ell)\}, \quad S_W^{red}(w) := \{p_{S_W^{ap}(w)}(x) \mid x \in S_W^\leftarrow(w)\}.$$

Theorem 14. *Let us consider the 2D cutting problem for the rectangle* $R = [L', L''] \times [W', W''']$ *with quality demands. Then, among all optimal patterns of this problem, there is an optimal pattern having only allocation points in* $S_L^{red}(\ell)$ *and* $S_W^{red}(w)$.

The proof is similar to that of Theorem 6.

The essential difference to the case without quality demands is again the fact that the yield of a rectangular region $R := [L', L''] \times [W', W''']$ of raw material depends on its position because of the varying quality. That means, the yield function used in a DP recursion is defined as follows:

$$\gamma(R) := \begin{cases} 0, & \text{if } I(R) = \emptyset, \\ \max\{\gamma_i \mid i \in I(R)\} & \text{otherwise,} \end{cases}$$

where $I(R) := \{i \in I \mid \exists k \in K_{q(i)} : R \subseteq A_k,\ L'' - L' \geq \ell_i,\ W'' - W' \geq w_i\}$.

Thus, a DP recursion to compute v which uses the sets of allocation points $S_L^{red}(\ell)$ and $S_W^{red}(w)$ is then as follows:

For all $R := [L', L''] \times [W', W''']$ with $L', L'' \in S_L^{red}(\ell)$ and $W', W''' \in S_W^{red}(w)$ set

$$\overline{v}(R) := \begin{cases} 0, & \text{if } L'' - L' < \ell_{min} \text{ or } W'' - W' < w_{min}, \\ \max\{\gamma(R), g(R), h(R)\} & \text{otherwise} \end{cases}$$

with

$$g(R) := \max\{\overline{v}(L', r, W', W'') + \overline{v}(r, L'', W', W''') \mid r \in S_L^{red}(\ell),\ L' < r < L''\},$$

$$h(R) := \max\{\overline{v}(L', L'', W', s) + \overline{v}(L', L'', s, W''')) \mid s \in S_W^{red}(w),\ W' < s < W'''\}.$$

The amount of time required to determine all $\overline{v}(R)$-values can be estimated as in Sect. 6.5.1.

Because of the dependence on the allocation areas, now $O(|S_L^{red}(\ell)|^2 \cdot |S_W^{red}(w)|^2)$ optimal values $\overline{v}(R)$ have to be computed with a DP approach. Due to this possibly large number, the usage of a B&B approach becomes more favorable. As upper bound for $v(R)$ one could simply use $u(L'' - L', W'' - W')$ but bounds which regard the allocation areas should be preferred.

Similar to the previous subsection, the number of small rectangles $R = [L', L''] \times [W', W''']$ used to define subproblems in a B&B algorithm can be further reduced. Therefore, for each R, the relevant allocation points should be computed as follows. Let $K_a(R), \ldots, K_d(R)$ denote those allocation areas which are relevant for allocating pieces into a rectangle R:

$$K_a(R) := \{k \in K \mid L' < a_k \le L'' - \ell_{min}, \ (W', W'') \cap [b_k, d_k] \ne \emptyset\},$$
$$K_b(R) := \{k \in K \mid W' < b_k \le W'' - w_{min}, \ (L', L'') \cap [b_k, c_k] \ne \emptyset\},$$
$$K_c(R) := \{k \in K \mid L' + \ell_{min} \le c_k < L'', \ (W', W'') \cap [a_k, d_k] \ne \emptyset\},$$
$$K_d(R) := \{k \in K \mid W' + w_{min} \le d_k < W'', \ (L', L'') \cap [a_k, c_k] \ne \emptyset\}.$$

The corresponding sets of allocation points are given by

$$\tilde{S}_L(R) := ((L' \oplus S(\ell)) \cup \{\textstyle\bigcup(a_k \oplus S(\ell)) \mid k \in K_a(R)\}) \cap [L', L''],$$
$$\tilde{S}_W(R) := ((W' \oplus S(w)) \cup \{\textstyle\bigcup(b_k \oplus S(w)) \mid k \in K_b(R)\}) \cap [W', W''].$$
$$\tilde{S}_L^{\leftarrow}(R) := ((L'' \ominus S(\ell)) \cup \{\textstyle\bigcup(c_k \ominus S(\ell)) \mid k \in K_c(R)\}) \cap [L', L''],$$
$$\tilde{S}_W^{\leftarrow}(R) := ((W'' \ominus S(w)) \cup \{\textstyle\bigcup(d_k \ominus S(w)) \mid k \in K_d(R)\}) \cap [W', W''].$$

Applying the Nicholson principle, we obtain the reduced sets of allocation points

$$\tilde{S}_L^{red}(R) := \{p_{\tilde{S}_L(R)}(x) \mid x \in S_L^{\leftarrow}(R)\}, \quad \tilde{S}_W^{red}(R) := \{p_{\tilde{S}_W(R)}(y) \mid y \in S_W^{\leftarrow}(R)\},$$

and we have

$$\tilde{S}_L^{red}(R) \subseteq S_L^{red}(\ell), \quad \tilde{S}_W^{red}(R) \subseteq S_W^{red}(w).$$

Finally, we get

Theorem 15. *We consider the 2D cutting problem for the rectangle $R = [L', L''] \times [W', W'']$ with quality demands. Then, among all optimal patterns of this problem, there is an optimal pattern which has only allocation points with coordinates in $\tilde{S}_L^{red}(R)$ and $\tilde{S}_W^{red}(R)$.*

Another option to reduce the number of subproblems in a B&B approach consists in replacing $S(\ell)$ and $S(w)$ by those sets which contain only combinations of lengths of items which can be feasibly placed within R.

6.5.3 Generalizations

As a generalization of the problem considered in the previous subsection, one may allow that the pieces have a variable size in one dimension and a fixed size for the other. Here we consider pieces with fixed widths and allow that the length ℓ_i of piece i can take any value in $[\underline{l}_i, \overline{l}_i]$. A related application occurs in hardwood cutting where, in general, the variable lengths are much larger than the fixed widths.

Although, from the theoretical point of view, it would be favorable to use a non-staged guillotine cutting technology, in practical application two-stage guillotine cutting (and probably, three-stage) is mostly used. We consider here an exact two-stage guillotine cutting (cf. [5]) with horizontal cuts, (i.e., in L-direction) in the first stage. No trimming is allowed. Different quality demands are represented again by allocation areas.

As a naive (basic) solution approach one can compute for each different width $\tilde{w}_j$ and each potential allocation point y the optimal value $v(y, \tilde{w}_j)$ for the part of raw material $[0, L] \times [y, y + \tilde{w}_j]$. In order to limit the computational amount the set of potential allocation points has to be defined appropriately, for instance as $S_W^{red}(w)$. The computation of $v(y, \tilde{w}_j)$ is in fact a 1D cutting problem whose input data are obtained as follows. For a piece $i \in I$ with $w_i = \tilde{w}_j$ and an allocation area $A_k = A_k(a_k, b_k, c_k, d_k)$ with $k \in K_{q(i)}$, the restriction of A_k to the strip $[0, L]x \times [y, y + \tilde{w}_j]$ determines the allocation interval $[a_k, c_k]$ for item i if $[y, y + \tilde{w}_i] \subseteq [b_k, d_k]$ and $l_i \leq c_k - a_k$.

Having computed all values $v(y, \tilde{w}_j)$ an optimal combination of the strips can be obtained by solving a 1D cutting problem in W-direction.

Depending on the real cutting technology, the positions of cross (vertical) and rip (horizontal) cuts can be restricted by reduced sets of allocation points in a similar way.

Moreover, practical requirements, such as a least distance between two cut (allocation) positions or a positive kerf, can be regarded within the proposed solution approaches or in the definition of sets of allocation points. If there are restrictions on how often a piece shall be placed, then an appropriate definition of the set $S(\ell)$ should be used, namely

$$\widehat{S}(\ell, u) := \left\{ \sum_{i \in I} \ell_i x_i \mid x_i \leq u_i, \, x_i \in \mathbb{Z}_+, \, i \in I \right\}.$$

6.6   Conclusions

Within this paper we considered one- and two-dimensional cutting and packing problems with additional placement constraints. Such constraints can be caused by defective parts of the raw material or by parts which satisfy different quality demands. We identified appropriate (reduced) sets of potential allocation points that do not depend on the profit coefficients. These sets either cover the set of jump discontinuities of the optimal value function of the allocation problem, or at least contain appropriate allocation points which still allow to compute an optimal pattern.

The proposed sets of potential allocation points strongly depend on the real data. If there are very small-sized pieces or many defects or different quality regions, the cardinality of these sets can be large but not greater than the corresponding size parameter of the raw material. In case of rather large pieces a high potential to save computational costs arises if the proposed sets of allocation points are used. Moreover, the cardinality of these sets does not change if the unit of measure is changed.

In the one-dimensional case, the explicit computation of a (reduced) set of potential allocation points may look as a meaningless expense, but the basic principle of its definition can be regarded directly within the solution approach as shown in the algorithms for 1D cutting problems with allocation intervals.

In the two-dimensional case, the construction of the (reduced) sets remains, in fact, a one-dimensional task since both dimensions can be handled independently. The use of the proposed sets of potential allocation points can lead to a significant reduction of the number of states, which have to be considered in a DP based approach, and, in a similar way, to a reduction of the number of subproblems which arise during a B&B based solution process. In particular, these reductions are of high importance in cases with complex quality demands.

An appropriate use of the profit coefficients of a problem might be helpful to further reduce the number of allocation points so that the computational effort for obtaining an optimal pattern can be reduced. This topic is left for future research. Moreover, we would like to mention that the definition of good allocation areas might become a non-trivial task if difficult quality demands have to be met, for example one may think of conditions on the number of knots within a certain area of a wooden board.

Acknowledgements This work is supported in a part by the German Research Foundation (DFG) in the Collaborative Research Center 912 "Highly Adaptive Energy-Efficient Computing."

References

1. Astrand, E., Rönnqvist, M.: Crosscut optimization of boards given complete defect information. For. Prod. J. **44**, 15–24 (1994)
2. Beasley, J.E.: An exact two-dimensional non-guillotine cutting tree search procedure. Oper. Res. **33**, 49–65 (1985)
3. Cintra, G.F., Miyazawa, F.K., Wakabayashi, Y., Xavier, E.C.: Algorithms for two-dimensional cutting stock and strip packing problems using dynamic programming and column generation. Eur. J. Oper. Res. **191**, 61–85 (2008)
4. Dowsland, K.A.: The three-dimensional pallet chart: An analysis of the factors affecting the set of feasible layouts for a class of two-dimensional packing problems. J. Oper. Res. Soc. **35**, 895–905 (1984)
5. Gilmore, P.C., Gomory, R.E.: Multistage cutting stock problems of two and more dimensions. Oper. Res. **13**, 94–120 (1965)
6. Hahn, S.G.: On the optimal cutting of defective sheets. Oper. Res. **16**, 1100–1114 (1968)
7. Herz, J.C.: Recursive computational procedure for two-dimensional stock cutting. IBM J. Res. Dev. **16**, 462–469 (1972)
8. Nicholson, T.A.J.: Finding the shortest route between two points in a network. Comp. J. **9**, 275–280 (1966)

9. Rönnqvist, M.: A method for the cutting stock problem with different qualities. Eur. J. Oper. Res. **83**, 57–68 (1995)

10. Rönnqvist, M., Astrand, E.: Integrated defect detection and optimization for cross cutting of wooden boards. Eur. J. Oper. Res. **108**, 490–508 (1998)

11. Scheithauer, G.: The solution of packing problems with pieces of variable length and additional allocation constraints. Optimization **34**, 81–96 (1995)

12. Scheithauer, G.: Zuschnitt- und Packungsoptimierung. Vieweg + Teubner, Wiesbaden (2008)

13. Scheithauer, G., Terno, J.: Guillotine cutting of defective boards. Optimization **19**, 111–121 (1988)

14. Sweeney, P.E., Haessler, R.W.: One-dimensional cutting stock decisions for rolls with multiple quality grades. Eur. J. Oper. Res. **44**, 224–231 (1990)

15. Sweeney, P.E., Paternoster, E.R.: Cutting and packing problems: A categorized application-oriented research bibliography. J. Oper. Res. Soc. **43**, 691–706 (1992)

16. Terno, J., Lindemann, R., Scheithauer, G.: Zuschnittprobleme und ihre praktische Lösung. Harri Deutsch, Thun and Frankfurt/Main (1987)

17. Wäscher, G., Haußner, H., Schumann, H.: An improved typology of cutting and packing problems. Eur. J. Oper. Res. **183**, 1109–1130 (2007)

Chapter 7
A Container Loading Problem MILP-Based Heuristics Solved by CPLEX: An Experimental Analysis

Stefano Gliozzi, Alessandro Castellazzo, and Giorgio Fasano

Abstract The issue of placing small boxes orthogonally, generally with the possibility of rotations, into a big box, maximizing the loaded volume, is usually referred to as the container loading problem. Despite its being notoriously of an NP-hard typology, a number of algorithms work out this problem very efficiently. The task becomes, nonetheless, even more challenging when additional conditions have to be taken account of. In such cases, a modeling-based approach is supposedly the most suitable and this definitely holds, in particular, when balancing requirements are posed. These, indeed, entail constraints of strong global impact that can hardly be coped with by sequential procedures, based on a step by step incremental loading of items.

MIP (Mixed Integer Programming) models relevant to the container loading problem or possible extensions of it are available in specialized literature. A dedicated MILP (Mixed Integer Linear Programming) formulation, supporting an overall heuristic approach, addressed to non-standard packing issues, is discussed in another chapter of this book. Hereinafter, some relevant computational aspects are looked into, restricting the consideration to the container loading problem, as per its classical statement. An ad hoc heuristics, derived from the above-mentioned overall approach, is outlined. The use of IBM ILOG CPLEX as an MILP optimizer is considered. Case studies concerning the solution of the MILP model tout court, when the instances involved are not of a large-scale nature, are reported first.

S. Gliozzi (✉)
IBM Italia S.p.A., Rome, Italy
e-mail: Stefano_gliozzi@it.ibm.com

A. Castellazzo
Altran Italia S.p.A. Consultant c/o Thales Alenia Space Italia S.p.A., Turin, Italy
e-mail: alessandro.castellazzo@external.thalesaleniaspace.com

G. Fasano
Thales Alenia Space Italia S.p.A., Turin, Italy
e-mail: giorgio.fasano@thalesaleniaspace.com

© Springer International Publishing Switzerland 2015
G. Fasano, J.D. Pintér (eds.), *Optimized Packings with Applications*, Springer
Optimization and Its Applications 105, DOI 10.1007/978-3-319-18899-7_7

157

Outcomes relevant to the ad hoc heuristics are further shown through a number of difficult instances. Examples of container loading issues, involving also balancing conditions, are additionally provided.

Keywords Container loading • Orthogonal packing with rotations • Mixed integer linear programming • MILP model • Heuristics • CPLEX • Computational results

7.1 Introduction

Container loading is a typical packing problem, concerning the orthogonal placement of small boxes (i.e., rectangular parallelepipeds) into a big box, maximizing the loaded volume. A very large number of specialist works are devoted to this subject and the reader is referred to the available literature for a wide-ranging overview (e.g., [1]). Hereinafter, we shall recall the modeling-based methodology discussed in depth in a dedicated chapter of this book [2].

This approach has been conceived to solve complex non-standard packing problems, allowing for tetris-like items inside convex domains, with additional conditions, such as balancing. The container loading problem, as per its classical formulation, represents a specific case addressed by the general MILP (Mixed Integer Linear Programming) mathematical model discussed in Fasano [2], Section 2. This can be utilized, directly, when a limited number of items are involved. Otherwise, when large-scale instances have to be coped with, the above-mentioned MILP model represents the basic "engine" of the overall heuristic approach outlined in Fasano [2], Section 6. Specific versions of this model are, in such cases, adopted to support all the relevant phases of the whole heuristic process, i.e.: *Initialization, Packing, Item-exchange, Hole-filling*. The present chapter focuses on what we currently consider the most promising solution strategies relevant to the modeling-based approach in question. These act at two different levels.

Firstly, an ad hoc strategy, delineated in Sect. 7.2.1, has been looked into for the above-mentioned overall heuristic procedure. As pointed out in Fasano [2], Section 6, indeed, the way the various modules (i.e., *Initialization, Packing, Item-exchange, Hole-filling*) are activated/executed actually determines a specific heuristics.

Secondly, dedicated MILP strategies have been studied to work out the general MILP model, as utilized in its different versions, i.e. either when the container loading problem is tackled tout court, or the various phases of the heuristic process have to be performed. As is well known, when managing an MILP model, the solution search effectiveness is strongly affected by the general features of the optimizer adopted, but even more by the way it is "driven." For instance, different branch and bound (B&B) strategies may yield very different outcomes, both in terms of solution quality and computational effort. In our research, IBM ILOG CPLEX [3] has been selected as the MILP solver and appropriate *drivers* set up to solve the MILP model, in its various versions, efficiently. This is the subject of Sect. 7.2.2.

A dedicated experimental analysis, covering non-trivial instances, has been carried out. Although the strategies proposed here are suitable for a number of non-standard packing issues, not limited to the classical container loading problem, our attention has been concentrated on it. In this case, the procedure put forward in the present chapter (as well as the modeling-based approach in general), being aimed at non-standard applications, is typically outperformed by most of the off-the-shelf algorithms, specific for the classical container loading problem (e.g., [4]). The choice of focusing on this problem in particular, however, aims to offer a useful reference in a standard test framework. Section 7.3.1 reports computational results relevant to the direct solution of the MILP model, with small-scale instances. Section 7.3.2 shows outcomes regarding demanding test cases, solved by the heuristics of Sect. 7.2.1. Insights concerning the presence of balancing conditions are additionally provided in Sect. 7.3.3, to draw the reader's attention to an application quite frequent in practice.

7.2 Solution Search Strategies

7.2.1 Heuristic Approach

A specific procedure deriving from the overall heuristic approach proposed in Fasano [2] is outlined in this section (in a streamlined form). Major interrelated concepts are those of *relative position* and *abstract configuration*. A *relative position* between two items expresses that one, with respect to the other, is located in compliance with one of the following conditions: on the left, on the right, in front, behind, above or below. An *abstract configuration*, relative to N items, is a set of $\frac{N(N-1)}{2}$ *relative positions*, one for each pair of items, that are feasible (i.e., all of them can be respected) in any unbounded domain. With a given *abstract configuration* items may be rotated and translated, keeping their *relative positions* unaltered.

The underlying idea of the overall heuristic approach is to generate a sequence of *abstract configurations* that allow the feasible placement (i.e., with no overlapping) of an increasing number of items in the given domain (i.e., the container), maximizing the loaded volume.

In the specific heuristics proposed in this chapter, the whole process is split into two macro-phases, i.e. the *main* and the *incremental* one, respectively, see Fig. 7.1. Both of them activate the modules of the overall heuristic approach (i.e., *Initialization, Packing, Item-exchange, Hole-filling*) sequentially. The macro-phases are executed recursively, performing a number of cycles. Items are added, time after time, following an overall *greedy* approach. At each module execution, the selection of items as candidates for loading (in addition to those previously accepted) is made on a larger-first priority criterion. This way, the procedure attempts to load items with the largest volumes whilst the domain is still quite unexploited. On the contrary, the smaller ones are tentatively introduced to fill the empty spaces, when

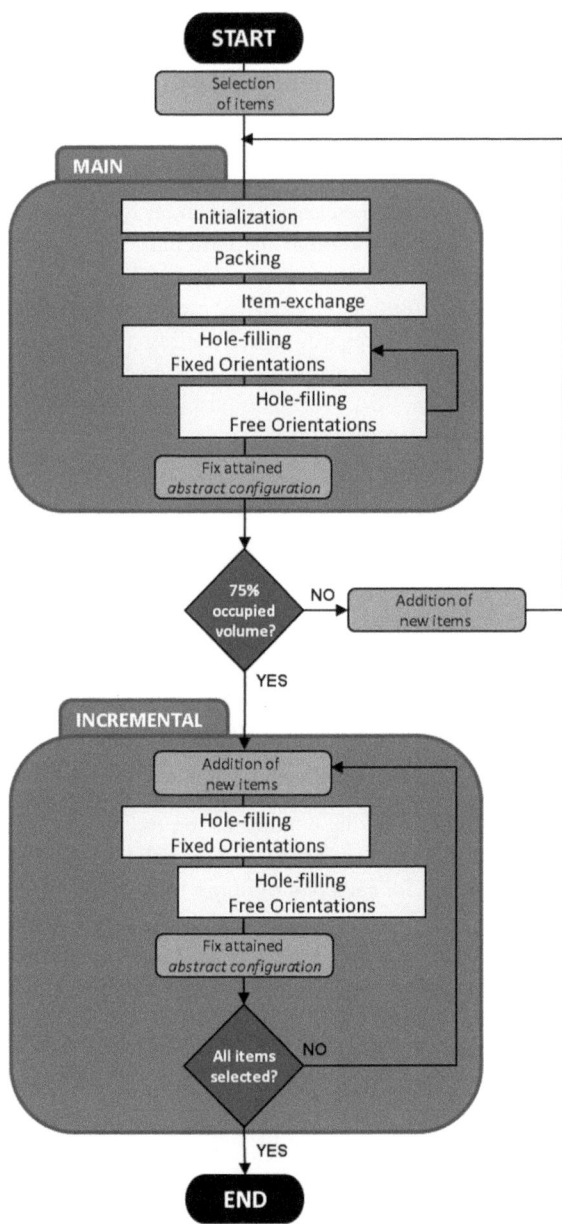

Fig. 7.1 Heuristics overall logic

high loading percentages have already been attained. At each module execution, the current *abstract configuration* is taken as input and an upgraded one is provided by it (if executed successfully).

7.2.1.1 *Main* Phase

A single cycle of the *main* phase consists of the prefixed sequence: *Initialization*, *Packing*, *Item-exchange*, and *Hole-filling*.

Their specific functionalities are summarized here below:

- the *Initialization* module generates a preliminary *abstract configuration*, for a subset of the items available, by means of a *relaxation* of the general MILP model (allowing item overlapping);
- the *Packing* module places items into the domain, in compliance with the current *abstract configuration* and maximizing the total volume loaded;
- the *Item-exchange* module attempts advantageous exchanges between (subsets of) non-loaded and loaded items (both are set free, with respect to the current *abstract configuration*, and a new one is correspondently generated, if the module execution has been successful);
- the *Hole-filling* module tries to add some of the unloaded items (that are set free, with respect to the current *abstract configuration*, and an upgraded one is correspondently generated, if the module execution has been successful). The *Hole-filling* module is first executed by fixing the orientation of all items involved. Afterwards, it is re-executed, if opportune, setting the item orientations free.

The *main* phase is carried on, by repeating single cycles, until either the loaded volume has attained 75 % of the domain's or a maximum time limit has been reached. The *abstract configuration* obtained at the end of this phase is handed over to the next.

7.2.1.2 *Incremental* Phase

A single cycle of the *incremental* phase consists of the prefixed sequence: *Hole-filling* and *Item-exchange* (at this level of volume exploitation, indeed, the first two modules are no longer effective, in particular *Initialization*, being based on a *relaxation* of the MILP model). Also in this case, the *Hole-filling* module is firstly executed by fixing the orientation of all items involved. Afterwards, if re-executed, these are set free. Here, the *Item-exchange* module has the role of performing backward iterations (to make up for possible previous inappropriate moves). The relevant functionalities of both modules employed are the same as described above. A number of single cycles are executed, until either all items have been processed or a maximum time limit has been reached.

7.2.2   Model Solving by CPLEX

The heuristics introduced in Sect. 7.2.1, and all of the tests reported in this chapter, were performed utilizing IBM ILOG CPLEX (see [3]) as the MILP optimizer. CPLEX carries out the optimization process by a branch & cut (B&C) algorithm, including several general purpose heuristics. It is also able to perform parallel optimization. Like most of the optimizers available to date, CPLEX has a default strategy for the MILP solution, which is flexible and adaptable to the model characteristics. Its level of sophistication is so advanced that a number of ad hoc optimizer parameters, able to outperform the default mode, can hardly be found. Moreover, the risk of "over engineering" the setting of the parameters, tuning them to a particular class of instances, rather than to the model intrinsic characteristics, cannot be neglected. Sometimes however, it can be useful to define a specific CPLEX optimization strategy. This holds, in particular, when the solution search is somehow time-boxed, and the proof of optimality is not necessary. This is the case of the two situations dealt with in Sect. 7.3, concerning either the solution of the MILP model directly or the execution of the heuristics of Sect. 7.2.1.

7.2.2.1   Direct Solution

When some difficult instances, albeit with a limited number of items, are tackled by solving the MILP model directly, i.e. in the first situation, the number of nodes generated by the B&C procedure really tends to "explode." In this circumstance, a specific strategy is needed, in order to reduce the node generation as much as possible and make the process spend more time at each B&B step, yielding (supposedly) better search choices. Another characteristic associated with the MILP model in question is that the LP-*relaxation* upper bound is usually coincident with the value of the optimal (integer) MILP solution. As a consequence, any strategy, aimed at generating cuts and improving the upper bound, results in being ineffective.

To cope with these difficult instances, an ad hoc approach was therefore devised. It is based on an intense employment of (CPLEX) heuristics, probing techniques (see [3]), a very limited use of cutting planes, and the "solution polishing" heuristics (see [3, 5]) that are activated when several solutions and at least 200 nodes have already been explored.

The priority order of the branching variables represents a further important feature of the approach studied. Since, from the model formulation and from the solution logic, some binary variables are supposedly able to induce a better separation in the search tree, they are provided with a higher priority in the process. Figure 7.2 illustrates the multiple CPLEX parameters (see [3]) that have been selected (the parameters not shown in the figure correspond to the default setting).

```
CPLEX Parameter File Version 12.6.0.1
#CPLEX Tuning for Direct Solution
CPX_PARAM_PRELINEAR 0
CPX_PARAM_MIPCBREDLP 0
CPX_PARAM_NODEFILEIND 3
CPX_PARAM_TILIM 3600
CPX_PARAM_POLISHAFTEREPAGAP 0.02
CPX_PARAM_POLISHAFTERNODE 200
CPX_PARAM_POLISHAFTERTIME 2400
CPX_PARAM_MIPEMPHASIS 1
CPX_PARAM_FLOWCOVERS 1
CPX_PARAM_MIRCUTS 1
CPX_PARAM_PRESLVND 3
CPX_PARAM_PROBE 3
CPX_PARAM_REPEATPRESOLVE 1
CPX_PARAM_RINSHEUR 5
CPX_PARAM_LBHEUR 1
CPX_PARAM_FRACCUTS -1
CPX_PARAM_LANDPCUTS 1
CPX_PARAM_SYMMETRY -1
```

Fig. 7.2 CPLEX parameter selection for the direct solution

```
CPLEX Parameter File Version 12.6.0.1
#CPLEX Tuning for Initialization
CPX_PARAM_TILIM               40
CPX_PARAM_PRELINEAR   0
CPX_PARAM_BRDIR              1
CPX_PARAM_POLISHAFTEREPGAP 0.1
```

```
CPLEX Parameter File Version 12.6.0.1
#CPLEX Tuning for
#Packing, Item-exchange and Hole-filling
CPX_PARAM_PRELINEAR 0
CPX_PARAM_MIPCBREDLP 0
CPX_PARAM_BRDIR 1
CPX_PARAM_OBJDIF 0.0001
```

Fig. 7.3 CPLEX parameter selection for the heuristic solution

7.2.2.2 Heuristic Solution

When the heuristics of Sect. 7.2.1 is utilized, i.e. in the second situation, it is of paramount importance to obtain quick, albeit sub-optimal, solutions for each module execution. Proof of optimality is not needed at all, although sometimes the heuristics performances can be biased by too many run interruptions (based on predefined-maximum-time limits). The priority order of the branching variable declaration is the same as in the direct solution situation. Figure 7.3 reports the CPLEX parameters adopted for *Initialization*, *Packing*, *Item-exchange*, and *Hole-filling*, respectively.

7.3 Experimental Analysis

The experimental analysis of this section is an extension of the previous, reported in Fasano [6]. The test campaign referred to hereinafter was performed using IBM CPLEX 12.6.0.1 (see [3]) as the optimizing engine, and IBM EasyModeler as the model generator. More precisely, the MILP solver available within CPLEX, statically linked to the C++ code generated by EasyModeler, was adopted, using the open source Coin-OR OSI 0.105.3 library as the interface between EasyModeler and the optimizer. The following computational supports were moreover utilized:

- platform: Lenovo Thinkpad W520 Laptop. with an Intel(R) Core (TM) i7-2620M at 2.7 GHz clock frequency (2 real core seen as 4 with Intel Hyperthreading) and 8 GB Ram available;
- operating system: Windows(R) 7 Professional OS.

All the tests were run using a parallel version of CPLEX. CPLEX 12.6 can execute the B&C in two different parallel flavors: Parallel Optimization on threads on the same (multi core) CPU, and Distributed Parallel with a messaging protocol among distinct CPUs. During the tests, the Parallel Optimization on Threads was employed; the number of Threads is defaulted to the number of cores seen by the OS, i.e. 4.

This section reports first a group of tests concerning the solution of the MILP model directly. Experimental results relevant to the use of the heuristics outlined in Sect. 7.2.1 are presented next. Additionally, instances with the balancing requirement are provided. All the case studies considered hereinafter involve box-shaped items and domains.

7.3.1 Direct Solution of Standard Instances

In order to test the MILP model for solving the container loading problem directly, we selected 5 fabricated instances (see [7]), whose optimal solutions were known a priori. Among them, 4 have a cube as a domain of 8, 9, 10, and 11 units, respectively. They are denoted in the following as: Cube-8, Cube-9, Cube-10, and Cube-11 tests. The domain of the further instance is a rectangular parallelepiped, obtained by merging two Cube-8 domains. The relevant test is referred to as: Double-cube-8. For all the tests considered in this section, no additional condition was posed. The instance data concerning the items available are reported, test by test, in the following Tables 7.1, 7.2, 7.3, 7.4, and 7.5.

A maximum time limit of 1 h was set for each test case. Two out of five (i.e., Cube-8 and Cube-9) were solved to optimality; in two cases (i.e., Cube-10 and Cube-11) only one item was rejected; in one (i.e., Double-cube-8) those not loaded were three. The relevant results are shown in Table 7.6 while Figs. 7.4 and 7.5 provide graphical views of the solutions obtained for Cube-8 and Double-cube-8.

Table 7.1 Cube-8 items

Item type	L1 side (units)	L2 side (units)	L3 side (units)	No. of items per type
A	4	4	4	1
B	2	3	5	6
C	1	3	6	6
D	1	2	6	6
E	1	3	3	6
F	1	2	2	6

Table 7.2 Cube-9 items

Item type	L1 side (units)	L2 side (units)	L3 side (units)	No. of items per type
A	5	5	5	1
B	2	4	6	6
C	1	3	7	6
D	1	2	7	6
E	1	3	4	6
F	1	2	2	6

Table 7.3 Cube-10 items

Item type	L1 side (units)	L2 side (units)	L3 side (units)	No. of items per type
A	6	6	6	1
B	2	5	7	6
C	1	3	8	6
D	1	2	8	6
E	1	3	5	6
F	1	2	2	6

Table 7.4 Cube-11 items

Item type	L1 side (units)	L2 side (units)	L3 side (units)	No. of items per type
A	7	7	7	1
B	2	6	8	6
C	1	3	9	6
D	1	2	9	6
E	1	3	6	6
F	1	2	2	6

Instances of the above tests, with all items pre-oriented (correspondently to the fabricated optimal solutions) were considered. Surprisingly enough, none of them was solved to optimality within 1 h. Further pre-oriented instances for Cube-8 and Cube-9 were hence taken into account, additionally. In such cases, the item pre-orientation was derived from the solutions reported in Table 7.6. The optimal solutions (or some equivalent) were re-obtained in almost half the time of the

Table 7.5 Double-cube-8 items

Item type	L1 side (units)	L2 side (units)	L3 side (units)	No. of items per type
A	4	4	4	2
B	2	3	5	12
C	1	3	6	12
D	1	2	6	12
E	1	3	3	12
F	1	2	2	12

Table 7.6 Direct solution tests

Test case	Max no. of items	No. of loaded items	Load factor (%)	Elapsed time (s)	Loaded items (%)	No. of nodes	Optimality proved
Cube-8	31	31	98.05	312	100.00	231	Yes
Cube-9	31	31	98.63	2,001	100.00	315	Yes
Cube-10	31	30	96.60	3,600	96.77	403	No
Cube-11	31	30	96.54	3,600	96.77	463	No
Double-cube-8	62	59	89.26	3,600	95.16	270	No

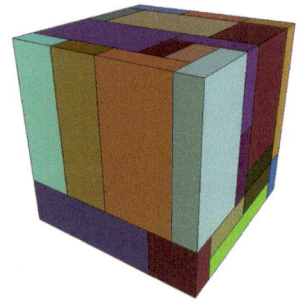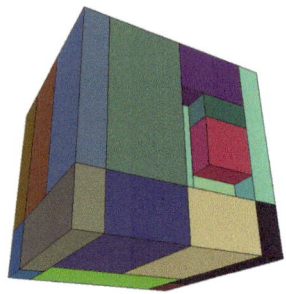

Fig. 7.4 Cube-8 graphical results

Fig. 7.5 Double-cube-8 graphical results

non-pre-oriented instances. Our interpretation in regard is that, since several *non-symmetric* solutions exist, the more "degrees of freedom" there are, the more effective the CPLEX heuristics (including the "solution polishing") results.

7.3.2   Heuristic Solution of Standard Instances

This section refers to 27 non-trivial test cases for the classical container loading problem, with no additional conditions. They are extracted from the reference: "Three Dimensional Cutting and Packing Data Sets - THPACK 1-7 BR" [8]: http://www.euro-online.org/web/ewg/25/esicup-euro-special-interest-group-on-cutting-and-packing. As is known, this test-bed consists of 7 sets of 100 test cases each. Among these, all those with an available number of items between 200 and 400 were selected. They are listed in Table 7.7 and, hereinafter, numbered sequentially, from 1 to 27.

The 27 selected test cases were solved using the heuristics of Sect. 7.2.1, with a maximum time limit of 1 h. The relevant results are summarized in Table 7.8 (the reported time elapses are often longer than 1 h, since the heuristics always finalized the last optimization steps, prior to performing the final housekeeping) (Fig. 7.6).

Load factors range from 57.45 to 87.07 %, with an average of 77.92 % and a standard deviation of 7.52 (graphical results relative to Test case 17 are illustrated in Fig. 7.7). It is interesting to note that the time spent in optimization is inversely correlated, as pointed out in Fig. 7.6. This appears as an indication that the heuristics' logic itself is more relevant than the optimizer speed. The instances are solved to a greater extent when the heuristics is able to generate easier sub-instances to solve. The heuristics' overall logic and its specific module features (including possible function extensions) are expected to represent the objective of further research.

Table 7.7 Selected test cases (THPACK 1-7 BR)

Set number	Test case
1	13,17,33,39,67,68,76,85,91,100
2	4,13,39,59,77,79,85,96
3	39,56,59,77
4	39,56,79
5	56
6	13

Table 7.8 Results for the selected test cases (THPACK 1-7 BR)

Test case	Max no. of items	No. of loaded items	Load factor (%)	Elapsed time	Loaded items (%)	Time spent in opti- mization	Time spent in opti- mization (%)	No. of sub- models solved
1	284	186	84.60	01:01:28	65.5	00:19:05	31.0	86
2	213	155	84.96	01:01:59	72.8	00:35:31	57.3	71
3	282	159	76.57	01:01:08	56.4	00:38:08	62.4	52
4	243	163	85.57	01:00:49	67.1	00:27:32	45.3	78
5	221	140	72.44	01:04:49	63.3	00:47:48	73.7	50
6	238	119	57.45	01:00:20	50.0	00:52:24	86.9	37
7	269	149	59.30	01:02:21	55.4	00:49:28	79.3	42
8	319	165	67.39	01:01:25	51.7	00:46:29	75.7	40
9	238	149	86.82	00:54:08	62.6	00:28:41	53.0	93
10	214	151	79.18	00:55:35	70.6	00:35:49	64.4	64
11	201	143	87.07	00:53:25	71.1	00:31:57	59.8	83
12	228	174	84.87	01:01:29	76.3	00:27:27	44.6	84
13	266	168	78.34	01:01:13	63.2	00:36:58	60.4	64
14	201	129	69.62	00:52:51	64.2	00:43:21	82.0	39
15	202	144	77.08	00:58:37	71.3	00:43:38	74.4	52
16	206	145	79.46	01:00:37	70.4	00:40:17	66.5	74
17	209	163	85.32	01:01:20	78.0	00:36:30	59.5	74
18	202	139	82.01	01:03:20	68.8	00:31:56	50.4	72
19	232	200	77.11	01:01:35	86.2	00:31:07	50.5	69
20	212	157	82.89	01:00:58	74.1	00:33:58	55.7	84
21	216	145	74.05	01:00:38	67.1	00:45:52	75.6	50
22	201	135	76.88	01:01:22	67.2	00:44:10	72.0	57
23	225	138	77.60	01:01:02	61.3	00:40:57	67.1	66
24	233	151	80.47	01:01:46	64.8	00:39:40	64.2	56
25	217	138	80.42	01:01:32	63.6	00:39:41	64.5	72
26	218	131	78.11	01:00:37	60.1	00:42:30	70.1	63
27	203	119	78.15	01:00:40	58.6	00:44:54	74.0	64

7.3.3 Heuristic Solution of Test Cases with Balancing Conditions

An extension of the classical loading problem is briefly discussed here. The (quite frequent in practice) balancing requirement, for which the overall center of mass (of the loaded container) must stay inside a convex domain is considered (see [6]). Each item (supposed to be of homogeneous density) is therefore represented by its side lengths and mass. In the following sections (for the sake of simplicity) no mass is associated with the container itself and the overall center of mass domain is assumed to be a (rectangular) parallelepiped.

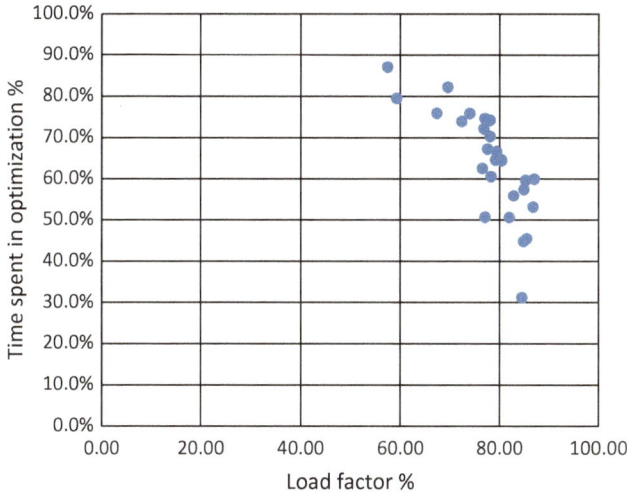

Fig. 7.6 Correlation between load factor and time spent in optimization

Fig. 7.7 Test case 17 graphical results

7.3.3.1   Items Having the Same Density

The 27 test cases of Sect. 7.3.2 were reconsidered assuming that all items had the same density. These test cases were run with a very tight restriction on the center of mass domain, which consisted of a cube of 2 units, centered with respect to the container. Since the container was a box of $587 \times 233 \times 220$ units, this meant a deviation from its center well below 0.5 % its side lengths. The relevant results are reported in Table 7.9.

Both from the load factor and computational performance viewpoints, the results are quite similar to the test cases reported in Sect. 7.3.2. A slight deviation from them can be noticed, consisting, essentially, of an average load factor decrement of a mere 1.14 %.

Table 7.9 Results for the selected test cases with items of constant density

Test case	Max no. of items	No. of loaded items	Load factor (%)	Elapsed time	Loaded items (%)
1	284	173	82.57	01:00:40	60.9
2	213	151	83.80	00:46:07	70.9
3	282	168	75.21	01:01:15	59.6
4	243	152	83.84	00:58:54	62.6
5	221	144	74.16	00:40:51	65.2
6	238	120	57.87	01:00:58	50.4
7	269	155	61.18	01:00:56	57.6
8	319	141	57.60	01:03:08	44.2
9	238	151	86.31	00:59:23	63.4
10	214	147	76.67	01:00:42	68.7
11	201	142	85.04	00:50:41	70.6
12	228	175	84.69	01:00:56	76.8
13	266	167	78.66	01:01:10	62.8
14	201	126	68.29	01:00:07	62.7
15	202	148	78.99	00:58:47	73.3
16	206	139	75.96	01:00:37	67.5
17	209	159	84.22	00:45:55	76.1
18	202	131	79.41	00:56:22	64.9
19	232	196	76.49	01:01:57	84.5
20	212	162	83.29	01:01:14	76.4
21	216	141	73.42	01:05:13	65.3
22	201	137	77.02	01:02:57	68.2
23	225	142	78.83	01:00:53	63.1
24	233	155	80.94	01:04:15	66.5
25	217	109	72.35	01:00:54	50.2
26	218	128	77.27	01:00:41	58.7
27	203	121	78.92	01:02:15	59.6

7.3.3.2 Items Having Different Densities

Three test cases, i.e. 4, 19 and 25, extracted from the set of 27 of Sect. 7.3.2 were considered, by providing the items with different densities (generated randomly). These are reported in Table 7.10 (referring to mass and volume units). For these three test cases the overall center of mass was requested to stay inside a slightly larger domain (roughly representing 5 % of tolerance over the length of each axis), centered with respect to the container. Table 7.11 shows the relevant results and Fig. 7.8 provides graphical views of the solution obtained for Test case 4.

Finally Test cases 4, 19 and 25 were considered, maintaining the same masses reported in Table 7.10, but with different conditions concerning the position of the center of mass domain (that continued to have the same dimension as before). This was placed in an off-centered position, inside the container. This request can occur

Table 7.10 Items with different densities

Test case	Average density	Standard deviation
4	1.5276	0.2373
19	1.5211	0.2269
25	1.5134	0.2317

Table 7.11 Results for Test cases 4, 19, and 25 with central balancing

Test case	Max no. of items	No. of loaded items	Load factor (%)	Elapsed time	Loaded items (%)
4	243	158	84.53	01:02:15	65.0
19	232	137	77.32	01:01:51	59.1
25	217	123	77.03	01:00:31	56.7

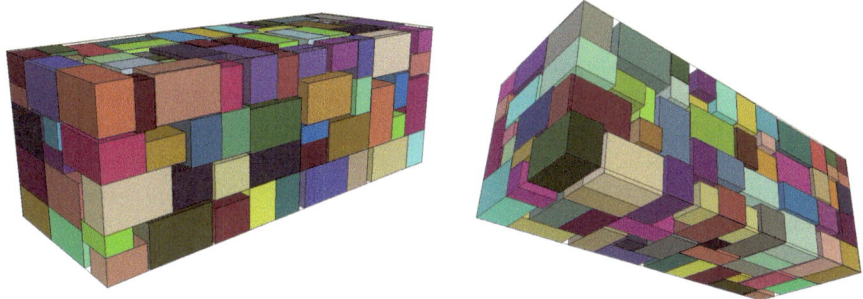

Fig. 7.8 Test case 4 (with balancing conditions) graphical results

Table 7.12 Center of mass domain off-centered locations

Domain dimensions			Center of mass coordinates		
x	y	z	x	y	z
587	233	220	293.5	116.5	55

Table 7.13 Results for Test cases 4, 19, and 25 with off-centered balancing

Test case	Max no. of items	No. of loaded items	Load factor (%)	Elapsed time	Loaded items (%)
4	243	87	51.31	01:06:35	34.4
19	232	73	47.36	01:03:10	31.5
25	217	46	36.77	01:01:54	21.2

in practice, for instance, for structural reasons (see [9]). Table 7.12 reports, for Test cases 4, 19, and 25 respectively, the positions of the relevant domain centers. The results obtained are shown in Table 7.13. Test case 4 solution is represented graphically in Fig. 7.9.

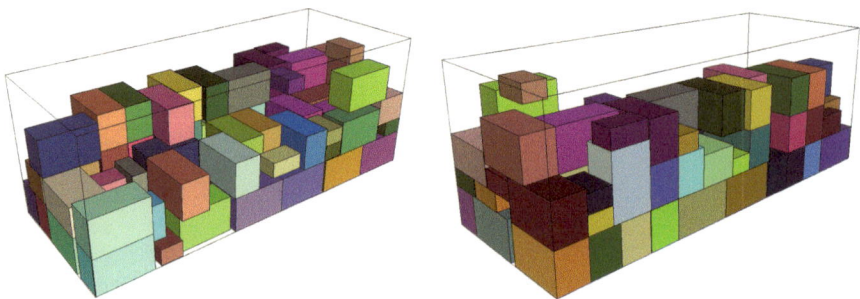

Fig. 7.9 Test case 4 (with off-centered balancing conditions) graphical results

7.4 Conclusive Remarks

This work focuses on experimental aspects relevant to the container loading problem, solved by a modeling-based heuristic approach. The relevant MILP model is discussed in depth in another chapter of this book and represents the reference framework for the underlying mathematical formulation. This approach is aimed at coping with complex non-standard packing problems, involving *tetris*-like items, non-box-shaped domains and additional conditions, such as balancing.

A standard context has, nonetheless, been targeted in this chapter, addressing the container loading problem, as per its classical statement. This concerns the placement of box-shaped items (with the possibility of rotation) into a box-shaped domain, with no additional conditions, maximizing the loaded volume.

The general MILP model has been tested to solve directly non-large-scale fabricated instances, whose optimal solutions were known a priori. Afterwards, a set of complex case studies have been studied and further examples involving the additional condition of balancing provided.

Although for the specific experimental context considered, the proposed approach is usually outperformed by most of the off-the-shelf container loading optimization methods, the authors deem that the results shown here provide a useful reference in a standard-based framework. Further research relevant both to the heuristics overall logics and its specific features, referred to a more general non-standard context, is in the pipeline for the near future.

Acknowledgements We wish to thank Janos D. Pintér for his accurate review of the manuscript. We are also grateful to Jane Evans for her support. Alessandro Castellazzo acknowledges the grant provided by Lagrange Project – Crt Foundation, Turin, Italy.

References

1. Bortfeldt, A., Wäscher, G.: Container Loading Problems – A State-of-the-Art Review. FEMM Working Papers 120007, Otto-von-Guericke University Magdeburg, Faculty of Economics and Management (2012)
2. Fasano, G.: A modeling-based approach for non-standard packing problems. In: Fasano, G., Pintér, J.D. (eds.) Optimized Packings and Their Applications. Springer Optimization and Its Applications. Springer Science + Business Media, New York (2015)
3. IBM: CPLEX 12.6.0 User Manual. http://www-01.ibm.com/support/knowledgecenter/ SSSA5P_12.6.0/ilog.odms.studio.help/Optimization_Studio/topics/COS_home.html?lang=en (2013)
4. Pisinger, D.: Heuristics for the container loading problem. Eur. J. Oper. Res. **141**(2), 382–392 (2002)
5. Rothberg, E.: An evolutionary algorithm for polishing mixed integer programming solutions. INFORMS J. Comput. **19**(4), 534–541 (2007)
6. Fasano, G.: Solving Non-standard Packing Problems by Global Optimization and Heuristics. SpringerBriefs in Optimization, Springer Science + Business Media, New York (2014)
7. Rietz, J.: Special Difficulties in the Three-Dimensional Container Loading Problem. Universidade do Minho, Braga (2008)
8. Bischoff, E.E., Ratcliff, M.S.W.: Issues in the development of approaches to container loading. OMEGA **23**(4), 377–390 (1995)
9. Fasano, G., Lavopa, C., Negri, D., Vola, M.C.: CAST: a successful project in support of the international space station logistics. In: Fasano, G., Pintér, J.D. (eds.) Optimized Packings and Their Applications. Springer Optimization and Its Applications. Springer Science + Business Media, New York (2015)

Chapter 8
Automatic Design of Optimal LED Street Lights

Balázs L. Lévai and Balázs Bánhelyi

Abstract The issue of light pollution, unnecessary lighting of outdoor areas, came into focus in the last 10 years. This is the reason why observatories should not be built in highly populated areas, it also disturbs the wild life, and it raises questions about energy conservation too. Based on its capabilities, LED technology offers a solution to this problem. Nowadays, travellers can visit many cities in developed countries and encounter LED street lights in streets as application of this technology spreading in public lighting. Designing orientation of LEDs in such street lights is a difficult problem as we need to use multiple LED packages to light an as large area as an incandescent light bulb can. Determining correct angles is a global optimization problem, a complex mathematical task related to the field of covering problems. In this chapter, we present an automatic designing method to construct LED configurations for street lights and a light pattern computation technique to evaluate these configurations. To speed up the whole designing process, a possible way of parallelization is also discussed.

Keywords Global optimization • Genetic algorithm • Covering problem • LED • Public lighting

8.1 Designing LED Street Lights

When we are looking at the view of a city at night, the first thought usually coming into our mind is how beautifully everything is lighted, admiring the luminous streets, buildings, and bridges. The last thing we realize is the price of shinning, the light pollution by name.

Nowadays, we reached a harmful level of light emission. It affects the wild life, especially the insects. Such a species as the fireflies, whose mating ritual essentially involves light signals, suffered a heavy drop in their numbers. Confused by artificial light, males and females cannot find each other. On a much global scale, light is

B.L. Lévai • B. Bánhelyi (✉)
Institute of Informatics, University of Szeged, 6701 Szeged, Hungary
e-mail: levaib@inf.u-szeged.hu; banhelyi@inf.u-szeged.hu

© Springer International Publishing Switzerland 2015   175
G. Fasano, J.D. Pintér (eds.), *Optimized Packings with Applications*, Springer
Optimization and Its Applications 105, DOI 10.1007/978-3-319-18899-7_8

also an important factor in animals' navigation and migration. Not just the timing of human created light is the problem, but its polarization too, because many animals use the natural polarization of sun light as information. The list of malicious effects on plants and animals could be continued, see [9].

Energy consumption is another relevant aspect. A rough estimation of 25 % of our total energy needs is required for lighting purposes. The introduction of daylight saving periods from April to October happened for a reason benefiting a large amount of energy saving every year. You can read more about this topic among other harmful effects of light pollution in the papers [2, 4, 12].

LED technology offers a possible solution to light pollution [5, 8, 10]. The light of LEDs is much more focusable [3] and can also be dimmed, even adaptively to traffic density [15]. LEDs have longer lifetime and consume less energy [14] than incandescent light bulbs, but there are drawbacks of this technology too. LEDs illuminate a relatively small area, therefore application of multiple LED packages in LED street lights is necessary to replace the currently operating public lighting. This fact leads us to the question of how LEDs should be directed in the housing of lamps.

Angles of LEDs in lamps have to be set carefully to distribute light emission equally on the target surface. Configuring LED directions is a complex task, and it depends on a lot of factors, dimensions of the street and the lamppost, the minimal and maximal allowed intensity of light, and so forth. The regulation of public lighting, the future surroundings of street lights, and the cost-effectiveness should be considered simultaneously. One may focus on only one aspect, while neglecting the others, to be able to manually create designs, but it is most likely that resulted configurations will not be competitive due to high cost, or large energy consumption, or something else. We have to consider everything at the same time and that is why automatic designing solutions are required.

The quality of lighting in public areas like roads, parks, etc. is regulated by law in protection of motorists. This means that the intensity and uniformity of light have to be in specified ranges. Considering the conical lighting characteristics of LEDs, the intersection of the target surface and the light cone cast by a single LED is an ellipse. Because intensities of different light sources simply add up, providing the required visibility can be interpreted as covering rectangle-shaped areas with ellipses while overlapping is allowed. Light intensity within the same ellipse varies depending on the lighting characteristic and the direction of the source LED, thus altering LED directions also changes their extent of contribution to the coverage. Even to the lay mind, LED configuration design for public lighting purposes is obviously not an ordinary covering problem. The complexity implies that there is no hope to handle successfully this type of task with direct and deterministic optimizer methods in reasonable time, therefore we decided to create a suitable genetic algorithm to search for acceptable LED configurations as the application of such heuristic methods proved to be a good strategy in similar situations [13].

To measure the goodness of configurations, we need to evaluate them based on their properties. Beside information already provided by manufacturers such as energy consumption or price, properties related to light quality are only available if we compute the light pattern generated by the studied configuration. This is the most important component of configuration evaluation as street lights violating the regulations cannot be deployed.

In a nutshell, the two cornerstones of automatic designing of LED street lights are the way how we construct new candidate configurations and how we determine their generated light pattern.

8.2   Light Pattern Computation

Light pattern computation means the determination of light intensity in given points on the surface we light. Regulation prescribes that these points must be the vertices of a grid with 1 m length of side. The height of lamppost and the overhanging of lamp are also necessary to proceed. Without loss of generality, we consider the housing of LEDs as a dimensionless point for simplicity because engineers can house LED sockets in a way that inserted LEDs will be directed through the same point. Lastly, LEDs are described by their lighting characteristic provided as intensities measured in different horizontal and vertical angles in fix distance from the light source following the format of EULUMDAT [1].

Algorithm 1 Compute Light Pattern

 1: generate evaluation points
 2: **for all** evaluation point p **do**
 3:         **for all** LED l **do**
 4:                calculate the direction vector LP pointing to p from l
 5:                calculate the angles of the direction pointing to p and the own direction of l
 6:                interpolate the base intensity towards p
 7:                determine the light intensity in p based on distance
 8:                increase the total intensity in p
 9:         **end for**
10: **end for**

The intensity of emitted light decreases quadratically by the distance measured from its source, and the effects of different LEDs simply add up, therefore the whole computation can be considered as repetition of an elementary subtask, calculating the intensity of light cast by a single LED in a single evaluation point.

Light intensities are only available in certain directions, therefore we have to determine which known values are the closest to the value we need. The first step is to calculate the direction from the light source to the evaluation point. Having the angles between this direction and the own direction of the LED, we are able to determine a base intensity towards the evaluation point. We applied bilinear interpolation for this purpose using four known intensity values. The final intensity can be obtained easily based on distance.

Algorithm 1 is a brief step-by-step pseudo code of the computation, but it will not give the correct pattern as it only takes into account the lamp which belongs to the target area, hence further adjustments are needed. Despite the high level control

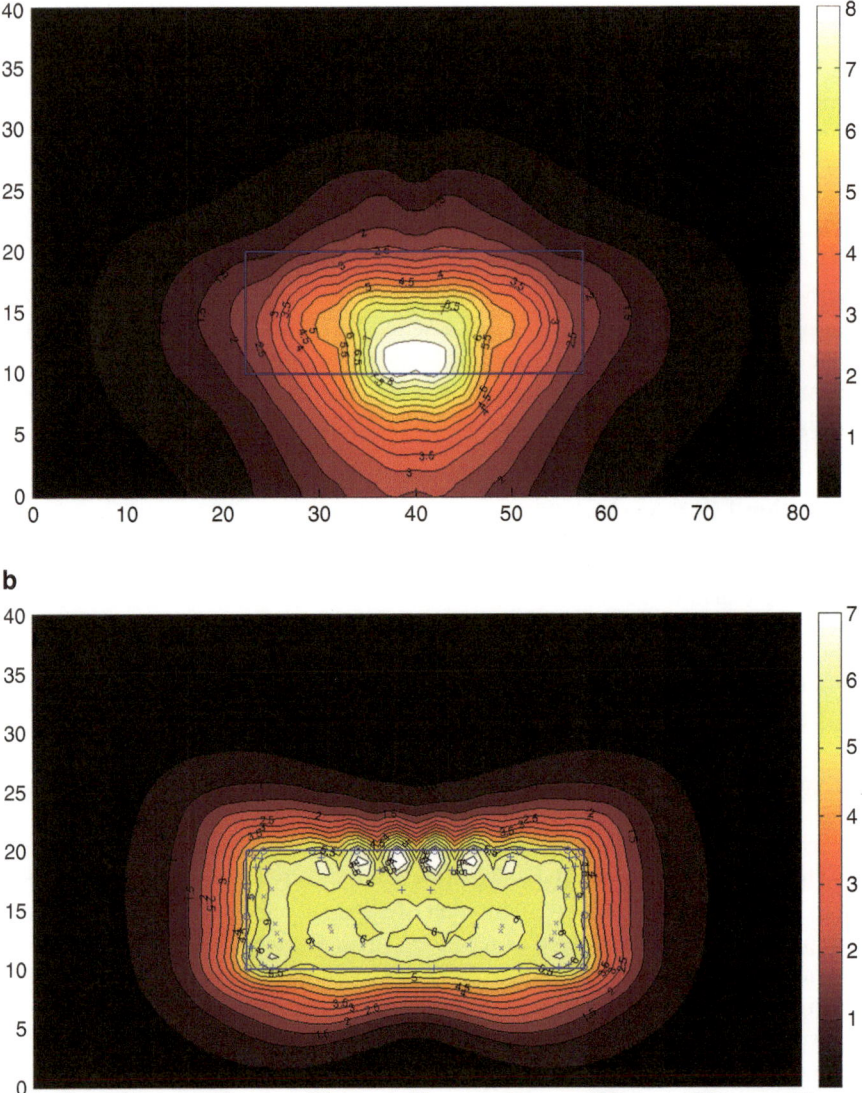

Fig. 8.1 Light patterns of street lights illuminating a *rectangle shaped* street section. *Brighter colours* denote higher intensities measured in LUX. (**a**) Regular incandescent street light. (**b**) Optimized LED street light

over the light of LEDs, we have to include the effects of neighbouring lamps to obtain a valid pattern because intensities do not drop to zero when we leave the borders of the target area as Fig. 8.1 shows. Light patterns also have at least one axis of symmetry, but 2 or 4 are also possible, depending on how the lamp will be deployed compared to the others. Therefore, we do not have to determine the intensity in every evaluation point. Exploiting symmetry, only half of points or even less need to be managed—significantly reducing the runtime.

All considered, 11 different light pattern settings are possible depending on lamppost deployment and pattern symmetries, summarized in Table 8.1, which cover most public lighting cases ranging from simple streets to parking lots.

8.3 Global Optimization by Genetic Algorithm

Constrained by complexity, we can only approximate the globally optimal LED configuration in acceptable time. Any designer tool has to be capable of combining different parts of configurations, which are already optimal at some level, to move towards better solutions while it also involves randomness to be able to leave local extremal points. Genetic algorithms seem to offer a suitable approach to handle our problem-type.

Researchers apply genetic algorithms in many areas ranging from optimization to machine learning to solve problems which cannot be handled by other means. The idea of genetic algorithms comes from natural evolution. The basic concept is to model the objects of a problem space as entities, or candidates in other words, of a population and let the rule of "the strong flourish, the weak perish" work out.

Table 8.1 The 11 different light pattern scenarios based on the number of axes of symmetry and the deployment of lampposts

				Not possible	Not possible	
			Not possible	Not possible	

Elaborating more, we repeat the following steps in subsequent iterations until some common stopping criteria are met:

1. Give a fitness value to every candidate based on its properties.
2. Take out some candidates from the population selecting more probably the ones whose fitness is low.
3. Mutate some candidates by slightly altering their properties.
4. Crossover entities mixing their properties somehow in the offspring to replace the ones you took out earlier.

This description may seem very intuitive and it also well shows how we let the principles of evolution help us to find whatever we are looking for represented as the best survivor in the population. The most important concepts are the genetic operators, the way we calculate fitness, the selecting strategy of survivors, and how the objects are distilled into candidates. For an in-depth study of genetic algorithms and applications, see the books [6, 7].

In our case, population naturally consists of different LED configurations. In more detail, each candidate solution contains vertical and horizontal angles, lighting characteristics, and power consumption data for every LED in the configuration. Determining and storing precise positions of LEDs in a lamp is omitted which the engineers designing the final physical product are responsible for.

Choosing the right operators was a more delicate decision. Three mutation operators are used to alter angles, or LED types, or to take out, and put back LEDs into configurations. Angle modification has the least impact on fitness function while the others concern not just the light pattern but every criterion too. Crossover, in contrast of mutation, has a much larger effect on configurations as it is tasked to introduce new approaches of lighting. As above mentioned, our intention was to combine configurations which light considerably well different parts of the target region, see Fig. 8.2. The following steps implement this idea:

1. Choose two parent configurations.
2. Generate a rectangle randomly in the target region with uniform distribution.
3. Select the LEDs pointing in the rectangle from one configuration and the LEDs pointing out of the rectangle from the other configuration. Switch the roles of parent configurations and repeat the process.
4. The two resulted sets of LEDs will compose the child configurations.

The above steps are simple and intuitive. We tested the operator with different parameters of rectangle generation to make it fit best to its intended goal. Finding the maximal and minimal allowed area of rectangles was the key. Too small rectangles result in an insignificant change of configurations while too large ones are likely to include poorly lighted areas as well, again not bringing improvement into the population.

This set of genetic operators provide a fine-grained tool set to create and modify candidate solutions in various levels.

The final element we have not discussed yet is the fitness function, the compass of designing for which we reasoned on behalf of automatic configuration design

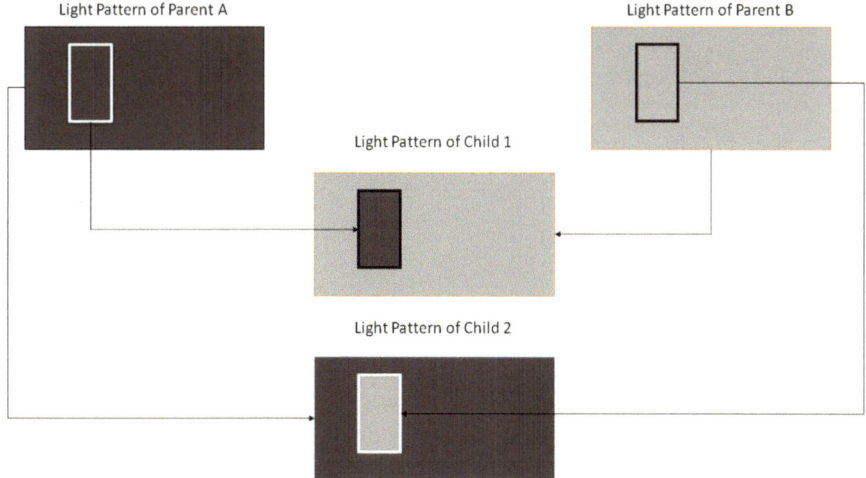

Fig. 8.2 Illustration of recombining two configurations based on light patterns

in the first place. The fitness value unites the goodness of configuration from the economic, energetic, and functional point of view simultaneously. We define fitness as the sum of several penalty terms derived by observing the following properties:

1. difference between the expected average intensity and the current values in evaluation points,
2. difference between the expected and current average of intensity,
3. total energy consumption,
4. the number of LEDs, and
5. difference between the expected and the current variance of intensity.

The first two expressions measure how far configurations are from regulations. If the intensities and the average are not in the allowed range, an additional penalty constant is also applied. Soundly, this forces the genetic algorithm to consider regulations first and everything else second. The user inputs are the strict bounds for term 1 and 2, and the weights expressing the relative importance compared to each other in case of the last three.

8.4   Results

The designing application has two components, a JAVA graphical user interface (GUI) and an optimizer written in c++. The GUI handles typical features as creating, opening, saving, etc. designing projects. When a new project is started, a step-by-step wizard guides the user to set optimization parameters, usable LED types, lamppost settings, and other user defined values. After everything is prepared, the

Table 8.2 Five test cases and their running time of automatic design

No.	Width (m)	Length (m)	LEDs	Intensity (lux)	Total CPU runtime (s)	Optimization runtime (s)
1.	30	10	27	6	299	269
2.	30	10	48	10	431	391
3.	35	20	38	6	705	651
4.	40	20	68	8	1,505	1,410
5.	50	25	100	8	2,849	2,673

GUI starts the optimizer. During optimization, the genetic algorithm frequently sends back the best solutions, whose light pattern and other describing numerical information are visualized in the GUI. The user can stop the optimization whenever he or she decides that the currently shown configuration fulfils the requirements. Otherwise, the process stops when all the characteristics of the best configuration found are within the allowed ranges and it does not change significantly over several iterations.

We implemented the genetic algorithm and the light pattern computation from scratch. Only the GUI relies on third party libraries to read and write XML files and to export LED configurations in PDF format. The program runs on Windows operating systems as this was the platform our industrial partner requested.

Assessing capabilities of the developed methodology was a difficult issue. First, LED street light manufacturers tend to keep their designing processes as well guarded secrets. Academic research groups and companies release improved designs from time to time but never designing tools. Programs available on the market are mainly concerned about visualizing light plans as realistic and fast as possible, but the burden of creating plans is left to the user. Unfortunately, this means that we were unable to compare our software to competitors as they have not presented their results yet.

Our only option was to compete with our industrial partner's engineers. Manual creation of even a single LED configuration takes long hours, therefore we could only ask for a few test cases from which several are shown on Table 8.2. We ran the tests on a simple laptop having an Intel Core I3-370M processor and 3 GB memory. On average, configurations our program found were at least twice better than manually created ones regarding the objective function. The largest difference appeared in the uniformity of light patterns as the automatically designed ones turned out to be much more smoother.

A practical feature that the industrial partner specially asked for is the possibility to add already configured LEDs to configurations in advance whose properties cannot be modified during design. This might seem a little bit odd at first glance. Why would anyone want to force such constraints to the algorithm by adding manually set elements to a lamp? Obviously no one would, but if we create a configuration for a certain lighting scenario, by this feature, we are able to adjust it with also automatically designed LEDs to fit another one. This allows us to produce

Table 8.3 Running time comparison of CPU and GPU based implementations in seconds

#	Total CPU time	Total GPU time	CPU optimization time	GPU optimization time
1.	299	269	144	111
2.	431	391	151	116
3.	705	651	184	134
4.	1,505	1,410	251	156
5.	2,849	2,673	334	181

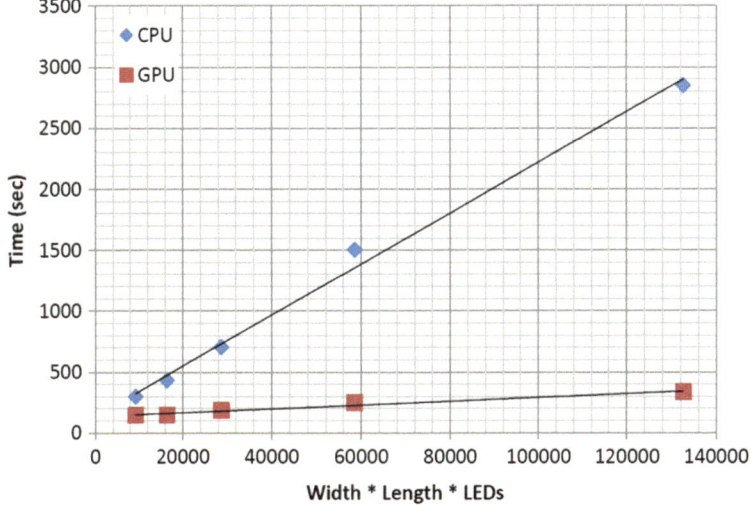

Fig. 8.3 The effect of parallelization

the same housing for different streets or roads. We only need to plug the right LEDs into the right sockets before deployment. This results in less product types to manage saving even more for the companies.

The stopping conditions of optimization were met after 3–4 h for typical design settings, and configurations fulfilling every hard condition already emerged after 20–30 min as Table 8.2 shows. After the first test runs, the industrial partner became interested in the reduction of optimization time assigning a new objective to us.

Profiling the software revealed that 80 % of executed operations are related to light pattern computation. As indicated before, most of these calculations can be executed independently, hence we decided to execute the intensity calculation of different evaluation points in parallel. We based the new implementation on NVIDIA's CUDA technology [11]. We repeated the optimization using the same seeds for random number generation, see Table 8.3. This happened on the same laptop we used earlier with an NVIDIA GeForce GT 335M video card.

As illustrated in Fig. 8.3, we analysed the runtime as a function of problem size, which is the product of the number of applied LEDs and the width and length of the target rectangle. Linear regression resulted in the following coefficients:

$$CPUruntime = 0.020{,}94 problemsize + 127,$$
$$GPUruntime = 0.001{,}58 problemsize + 134.$$

(8.1)

Dividing the steepness' of (8.1) by each other, we obtain a 13 times speedup limit in runtime. Although this growth in performance can truly be harnessed when larger problems are encountered, parallelization significantly reduces runtime in every case.

8.5 Conclusion

Designing optimal LED configurations for public lighting purposes is far more complex, even for expert engineers, than to be handled manually. This global optimization problem belongs to the classic field of covering problems; however, it can only be approached by stochastic optimization methods due to high dimensionality and special constraints.

We developed a software solution which is capable of designing LED configurations automatically while it considers every relevant factor during the process. Our approach is to handle configuration construction by a genetic algorithm which combines configurations based on partially good light patterns using crossover to obtain better candidate solutions whom mutation refines further. The objective function is the weighted sum of different penalty terms measuring the goodness of energy consumption, quality of lighting, and total cost of applied light sources.

As light pattern related operations put out the bulk of required computation during design, we took advantage of any axial symmetry present in the problems to reduce light intensity evaluation to the most necessary level. After finishing the first prototype, we reimplemented light pattern computation using NVIDIA's CUDA technology to make the optimization even faster by handling effects of different LEDs simultaneously. The result of this effort is a 13 times speedup in limit.

In our experience, automatic design can lead to at least twice better configurations than manual design. The most outstanding difference comes out in the uniformity of light intensities revealing the main strength of our algorithm. The presented test cases prove that the developed optimization technique can truly help the work of engineers reducing designing time and other costs.

References

1. EULUMDAT specification: http://www.helios32.com/Eulumdat.htm (2014). Accessed 12 Nov 2014
2. Falchia, F., Cinzanoa, P., Elvidgeb, C.D., Keithc, D.M., Haimd, A.: Limiting the impact of light pollution on human health, environment and stellar visibility. J. Environ. Manag. **92**, 2714–2722 (2011)
3. Fournier, F., Cassarly, W.: SOFTWARE & COMPUTING: Freeform optics design advances lighting and illumination. Laser focus world. http://www.laserfocusworld.com/articles/print/volume-47/issue-3/columns/software-computing/freeform-optics-design-advances-lighting-and-illumination.html (2011). Accessed 12 Nov 2014
4. Gallaway, T., Olsen, R.N., Mitchell, D.M.: The economics of global light pollution. Ecol. Econ. **69**, 658–665 (2011)
5. Gereffi, G., Dubay, K., Lowe, M.: Manufacturing Climate Solutions Carbon Reducing Technologies and U.S. Jobs. Center on Globalization, Governance & Competitiveness, Duke University, Durham (2008)
6. Goldberg, D.E.: Genetic Algorithms in Search, Optimization and Machine Learning. Addison-Wesley Professional, Boston (1989)
7. Lawrence, D.: Handbook of Genetic Algorithms. Van Nostrand Reinhold, New York (1991)
8. Lee, H., et al.: High-performance LED street lighting using microlens arrays. Opt. Express **21**, 10612–10621 (2013)
9. Longcore, T., Rich, C.: Ecological light pollution. Front Ecol. Environ. **2**, 191–198 (2004)
10. Nuttall, D.R., Shuttleworth, R., Routledge, G.: Design of a LED street lighting system. In: 4th IET International Conference on Power Electronics, Machines and Drives, pp. 436–440 (2008)
11. NVIDIA's CUDA homepage: http://www.nvidia.com/object/about-nvidia.html (2014). Accessed 12 Nov 2014
12. Riegel, K.W.: Light pollution: Outdoor lighting is a growing threat to astronomy. Science **179**, 1285–1291 (1973)
13. Szabó, P.G., Csendes, T., Casado, L.G., García, I.: Lower bounds for equal circles packing in a square problem using the TAMSASS-PECS stochastic algorithm. In: Abstracts of GO. '99, Firenze, pp. 122–126 (1999)
14. White, J.J.: No-cost LED street lighting modernization. EATON report (2013)
15. Wu, Y., Shi, C., Zhang, X., Yang, W.: Design of new intelligent street light control system. In: 2010 8th IEEE International Conference on Control and Automation (ICCA), pp. 1423–1427 (2010)

Chapter 9
Approximate Packing: Integer Programming Models, Valid Inequalities and Nesting

Igor Litvinchev, Luis Infante, and Lucero Ozuna

Abstract Using a regular grid to approximate a container, packing objects is reduced to assigning objects to the nodes of the grid subject to non-overlapping constraints. The packing problem is then stated as a large scale linear 0-1 optimization problem. Different formulations for non-overlapping constraints are presented and compared. Valid inequalities are proposed to strengthening formulations. This approach is applied for packing circular and L-shaped objects. Circular object is considered in a general sense as a set of points that are all the same distance (not necessary Euclidean) from a given point. Different shapes, such as ellipses, rhombuses, rectangles, octagons, etc., are treated similarly by simply changing the definition of the norm used to define the distance. Nesting objects inside one another is also considered. Numerical results are presented to demonstrate the efficiency of the proposed approach.

Keywords Packing problems • Integer programming • Large-scale optimization

9.1 Introduction

Packing problems generally consist of packing a set of items of known dimensions into one or more large objects or containers to minimize a certain objective (e.g. the unused part of the container or waste). Packing problems constitute a family of natural combinatorial optimization problems applied in computer science, industrial engineering, logistics, manufacturing and production processes (see, e.g., [1–4] and the references therein).

I. Litvinchev (✉)
Complex Systems Department, Computing Center, Russian Academy of Sciences, Moscow, Russia
e-mail: igorlitvinchev@gmail.com

L. Infante • L. Ozuna
Faculty of Mechanical and Electrical Engineering, Nuevo Leon Sate University, Monterrey, Mexico
e-mail: luisinfanterivera@gmail.com; luceroozuna@gmail.com

© Springer International Publishing Switzerland 2015   187
G. Fasano, J.D. Pintér (eds.), *Optimized Packings with Applications*, Springer
Optimization and Its Applications 105, DOI 10.1007/978-3-319-18899-7_9

Along with industrial applications one may find packing problems in healthcare issues (e.g., [5, 6]). Wang [6] considered automated radiosurgical treatment planning for treating brain and sinus tumours. Radiosurgery uses the gamma knife to deliver a set of extremely high dose ionizing radiation, called "shots" to the target tumour area. For large target regions multiple shots of different intensity are used to cover different parts of the tumour. However, this procedure may result in large doses due to overlap of the different shots. Optimizing the number, positions and individual sizes of the shots can reduce the dose to normal tissue and achieve the required coverage.

Packing problems for regular shapes (circles and rectangles) of objects and/or containers are well studied (see, e.g., a review by [7] for circle packing). In circle packing problem the aim is to place a certain number of circles, each one with a fixed known radius inside a container. The circles must be totally placed in the container without overlapping. The shape of the container may vary from a circle, a square, a rectangular, etc. For the rectangular container there are two principal types of objectives [8, 9]: (a) regarding the circles (not necessary equal) as being of fixed size and the container as being of variable size and (b) regarding the circles and the container as being of fixed size and minimize "waste". Examples of the first approach include: minimize the perimeter or the area of the rectangle; considering one dimension of the rectangle as fixed, minimize the other dimension (strip packing or open dimension problem). For the second approach various definitions of the waste can be used. The waste can be defined in relation to circles not packed or introducing a value associated with each circle that is packed (e.g., area of the circles packed).

Many variants of packing circular objects have been formulated as nonconvex (continuous) optimization problems with decision variables being coordinates of the centres [7]. Non-overlapping typically is assured by nonconvex constraints representing that the Euclidean distance separating the centres of the circles is greater than a sum of their radii. The nonconvex problems can be tackled by available nonlinear programming (NLP) solvers, however most NLP solvers fail to identify global optima and global optimization techniques have to be used [2, 10]. The nonconvex formulations of circular packing problem give rise to a large variety of algorithms which mix local searches with heuristic procedures in order to widely explore the search space. We will refer the reader to review papers presenting the scope of techniques and applications for regular packing problem (see, e.g., [8, 9, 11–13] and the references therein).

Irregular packing problems involve non-standard shapes of objects and/or containers. Irregular shapes are those that require non-trivial handling of the geometry [14, 31]. One of the most common representations for irregular shape is a polyhedral domain which may by nonconvex or multi-connected. Heuristic and metaheuristic algorithms are the basis for the solution approaches (see [3, 15] and the references therein).

Discrete approximations of objects by tetris-like items [3] and containers by grids [15–20] were recently used to simplify packing problems. This approach allows handling irregular shapes and reduces (approximately) packing problems to discrete optimization problems. To the best of our knowledge, the proposal to use a grid was first applied by Beasley [21] in the context of cutting problems.

This work is a continuation of Litvinchev and Ozuna [17]. Using a regular grid to approximate the container, packing is reduced to assigning the objects to the nodes of the grid subject to non-overlapping constraints. Different formulations for non-overlapping are considered and compared. Valid inequalities are proposed to strengthening formulations. This approach is applied for packing circular and L-shaped objects. Circular object is considered as a set of points that are all the same distance (not necessary Euclidean) from a given point. This way different shapes, such as ellipses, rhombuses, rectangles, octagons, etc. can be treated by simply changing the norm used to define the distance. Nesting objects inside one another is also considered. Numerical results are presented to demonstrate efficiency of the proposed approach.

The rest of the work is organized as follows. In Sect. 9.2 integer programming approximation of the packing problem is presented along with different formulations for non-overlapping. In Sect. 9.3 the proposed approach is applied to packing circular objects. Experimental results for packing different circular shapes are provided to demonstrate usefulness of valid inequalities proposed in Sect. 9.2. L-shaped objects and containers are considered in Sect. 9.4, while Sect. 9.5 presents concluding remarks and directions for the future research.

9.2 Basic Constructions

Suppose we have non-identical objects G_k, $k \in K = \{1, 2, \ldots K\}$ which have to be packed in a container G. In what follows we will use the same notation G_k, G for the domain in $\mathbb{R}^n$ and for its boundary assuming that it is easy to understand from the context what do we mean. It is assumed that no two objects overlap with each other and each packed object lies entirely in the container. Denote by S_k the area of G_k. Let at most M_k objects G_k are available for packing and at least m_k of them have to be packed. Denote by p_i, $i \in I = \{1, 2 \ldots, n\}$ the nodes of a grid covering the container, $p_i \in G$. It is assumed that the position of the object in the container is completely characterized by the position of its reference point. Define binary variables $x_i^k = 1$ if the reference point of the object G_k is assigned to the node i; $x_i^k = 0$ otherwise. In what follows we will say that the object is assigned to the node i if the corresponding reference point is assigned to that node and will denote this as G_k^i. For fixed i, k let

$$N_{ik} = \left\{ j, l : i \neq j \text{ such that } G_k^i \text{ overlaps with } G_l^j \right\}.$$

Let n_{ik} be the cardinality of $N_{ik} : n_{ik} = |N_{ik}|$. Then the problem of maximizing the area covered by the objects can be stated as follows:

$$\max \sum_{i \in I} \sum_{k \in K} S_k x_i^k \tag{9.1}$$

subject to

$$m_k \le \sum_{i \in I} x_i^k \le M_k, \quad k \in K, \tag{9.2}$$

$$\sum_{k \in K} x_i^k \le 1, \quad i \in I, \tag{9.3}$$

$$x_i^k = 0 \text{ for } G_k^i \setminus \left(G \cap G_k^i \right) \ne \varnothing \quad \text{for } i \in I, \ k \in K, \tag{9.4}$$

$$x_i^k + x_j^l \le 1, \text{ for } i \in I, \ k \in K, \ (j, l) \in N_{ik}, \tag{9.5}$$

$$x_i^k \in \{0, 1\}, \quad i \in I, \ k \in K. \tag{9.6}$$

Constraints (9.6) ensure that the number of objects packed is between m_k and M_k; constraints (9.3) that at most one object is assigned to any node; constraints (9.4) that G_k cannot be assigned to the node i if G_k^i is not totally placed inside G; pairwise constraints (9.5) guarantee that there is no overlapping between the objects; constraints (9.6) represent the binary nature of variables.

Remark 2.1 Linear non-overlapping constraints (9.5) are equivalent to a single quadratic constraint

$$Q(x) \equiv \sum_{i,k} x_i^k \sum_{j,l \in N_{ik}} x_j^l = 0 \ (\le 0). \tag{9.7}$$

If (9.7) holds, then for $x_i^k = 1$ we have $\sum_{j,l \in N_{ik}} x_j^l = 0$ yielding $x_j^l = 0$, $(j, l) \in N_{ik}$, and if $x_j^l = 1$ at least for one pair $(j, l) \in N_{ik}$, then $x_i^k = 0$. Thus (9.5) can be considered as a specific linearization of (9.7). Other linearizations and relaxations of (9.7), e.g. used for the quadratic assignment problem [22] can also be considered.

Below we present different formulations for the non-overlapping constraints (9.5) which remain valid for the general definition of N_{ik}.

By the definition of N_{ik} if $(j, l) \in N_{ik}$, then $(i, k) \in N_{jl}$. Thus a half of the constraints in (9.5) are redundant since we have:

$$x_i^k + x_j^l \le 1, \text{ for } i \in I, \ k \in K, \ (j, l) \in N_{ik},$$

$$x_j^l + x_i^k \le 1, \text{ for } j \in I, \ l \in K, \ (i, k) \in N_{jl}.$$

We may eliminate any (none) of these two constraints to get the *reduced* equivalent formulation. This can be represented by multiplying constraints (9.5) by a fixed $\lambda_j^l \in \{0, 1\}$:

$$x_i^k \lambda_j^l + x_j^l \lambda_j^l \le \lambda_j^l, \text{ for } i \in I, \ k \in K, \ (j, l) \in N_{ik}, \tag{9.8}$$

subject to $\lambda_j^l + \lambda_i^k \geq 1$. This way either one of the redundant constraints is eliminated $(\lambda_j^l + \lambda_i^k = 1)$ or no-one $(\lambda_j^l + \lambda_i^k = 2)$. Since eliminating redundant constraints does not affect the feasible set, the problem (9.1)–(9.6) is equivalent to (9.1)–(9.4), (9.6), (9.8) for any λ fulfilling the normalized condition

$$\lambda \in \Lambda = \left\{ \lambda_j^l \in \{0,1\} : \lambda_j^l + \lambda_i^k \geq 1, \ (j,l) \in N_{ik} \right\}.$$

Similar to plant location problems [23] we can state non-overlapping conditions in a more *compact* form. Summing up constraints (9.7) over $(j,l) \in N_{ik}$ we get

$$x_i^k \sum_{(j,l) \in N_{ik}} \lambda_j^l + \sum_{(j,l) \in N_{ik}} \lambda_j^l x_j^l \leq \sum_{(j,l) \in N_{ik}} \lambda_j^l, \text{ for } i \in I, k \in K. \tag{9.9}$$

Proposition 2.1 *For any $\lambda \in \Lambda$ constraints (9.5), (9.6) are equivalent to constraints (9.6), (9.9).*

Proof If constraints (9.5) are fulfilled, then obviously constraints (9.9) hold by construction. Now let constraints (9.9) are fulfilled. Define

$$N_{ik}^1 = \left\{ (j,l) \in N_{ik} : \lambda_j^l = 1 \right\}, \ N_{ik}^0 = \left\{ (j,l) \in N_{ik} : \lambda_j^l = 0 \right\}, \ N_{ik}^1 \cup N_{ik}^0 = N_{ik},$$
$$\left| N_{ik}^1 \right| = n_{ik}^1, \ \left| N_{ik}^0 \right| = n_{ik}^0.$$

By (9.9) we have

$$x_i^k n_{ik}^1 + \sum_{(j,l) \in N_{ik}^1} x_j^l \leq n_{ik}^1$$

and hence,

$$\text{if } x_i^k = 1, \text{ then } x_j^l = 0 \text{ for } (j,l) \in N_{ik}^1. \tag{9.10}$$

By the definition, if $(j,l) \in N_{ik}$, then $(i,k) \in N_{jl}$. Thus by (9.9) we have

$$x_j^l \sum_{(i,k) \in N_{jl}} \lambda_i^k + \sum_{(i,k) \in N_{jl}} \lambda_i^k x_i^k \leq \sum_{(i,k) \in N_{jl}} \lambda_i^k \text{ for } j \in I, \ l \in K. \tag{9.11}$$

In particular, (9.11) is fulfilled for $(j,l) \in N_{ik}^0$. Since $\lambda_j^l + \lambda_i^k \geq 1$, then for $(j,l) \in N_{ik}^0$ all λ_i^k in (9.11) are positive $(\lambda_i^k = 1)$. Then by (9.11) we have:

$$\text{if } x_j^l = 1 \text{ for at least one } (j,l) \in N_{ik}^0, \text{ then } x_i^k = 0. \tag{9.12}$$

Note that constraints (9.5) can be interpreted in two ways. First if $x_i^k = 1$, then $x_j^l = 0$ for all $(j,l) \in N_{ik}$. Second, if $x_j^l = 1$ for at least one $(j,l) \in N_{ik}$, then $x_i^k = 0$. Combining (9.10) and (9.12) we may conclude that if constraints (9.9) are fulfilled, then (9.5) hold. $\qquad\square$

Remark 2.2 In Galiev and Lisafina [16] the compact formulation

$$x_i n_i + \sum_{j \in N_i} x_j \leq n_i \text{ for } i \in I \tag{9.13}$$

was used to represent non-overlapping constraints for the case of packing identical circles. This corresponds to a singleton set K and all multipliers λ equal to 1 in (9.9).

Remark 2.3 Proposition 2.1 remains true for nonnegative (not necessary binary) multipliers λ subject to $\lambda_j^l + \lambda_i^k \neq 0$. The proof is similar.

As follows from Proposition 2.1, the non-overlapping constraints can be stated in different forms (see [20] for an illustrative example). We have a family of formulations equivalent to (9.5) and obtained for different multipliers λ in (9.9). To compare equivalent formulations, let

$$P_1 = \left\{ x \geq 0 : x_i^k + x_j^l \leq 1, \text{ for } i \in I, \ k \in K, (j, l) \in N_{ik} \right\},$$

$$P_2 = \left\{ x \geq 0 : x_i^k \sum_{(j,l) \in N_{ik}} \lambda_j^l + \sum_{(j,l) \in N_{ik}} \lambda_j^l x_j^l \leq \sum_{(j,l) \in N_{ik}} \lambda_j^l, \ i \in I, \ k \in K \right\},$$

where multipliers λ in P_2 fulfil the normalizing condition stated in Proposition 2.1.

Proposition 2.2 $P_1 \subset P_2$.

Proof Since constraints of P_2 are a linear combination of those in P_1 with nonnegative multipliers λ, then $P_1 \subseteq P_2$. To show that $P_1 \subset P_2$ we need to find a point in P_2 that is not in P_1.

This point can be constructed as follows. Choose $(i, k) \in N_{jl}$ and $(j, l) \in N_{ik}$ such that $\sum_{(j,l) \in N_{ik}} \lambda_j^l, \ \sum_{(i,k) \in N_{jl}} \lambda_i^k \geq 2$. Set to zero all the variables except x_i^k, x_j^l. Obviously all constraints in P_2 corresponding to zero variables are fulfilled. Define x_i^k, x_j^l to fulfil the two remaining constraints as equalities:

$$x_i^k \sum_{(j,l) \in N_{ik}} \lambda_j^l + x_j^l = \sum_{(j,l) \in N_{ik}} \lambda_j^l, \quad x_j^l \sum_{(i,k) \in N_{jl}} \lambda_i^k + x_i^k = \sum_{(i,k) \in N_{jl}} \lambda_i^k.$$

Denote $\bar{n}_{ik} = \sum_{(j,l) \in N_{ik}} \lambda_j^l, \ \bar{n}_{jl} = \sum_{(i,k) \in N_{jl}} \lambda_i^k$ with $\bar{n}_{ik}, \bar{n}_{jl} \geq 2$. The corresponding solution of the two equations above is

$$x_i^k = \frac{\bar{n}_{jl}(\bar{n}_{ik} - 1)}{\bar{n}_{jl}\bar{n}_{ik} - 1} < 1, \quad x_j^l = \frac{\bar{n}_{ik}(\bar{n}_{jl} - 1)}{\bar{n}_{jl}\bar{n}_{ik} - 1} < 1$$

with

$$x_i^k + x_j^l = 1 + \frac{1 + \bar{n}_{jl}\bar{n}_{ik} - \bar{n}_{jl} - \bar{n}_{ik}}{\bar{n}_{jl}\bar{n}_{ik} - 1} > 1.$$

This point violates corresponding constraint in P_1 and hence $P_1 \subset P_2$ as desired.   □

As follows from Proposition 2.2, the pairwise formulation (9.1)–(9.6) is stronger than the compact one (9.1)–(9.4), (9.6), (9.9) in the sense of Wolsey [23].

In general, checking if the object is not totally placed inside the container is tricky. However, for a convex container and a polygonal object this problem can be simplified as stated below.

Proposition 2.3 *Let G be a convex set and G_k be a (not necessary convex) polygon. Let v_{ki}^t, $t = 1, \ldots, T_k$ be all vertices of G_k^i. Then $G_k^i \subseteq G$ iff $v_{ki}^t \in G$, $t = 1, \ldots, T_k$.*

Proof If $G_k^i \subseteq G$, then obviously all vertices of G_k^i are in G. Let now $v_{ki}^t \in G$, $t = 1, \ldots, T_k$. Consider the convex hull of G_k^i, $conv\left(G_k^i\right) = \left\{ y : y = \sum_t \alpha_t v_{ki}^t, \sum_t \alpha_t = 1, \alpha_t \geq 0 \right\}$. Since all vertices of G_k^i are in G, then by convexity of G any convex linear combination of vertices also belongs to G and hence $conv\left(G_k^i\right) \subseteq G$. By the definition of convex hull, $G_k^i \subseteq conv\left(G_k^i\right)$ and hence $G_k^i \subseteq G$ as desired.   □

Good upper (dual) bounds are very important to solve integer programming problems. We may expect that the upper bound obtained by the linear programming relaxation of the problem (9.1)–(9.6) provides a poor upper bound for the optimal objective. For example, for packing equal circles in a rectangular container the objective value of the LP-relaxation grows linearly with respect to the number of grid nodes (see [20] for details).

To tightening the LP-relaxation we consider valid inequalities ensuring that no grid node is covered by two objects. To present this family, define matrix $[\alpha_{ij}^k]$ as follows. Let $\alpha_{ij}^k = 1$ if G_k covers a node j, $\alpha_{ij}^k = 0$ otherwise. The following constraints ensure that no nodes of the grid can be covered by two objects:

$$\sum_{k \in K} \sum_{j \in I} \alpha_{ij}^k x_j^k \leq 1, \quad i \in I. \tag{9.14}$$

Note that (9.14) is not equivalent to the non-overlapping constraints (9.5).

9.3 Circular Objects

Define a circular object C_k as a set of points that all are at most the distance R_k from a given point called centre, $C_k = \{y : \|y - y_{0k}\| \leq R_k\}$. Here the norm used to define the object is not necessary the Euclidean [32]. Let d_{ij} be the distance between node points i, j in the sense of the norm used to define the circular object.

The set N_{ik} in (9.5) is now defined as follows: $N_{ik} = \{j, l : i \neq j, \ d_{ij} < R_k + R_l\}$. For matrix $[\alpha_{ij}^k]$ in (9.14) we have $\alpha_{ij}^k = 1$ for $d_{ij} < R_k$, $\alpha_{ij}^k = 0$ otherwise.

Using different norms we can use constructions of the previous section for packing different geometrical objects of the same shape. For example, a circular object in the maximum norm $\|y\|_\infty := \max_r \{|y_r|\}$ is represented geometrically by a square, taxicab norm $\|y\|_1 := \sum_r |y_r|$ yields a rhombus. In a similar way we may handle rectangles, ellipses, etc. Using a superposition of norms, we can consider more complex circular objects. For

$$\|y\| := \max_r \left\{|y_r|, \gamma \sum_r |y_r|\right\}$$

and a suitable $0.5 < \gamma < 1$ we get an octagon, an intersection of a square and a rhombus.

A numerical experiment was designed to evaluate the performance of different non-overlapping formulations and to see the impact of the valid inequalities for packing circular objects in a rectangular container.

In the first part of the experiment the test bed set of 9 instances from ([16], Table 3) was used for packing maximal number of circles into a rectangle of width 3 and height 6. A rectangular uniform grid of size Δ along both sides of the container was used. It was assumed that the supply of the objects is unlimited and constraints (9.2) were relaxed. Similar to [16] the nodes located too close (close than a radius) to the boundary were eliminated from consideration and thus constraints (9.4) were omitted. In all experiments optimization problems were solved by the system CPLEX 12.6 [24]. The runs were executed on a desktop computer with CPU AMD FX 8350 8-core processor 4 GHz and 32 GB RAM.

The following four formulations were compared: pairwise formulation (9.1)–(9.6) (Cmpl), reduced formulation (9.1)–(9.6) without redundant constraints (CmplH), compact formulation (9.13) as in Galiev and Lisafina [16] (Cmpct), and compact formulation obtained by summing up constraints in the reduced formulation (9.1)–(9.6) (CmpctH). All these four formulations were combined with valid inequalities (cuts) (9.14), the corresponding formulations are denoted by CmplC, CmplHC, CmpctC, CmpctHC. The results of the numerical experiment are given in Table 9.1. Here the first three columns present instance number, circle radius, and grid size Δ. The last columns give CPU time (in seconds) for different formulations. For all problem instances *mipgap* = 0 was set for running CPLEX. In this table asterisk indicates that the computation was interrupted after

Table 9.1 CPU-time for circles (*gap* 0 %)

#	R	Δ	Cmpl	CmplC	CmplH	CmplHC	Cmpct	CmpctC	CmpctH	CmpctHC
1	0.5	0.125	2	2	1	1	276	4	5	4
2	0.625	0.078125	71	15	41	11	1,040	35	50	12
3	0.5625	0.0625	337	82	186	75	11,666	87	831	72
4	0.375	0.09375	6	9	4	4	2,698	29	169	92
5	0.3125	0.078125	96	163	114	189	*	819	*	1,027
6	0.4375	0.546875	17,437	1,392	17,654	1,379	*	39,347	*	*
7	0.25	0.0625	*	3,531	*	3,178	*	*	*	*
8	0.275	0.06875	132	87	177	87	*	2,523	*	2,860
9	0.1875	0.046875	*	17,437	*	*	*	*	*	*

Table 9.2 LP-relaxations

#	n_Δ	LP	O	LPC	R	LPC	C	LPC	E	LPC
1	697	348.5	18	19	28	33.43	18	19	34	36
2	1,403	701.5	9	10	15	16.87	10	10	21	25
3	2,449	1,224.5	12	14.0743	20	22.25	13	14.07	27	29.91
4	1,425	712.5	26	30.9485	39	41.37	32	36.33	59	68.86
5	2,139	1,069.5	41	53.4043	76	94.76	45	53.4	99	110
6	3,666	1,833.5	20	22.5537	35	39.72	21	23.86	43	49.787
7	3,649	1,824.5	72	90.9767	127	157.96	74	90.98	137	182
8	2,880	1,440	50	59.014	75	79.53	61	72	108	134.56
9	6,897	3,448.5	106	134.342	167	182.28	140	162	261	273.61

the computation time exceeded 12-h CPU. Number of binary variables and optimal packings are presented in Table 9.2 in columns (n_Δ) and (C), correspondingly.

As we can see from Table 9.1, CPU time for complete formulations is lower than for the compact, especially for large instances. Eliminating redundant constraints typically (but not always) reduces CPU time. Although eliminating redundancy does not change corresponding LP-relaxation, it may affect the path selected by branch and bound technique and thus result in increase/decrease of CPU time.

Introducing valid inequalities decreases CPU time for all problem instances and for all problem formulations. Although introducing valid inequalities slightly increases time to solve the LP-relaxation, the effect of improving quality of the LP-bound becomes more important for the convergence of the overall branch and bound scheme. That is why CPU time decreases significantly for hard instances 6, 7, 9, while for "easy" instances the decrease may be relatively modest. Moreover, with valid inequalities CPU time necessary to get provably optimal solution (*mipgap* = 0) is comparable with that reported in Galiev and Lisafina [16] for their heuristic approach.

Table 9.2 presents values of the LP-relaxations with/without valid inequalities for packing equal circles (C), ellipses (E), rhombuses (R) and octagons (O) into the

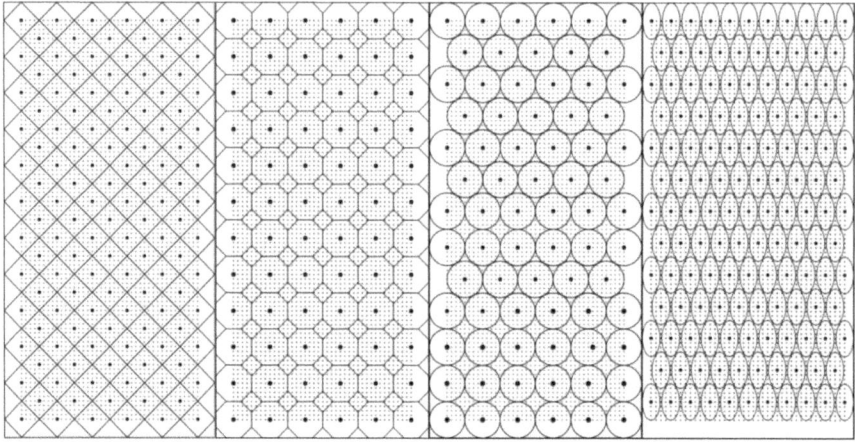

Fig. 9.1 Packing equal circular objects for instance 7

same 3×6 rectangle using the same values Δ for the grid. The standard Euclidean and taxicab norms were used to define circles and rhombuses, while norms

$$\|y\| := \left(2y_1^2 + y_2^2\right)^{1/2} \text{ and } \|y\| := \max\left\{|y_1|, \ |y_2|, \ \left(1/\sqrt{2}\right)(|y_1| + |y_2|)\right\}$$

were used for ellipses and octagons. The same values of radii as in Table 9.1 were used to define circular objects. In Table 9.2 the first three columns present instance number, number of binary variables (n_Δ) and value of the LP-relaxation without valid inequalities (LP). For all circular objects the optimal value of the LP-relaxation was $0.5n_\Delta$ (all variables equal to 0.5). The last eight columns give the value of the optimal integer solution (in bold) and the value LPC of the LP-relaxation improved by the valid inequalities (next to bold). We see that introducing valid inequalities improves significantly the quality of the LP bound for all shapes of the objects. The detailed study of this subject for the case of circles one can find in Litvinchev et al. [20] for the same test bed instances. Packings for the instance 7 are presented in Fig. 9.1.

In many applied problems packing smaller objects inside a larger one is permitted. For example, in tube industry the tubes are produced in a continuous extract machine and cut to the length of the container used for shipping. Before being placed in the container they may be inserted inside other, thicker tubes, so that usage of container space is maximized. Since all the tubes have the same length, maximizing container load is equivalent to maximizing the area filled with circles (rings) in a section of the container. Similar problems arise, e.g. in stacking up different containers to form a tower [25] and in visualization of large hierarchical data by 3D nested cylinders [26]. In tube industry the process is usually named telescoping [27], in optimized packing context the terms nesting [2] or recursive packing [28] are used. Although the term nesting is also used for packing irregular objects [15], we will use nesting for packing smaller objects inside larger ones assuming that it is easy to understand from the context what do we mean.

To consider nesting circular objects inside one another, we only need to modify the non-overlapping constraints. In order to C_k^i be non-overlapping with other objects being packed (including objects placed inside C_k^i), it is necessary that $x_j^i = 0$ for $j \in I,\ l \in K$, such the $R_k - R_l < d_{ij} < R_k + R_l$ for $R_k > R_l$. Let

$$\Omega_{ik} = \left\{ j, l : i \neq j,\ R_k - R_l < d_{ij} < R_k + R_l,\ R_k > R_l \right\}.$$

Then the non-overlapping constraints for packing circular objects with nesting can be stated as

$$x_i^k + x_j^l \leq 1,\ \text{for } i \in I; k \in K; (j, l) \in \Omega_{ik}. \tag{9.15}$$

Constraints (9.3) have to be omitted in case of nesting.

If nesting is permitted it may be necessary to take into account the difference between external and internal sizes of the object, i.e. consider the object as a circular ring (a region bounded by two concentric circular objects) having a positive thickness. To consider nesting-subject-to-thickness we need only to redefine the set Ω_{ik}. Let g_k be the thickness of the circle C_k. For Ω_{ik} defined as

$$\Omega_{ik} = \left\{ j, l : i \neq j,\ R_k - g_k - R_l < d_{ij} < R_k + R_l,\ R_k - g_k > R_l \right\}$$

we get non-overlapping constraints similar to (9.15).

The results for packing two different octagons in a square 30×30 container maximizing the total area of the packed objects are presented in Table 9.3. Here the first three columns give instance number, radii, and a number of grid nodes (integer variables). The last columns give the total area without nesting (N$-$), with nesting (N$+$) and with nesting and thickness (N$+$T), number of small (O1) and large (O2) objects packed, as well as corresponding CPU time in sec. The thickness g_k was defined as $0.1R_k$. The packings obtained for the instance 1 are presented in Fig. 9.2.

Table 9.3 Packing 2 different octagons

#	R_1, R_2	$n_\triangle$	N-	O1, O2	CPU	N+	O1, O2	CPU	N+T	O1, O2	CPU
1	0.6, 6.3	441	627.48	85, 4	1	842.21	265, 4	1	804.37	233, 4	1
2	0.6, 6.3	961	699.06	145, 4	6	971.05	373, 4	3	910.209	322, 4	5
3	1, 5.3	441	699.35	41, 6	1	952.82	119, 6	1	922.99	110, 6	1
4	1, 5.3	961	750.09	114, 4	57	1,158.27	181, 6	129	1,019.1	139, 6	49

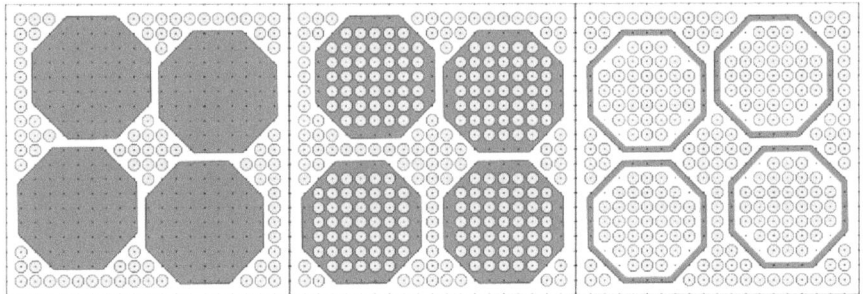

Fig. 9.2 Packing two octagons for instance 1

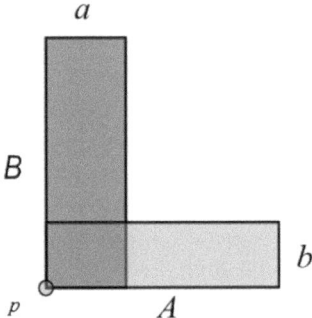

Fig. 9.3 L-object

9.4 L-shaped Objects and Containers

In this section we consider packing L-shaped objects. These shapes appear, e.g. in packing interpretations of scheduling with non-constant operational cycles [3]. Let L-object (see Fig. 9.3) be a superposition of rectangles $(A \times b)$ and $(a \times B)$ with edges parallel to the principal axes, $A > a > 0$, $B > b > 0$ and the principal corner considered as a reference point.

To state the problem (9.1)–(9.6) we need to specify constraints (9.4), (9.5), i.e. present a constructive way to check if the object is totally placed inside the container and if the objects overlap. Suppose we have two L-objects, L_i and L_j, with the reference points located at (y_{1i}, y_{2i}) and (y_{1j}, y_{2j}). Introducing binary variables $z_i, z_j \in \{0, 1\}$ these objects can be represented as follows:

$$L_i = \Big\{ (y_1, y_2, z_i) : z_i \in \{0, 1\}, \ \ 0 \leq y_1 - y_{1i} \leq A_i + z_i (a_i - A_i),$$

$$0 < y_2 - y_{2i} \leq b_i + z_i (B_i - b_i) \Big\},$$

$$L_j = \left\{ (y_1, y_2, z_j) : z_j \in \{0, 1\}, \quad 0 \leq y_1 - y_{1j} \leq A_j + z_j \left(a_j - A_j \right), \right.$$

$$\left. 0 \leq y_2 - y_{2j} \leq b_j + z_j \left(B_j - b_j \right) \right\}.$$

We wonder if $L_i \cap L_j \neq \emptyset$. This holds if the system of inequalities

$$\max \left\{ y_{1i}, y_{1j} \right\} \leq y_1 \leq \min \left\{ y_{1i} + A_i + z_i \left(a_i - A_i \right), \ y_{1j} + A_j + z_j \left(a_j - A_j \right) \right\},$$

$$\max \left\{ y_{2i}, y_{2j} \right\} \leq y_2 \leq \min \left\{ y_{2i} + b_i + z_i \left(B_i - b_i \right), \ y_{2j} + b_j + z_j \left(B_j - b_j \right) \right\}$$

is consistent at least for one combination of binary z_i, z_j. This can be verified by inspection.

Substituting $z_i = z_j = 0$ yields

$$\max \left\{ y_{1i}, y_{1j} \right\} \leq \min \left\{ y_{1i} + A_i, y_{1j} + A_j \right\}, \quad \max \left\{ y_{2i}, y_{2j} \right\} \leq \min \left\{ y_{2i} + b_i, y_{2j} + b_j \right\}.$$

For $z_i = z_j = 1$ we have

$$\max \left\{ y_{1i}, y_{1j} \right\} \leq \min \left\{ y_{1i} + a_i, y_{1j} + a_j \right\}, \quad \max \left\{ y_{2i}, y_{2j} \right\} \leq \min \left\{ y_{2i} + B_i, y_{2j} + B_j \right\}.$$

Substituting $z_i = 1, \ z_j = 0$ yields

$$\max \left\{ y_{1i}, y_{1j} \right\} \leq \min \left\{ y_{1i} + a_i, y_{1j} + A_j \right\}, \quad \max \left\{ y_{2i}, y_{2j} \right\} \leq \min \left\{ y_{2i} + B_i, y_{2j} + b_j \right\}.$$

And finally for $z_i = 0, \ z_j = 1$ we get

$$\max \left\{ y_{1i}, y_{1j} \right\} \leq \min \left\{ y_{1i} + A_i, y_{1j} + a_j \right\}, \quad \max \left\{ y_{2i}, y_{2j} \right\} \leq \min \left\{ y_{2i} + b_i, y_{2j} + B_j \right\}.$$

Thus if at least one pair of inequalities above hold, then $L_i \cap L_j \neq \emptyset$. In a similar way we can check overlapping for the other composite objects, e.g., for star-shapes represented as a superposition of a square and a rhombus.

To check if L-object is totally placed inside a convex container we can use Proposition 2.3 since all vertices of the object are easily identified. However, for rectangular and L-shaped containers with all edges parallel to the principal axes we can state constraints (9.4) based on simple geometrical considerations.

Below we present results of a numerical experiment for packing L-objects in rectangular and L-shaped containers. The normalized objective was defined as the total area of the objects divided over the area of the smallest object. For the case of equal objects the normalized objective coincides with the number of objects.

In the first part of the experiment the test bed set of 6 instances was used for packing maximal number of equal L-objects into a rectangular container of width 3 and height 6. Two types of the objects were considered with the shapes corresponding to $A = B = 2R$ and $B = 0.5A = 2R$. The thickness of

Table 9.4 Equal L-objects

#	R	n_Δ	z	LP	LP+C	T	T+C	Z	LP	LP+C	T	T+C
1	0.5	3,321	37	1,163	135.5	16	11	22	1,029	117.2	25	14
2	0.625	2,145	20	680.7	107.7	4	3	10	580.1	79.08	4	4
3	0.5625	2,556	25	851.7	127.5	8	5	14	739.8	92.45	8	8
4	0.375	5,778	75	2,223	271.6	150	150	42	2,036	195.6	610	330
5	0.3125	8,385	116	3,379	384.9	1,867	620	71	3,144	275.4	3,581	1,930
6	0.4375	4,186	48	1,537	201.5	76	76	29	1,383	146.5	73	55

Table 9.5 Packing 2 different L-objects

#	R_1, R_2	n_Δ	z	L1	L2	T	LR	LR+C	L1+	L2+	T+
1	0.6, 6.3	3,969	750	309	4	* (12 %)	78, 384	5,980	681	4	* (7.8 %)
2	0.6, 6.3	3,969	447.5	227	2	530	51, 404	4,931	400	2	* (6.0 %)
3	1, 5.3	1,369	271.2	215	2	17	8, 458	1,352	197	6	* (5.7 %)
4	1, 5.3	1,369	147.2	91	2	16	6, 221	1,107	114	3	440

L-object was defined as $a = b = 0.3R$ in both cases. A rectangular uniform grid of size $\Delta = 0.15R$ (a half of the thickness) was used. The results of the numerical experiment are given in Table 9.4. The first three columns present instance number, value of R and a number of binary variables n_Δ. The next five columns present indicators for the case $A = B$: the optimal value of integer solution z; value of the LP-relaxation without and with valid cuts, LP and LP+C; CPU time in sec. to get integer solution without and with valid cuts, T and T+C. The last five columns present similar indicators for the case $B = 0.5A$. For all problem instances $mipgap = 0$ was set for running CPLEX.

As we can see from Table 9.4 introducing valid inequalities improves significantly the LP-bound and reduces CPU-time, especially for hard instances.

In the second part of the experiment two different L-objects were packed in a square 30×30 container maximizing the total normalized area of the packed objects. Four instances were considered according to the shape of the objects. For instances 1 and 3 $A = B = 2R$ and for instances 2 and 3, $B = 0.5A = 2R$. In all cases $a = b = R$. Two values of R were considered and for $R = R_2$ (large object) the minimal number of the objects to be packed was set to two, $m_2 = 2$ in (9.2).

The results are presented in Table 9.5. Here the first four columns give instance number, radii R_1, R_2, number n_Δ of grid nodes (integer variables) and the value z of the optimal solution. Columns 5 and 6 give the number of small (L1) and large (L2) objects in the optimal solution, while column 7 indicates corresponding CPU time in sec. for the case of using the valid inequalities (9.14). Asterisk indicates that the computation was interrupted after the computation time exceeded 1,800 s. CPU time and the value in parenthesis gives the corresponding *mipgap*. Columns 8 and 9 present the value of the LP-relaxation without (LR) and with (LR+C) valid inequalities. The last three columns give the number of objects packed (L1+, L2+)

Fig. 9.4 Instances 1, 2

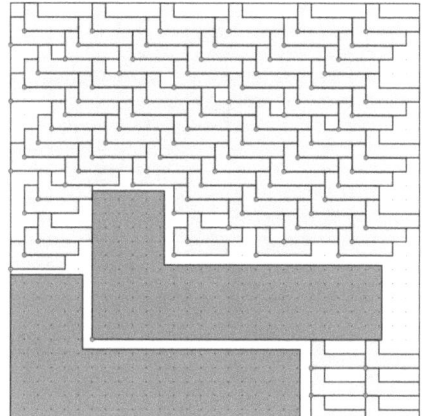

Fig. 9.5 Instances 3, 4

and CPU time (T+) for the case of nesting allowed. Optimal packings for instances 1–4 are presented in Figs. 9.4 and 9.5 for the case without nesting and in Figs. 9.6 and 9.7 for nesting allowed.

In the final part of experimentation L-shaped container was considered for $A = B = 30$, $a = b = 12$. Two instances were considered according to the shape of the two different L-objects. For the first instance $A = B = 2R$ for both objects and for the second $B = 0.5A = 2R$. In all cases $a = b = R$. Two values of R were used, $R_1 = 1$, $R_2 = 5.3$ and for $R = R_2$ we set $m_2 = 2$ in (9.2). A rectangular uniform grid of size $\Delta = \min\{R_1, R_2\} = 1$ was used giving $n_\Delta = 637$ grid nodes in the L-container. The optimal solution was obtained in less than 1 s. CPU time. For the first instance ($A = B = 2R$) the optimal solution gives 108 small and 2 large

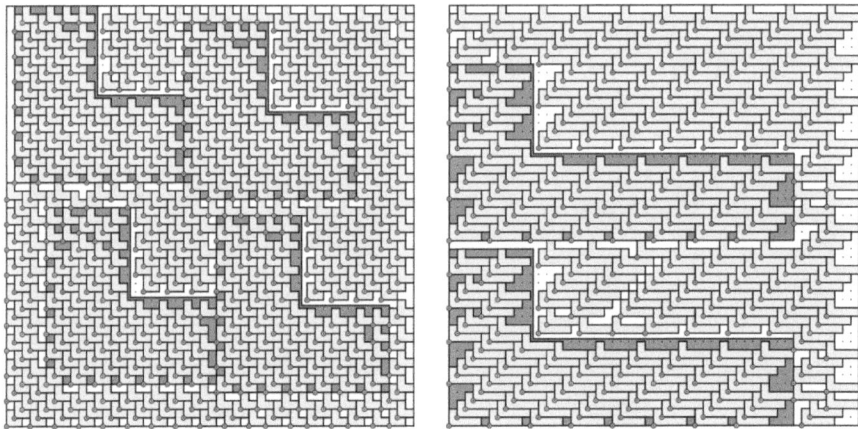

Fig. 9.6 Instances 1, 2 with nesting

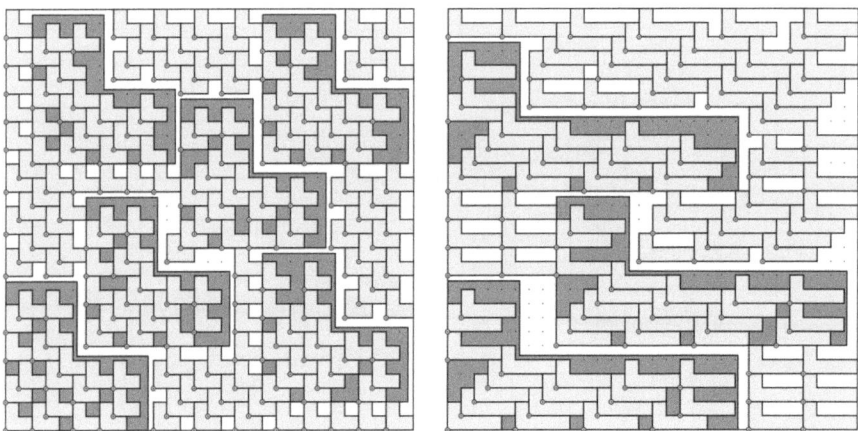

Fig. 9.7 Instances 3, 4 with nesting

L-objects without nesting and (120, 4) for nesting allowed. For the second instance ($B = 0.5A = 2R$) we get (33, 2) and (71, 2) objects, respectively. The optimal packings are presented in Figs. 9.8 and 9.9.

9.5   Conclusions

Integer programming formulations were considered for approximated packing objects in a container. Using a grid approximation of the container packing problems can be transformed into optimal assignment of the objects (reference points) to nodes of the grid subject to non-overlapping constraints. In this work we used

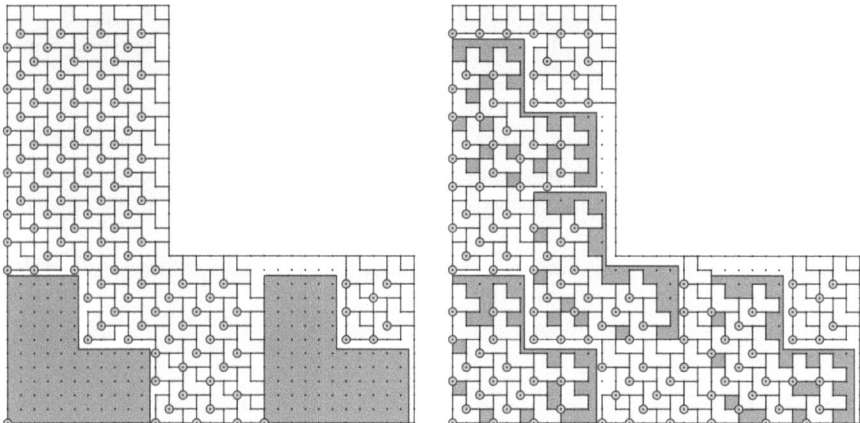

Fig. 9.8 Instance 1

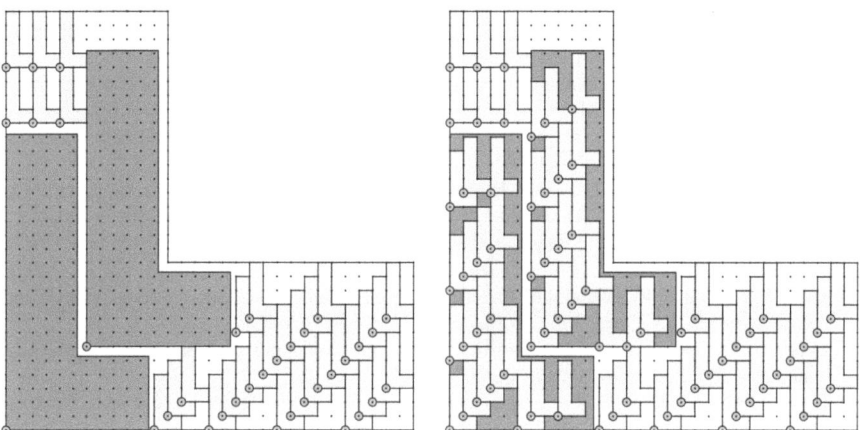

Fig. 9.9 Instance 2

linear non-overlapping constraints. However, as noted in Remark 2.1, the problem (9.1)–(9.6) is closely related to the quadratic assignment problem and corresponding approaches can be also used for packing problems. Some results in this direction are in course.

Valid inequalities (9.14) were proposed to strengthening the formulation and our numerical experiments demonstrate that the value of the LP-relaxation can be tightened significantly by (9.14). Moreover, aggregating valid cuts not only improve the value of the relaxation, but also change the structure of the optimal LP-solution. A simple LP-based heuristic is proposed in Litvinchev et al. [29] for packing circular objects.

Grid approximation of the container results in a large-scale integer optimization problem. Using decomposition and/or aggregation techniques [30] to split the nodes of the grid into smaller subsets (container decomposition) and/or creating "macro nodes" (nodes aggregation) may be helpful to cope with high dimension. Some results in this direction are in course.

A critical question in grid approximation is how to choose parameters of the grid, e.g. shape and number of nodes, to get a reasonable trade-off between computational burden and proximity to the true optimal packing. The use of non-uniform and/or adaptive grids seems to be interesting direction for the future research.

Acknowledgements This work was partially supported by Grants from RFBR, Russia (12 01 00893 a), and CONACYT, Mexico (167019).

References

1. Baltacioglu, E., Moore, J.T., Hill, R.R.: The distributor's three-dimensional pallet-packing problem: a human-based heuristical approach. Int. J. Oper. Res. **1**, 249–266 (2006)
2. Castillo, I., Kampas, F.J., Pinter, J.D.: Solving circle packing problems by global optimization: numerical results and industrial applications. Eur. J. Oper. Res. **191**, 786–802 (2008)
3. Fasano, G.: Solving Non-standard Packing Problems by Global Optimization and Heuristics. Springer-Verlag, Berlin (2014)
4. Frazer, H.J., George, J.A.: Integrated container loading software for pulp and paper industry. Eur. J. Oper. Res. **77**, 466–474 (1994)
5. Stevenson, D., Searchfield, G., Xu, X.: Spatial design of hearing aids incorporating multiple vents. Trends Hear. **18** (2014). doi:10.1177/2331216514529189
6. Wang, J.: Packing of unequal spheres and automated radiosurgical treatment planning. J. Comb. Optim. **3**, 453–463 (1999)
7. Hifi, M., M'Hallah, R.: A literature review on circle and sphere packing problems: models and methodologies. Adv. Oper. Res. (2009). doi:10.1155/2009/150624
8. Lopez, C.O., Beasley, J.E.: A heuristic for the circle packing problem with a variety of containers. Eur. J. Oper. Res. **214**, 512–525 (2011)
9. Lopez, C.O., Beasley, J.E.: Packing unequal circles using formulation space search. Comput. Oper. Res. **40**, 1276–1288 (2013)
10. Pinter, J.D., Kampas, F.J.: Nonlinear optimization in Mathematica with MathOptimizer Professional. Math. Educ. Res. **10**, 1–18 (2005)
11. Akeb, H., Hifi, M.: Solving the circular open dimension problem using separate beams and look-ahead strategies. Comput. Oper. Res. **40**, 1243–1255 (2013)
12. Birgin, E.G., Gentil, J.M.: New and improved results for packing identical unitary radius circles within triangles, rectangles and strips. Comput. Oper. Res. **37**, 1318–1327 (2010)
13. Stoyan, Y.G., Yaskov, G.N.: Packing congruent spheres into a multi-connected polyhedral domain. Int. Trans. Oper. Res. **20**, 79–99 (2013)
14. Bennel, J.A., Olivera, J.F.: A tutorial in irregular shape packing problems. J. Oper. Res. Soc. **60**, 93–105 (2009)
15. Toledo, F.M.B., Carravilla, M.A., Ribero, C., Oliveira, J.F., Gomes, A.M.: The dotted-board model: a new MIP model for nesting irregular shapes. Int. J. Prod. Econ. **145**, 478–487 (2013)
16. Galiev, S.I., Lisafina, M.S.: Linear models for the approximate solution of the problem of packing equal circles into a given domain. Eur. J. Oper. Res. **230**, 505–514 (2013)
17. Litvinchev, I., Ozuna, L.: Packing circles in a rectangular container. Paper presented at the 1st international congress on logistics and supply chain, Mexican Institute of Transportation, Queretaro, Mexico, 24–25 October 2013

18. Litvinchev, I., Ozuna, L.: Integer programming formulations for approximate packing circles in a rectangular container. Math. Probl. Eng. (2014). Article ID 317697, doi:10.1155/2014/317697
19. Litvinchev, I., Ozuna, L.: Approximate packing circles in a rectangular container: valid inequalities and nesting. J. Appl. Res. Technol. **12**, 716–723 (2014)
20. Litvinchev, I., Infante, L., Ozuna, L.: Approximate circle packing in a rectangular container: integer programming formulations and valid inequalities. Lect. Notes Comput. Sci. **8760**, 47–61 (2014)
21. Beasley, J.E.: An exact two-dimensional non-guillotine cutting tree search procedure. Oper. Res. **33**, 49–64 (1985)
22. Burkard, R., Dell'Amico, M., Martello, S.: Assignment Problems, Revised Reprint. SIAM (2012)
23. Wolsey, L.A.: Integer Programming. Wiley, New York (1999)
24. ILOG CPLEX, Mathematical programming optimizers. Version 12.6 (2013)
25. Bortfeldt, A., Wäscher, G.: Constraints in container loading—a state-of-the-art review. Eur. J. Oper. Res. **229**, 1–20 (2013)
26. Wang, W., Wang, H., Dai, G., Wang, H.: Visualization of large hierarchical data by circle packing. CHI '06 Proceedings of the SIGCHI Conference on Human Factors in Computing Systems, April 22–27, Montreal, Canada, pp. 517–520 (2006)
27. George, J.A.: Multiple container packing: a case study of pipe packing. J. Oper. Res. Soc. **47**, 1098–1109 (1996)
28. Pedroso, J.P., Cunha, S., Tavares, J.N.: Recursive circle packing problems. Int. Trans. Oper. Res. (2014). doi:10.1111/itor.12107
29. Litvinchev, I., Infante, L., Ozuna, L.: LP-based heuristic for packing circular-like objects in a rectangular container. Math. Probl. Eng. (to appear)
30. Litvinchev, I., Tsurkov, V.: Aggregation in Large Scale Optimization. Kluwer, Boston (2003)
31. Burke, E.K., Hellier, R.S., Kendall, G., Whitwell, G.: Irregular packing using the line and arc no-fit polygon. Oper. Res. **58**, 948–970 (2010)
32. Litvinchev, I., Infante, L., Ozuna, L.: Packing circular-like objects in a rectangular container. J. Comput. Syst. Sci. Int. **54**, 259–267 (2015)

Chapter 10
Exploiting Packing Components in General-Purpose Integer Programming Solvers

Jakub Mareček

Abstract The problem of packing boxes into a large box is often only a part of a complex problem. For example in furniture supply chain applications, one needs to decide what trucks to use to transport furniture between production sites and distribution centres and stores, such that the furniture fits inside. Such problems are often formulated and sometimes solved using general-purpose integer programming solvers.

This chapter studies the problem of identifying a compact formulation of the multi-dimensional packing component in a general instance of integer linear programming, reformulating it using the discretisation of Allen–Burke–Mareček, and solving the extended reformulation. Results on instances of up to 10,000,000 boxes are reported.

Keywords Packing • Multi-dimensional packing • Integer programming • Structure exploitation

10.1 Introduction

It is well known that one problem may have many integer linear programming formulations, in various dimensions, whose computational behaviour differs widely. The problem of packing three-dimensional boxes into a larger box is a particularly striking example. The trivial question of how many unit cubes can be packed into a $(k \times 1 \times 1)$ box is impossible to answer for $k = 12$ within an hour using the widely known formulation of Chen/Padberg/Fasano and state-of-the-art solvers (IBM ILOG CPLEX, FICO XPress MP, Gurobi Solver), despite the fact that after pre-solve, there are only 616 rows and 253 columns and 4,268 non-zeros. In contrast, a discretised ("space-indexed") formulation of [3] makes it possible to solve the instance with $k = 10,000,000$ within an hour, where the instance

J. Mareček (✉)
IBM Research – Ireland, B3 F14 Damastown Campus, Dublin 15, The Republic of Ireland
e-mail: jakub.marecek@ie.ibm.com

© Springer International Publishing Switzerland 2015
G. Fasano, J.D. Pintér (eds.), *Optimized Packings with Applications*, Springer
Optimization and Its Applications 105, DOI 10.1007/978-3-319-18899-7_10

of linear programming had 10,000,002 rows, 10,000,000 columns, and 30,000,000 non-zeros. This is due to the fact that the discretised linear programming relaxation provides a particularly strong bound. On a large-scale benchmark of randomly generated instances, the values of the linear programming relaxations at the root node are 10.49 % and 0.37 % away from the respective integer optima for the Chen/Padberg/Fasano and the formulation of [3], respectively.

The Allen–Burke–Marecek formulation with adaptive discretisations is, however, often rather hard to formulate in an algebraic modelling language. First, the choice of the discretisation of the larger box is hard in any case; in terms of computational complexity, finding the best possible discretisation is Δ_2^p-Hard. Second, algebraic modelling languages such as AMPL, GAMS, MOSEL, and OPL are ill-suited to the dynamic programming required by efficient discretisation algorithms. Finally, non-uniform grids pose a major challenge in debugging the formulation and analysis of the solutions. Within a modelling language, the discretisation is hence often chosen in an ad hoc manner, without considerations of optimality.

Instead, this chapter studies the related problems of identifying a packing component in the Chen/Padberg/Fasano formulation in a larger instance of integer linear programming, automating the reformulation of Chen/Padberg/Fasano formulation into the Allen–Burke–Marecek formulation, obtaining a reasonable discretisation in the process, and translating the solutions obtained by solving the Allen–Burke–Marecek formulation back to the original one. The main contributions are:

1. An overview of integer linear programming formulations of packing boxes into a larger box, covering both the formulation of Chen/Padberg/Fasano and the discretisation of [3].
2. A formal statement of the problems of extracting the component and row-block, respectively, which correspond to the formulation of Chen/Padberg/Fasano for packing boxes into a larger box, from a general integer linear programming instance. Surprisingly, we show that the extraction of the row-block is solvable by polynomial-time algorithms.
3. A formal statement of the problem of finding the best possible discretisation, i.e., the best possible reformulation of Chen/Padberg/Fasano to the formulation of [3]. We show that the problem is Δ_2^p-Hard, but that there are very good heuristics.
4. A novel computational study of the algorithms for the problems above. There, we take the integer linear programming instance, extract the row-block corresponding to the Chen/Padberg/Fasano formulation, perform the discretisation, write out the integer linear programming instance using the Allen–Burke–Marecek formulation, and solve it.

A general-purpose integer linear programming solver using the above results could solve much larger instances of problems involving packing components, compared with the state-of-the-art solvers using the Chen/Padberg/Fasano formulation.

10.2   Background and Definitions

In order to motivate the study of packing components, let us consider:

Problem 1.1. The Precedence-Constrained Scheduling (PCS): Given integers $r, n \geq 1$, amounts $a_i > 1$ of resources $i = 1, 2, \ldots, r$ available, resource requirements of n jobs represented by $D \in R^{n \times r}$, and p pairs of numbers $P \subseteq \{(i, j) \mid 1 \leq i < j \leq n\}, |P| = p$ expressing job i should be executed prior to executing j, find the largest integer k so that k jobs can be executed using the resources available.

This problem on its own has numerous important applications, notably in extraction of natural resources [7, 28], where it is known as the open-pit mine production scheduling problem, problems in supply chain applications, where both weight and volume of the load is considered (e.g., [18, 21]) and each order can consist of multiple boxes, and further problems in aerospace engineering [5, 17], where one needs to pack the load of an aircraft, satellite, or similar, and there are similar constraints.

Clearly, there is a packing component to Precedence-Constrained Scheduling (Problem 1.1). Let us fix the order of six allowable rotations in dimension three arbitrarily and define:

Problem 1.2. The Container Loading Problem (CLP): Given dimensions of a large box ("container") $x, y, z > 0$ and dimensions of n small boxes $D \in R^{n \times 3}$ with associated values $w \in R^n$, and specification of the allowed rotations $r = \{0, 1\}^{n \times 6}$, find the greatest $k \in R$ such that there is a packing of small boxes $I \subseteq \{1, 2, \ldots, n\}$ into the container with value $k = \sum_{i \in I} w_i$. The packed small boxes I may be rotated in any of the allowed ways, must not overlap, and no vertex can be outside of the container.

Similarly:

Problem 1.3. The Van Loading Problem (VLP): Given dimensions of a large box ("van") $x, y, z > 0$, maximum mass $p \geq 0$ it can hold ("payload"), dimensions of n small boxes $D \in R^{n \times 3}$ with associated values $w \in R^n$, mass $m \in R^n$, and specification of the allowed rotations $r = \{0, 1\}^{n \times 6}$, find the greatest $k \in R$ such that there is a packing of small boxes $I \subseteq \{1, 2, \ldots, n\}$ into the container with value $k = \sum_{i \in I} w_i$ and mass $\sum_{i \in I} m_i \leq p$. The packed small boxes I may be rotated in any of the allowed ways, must not overlap, and no vertex can be outside of the container.

Such problems are particularly challenging. Ever since the work of [19], there has been much research on extended formulations of 2D packing problems using the notion of patterns, e.g. [25]. See [4] for an excellent survey. Only in the past decade or two has the attention focused to exact solvers for 3D packing problems [27], where even the special case with rotations around combinations of axes in multiples of 90° is NP-Hard to approximate [9]. Although there are a number of excellent heuristic solvers, the progress in exact solvers for the CLP has been limited, so far.

Table 10.1 Notation used in this chapter, which matches [3]

Symbol	Meaning
n	The number of boxes
H	A fixed axis, in the set $\{X, Y, Z\}$
α	An axis of a box, in the set $\{1, 2, 3\}$
$L_{\alpha i}$	The length of axis α of box i
$l_{\alpha i}$	The length of axis α of box i halved
D_H	The length of axis H of the container
w_i	The volume of box i in the CLP

The Formulation of Chen/Padberg/Fasano Chen et al. [8] introduced an integer linear programming formulation using the relative placement indicator:

$$\lambda_{ij}^H = \begin{cases} 1 & \text{if box } i \text{ precedes box } j \text{ along axis } H \\ 0 & \text{otherwise} \end{cases},$$

$$\delta_{\alpha i}^H = \begin{cases} 1 & \text{if box } i \text{ is rotated so that axis } \alpha \text{ is parallel to fixed } H \\ 0 & \text{otherwise} \end{cases},$$

$$x_i^H = \text{absolute position of box } i \text{ along axis } H.$$

Using the notation of Table 10.1 and implicit quantification, it reads:

$$\max \sum_{i=1}^{n} \sum_{H} w_i \delta_{1i}^H \tag{10.1}$$

$$\text{s.t.} \sum_{H} \delta_{2i}^H = \sum_{H} \delta_{1i}^H \tag{10.2}$$

$$\sum_{H} \delta_{1i}^H = \sum_{\alpha} \delta_{\alpha i}^H \tag{10.3}$$

$$L_{1j(i)} \lambda_{j(i)i}^H + \sum_{\alpha} l_{\alpha i} \delta_{\alpha i}^H \leq x_i^H \tag{10.4}$$

$$x_i^H \leq \sum_{\alpha} (D_H - l_{\alpha i}) \delta_{\alpha i}^H - L_{1j(i)} \lambda_{ij(i)}^H \tag{10.5}$$

$$D_H \lambda_{ji}^H + \sum_{\alpha} l_{\alpha i} \delta_{\alpha i}^H - \sum_{\alpha} (D_H - l_{\alpha j}) \delta_{\alpha j}^H \leq x_i^H - x_j^H \tag{10.6}$$

$$x_i^H - x_j^H \leq \sum_{\alpha} (D_H - l_{\alpha i}) \delta_{\alpha i}^H - \sum_{\alpha} l_{\alpha j} \delta_{\alpha j}^H - D_H \lambda_{ij}^H \tag{10.7}$$

$$\sum_{H} (\lambda_{ij}^H + \lambda_{ji}^H) \leq \sum_{H} \delta_{1i}^H \tag{10.8}$$

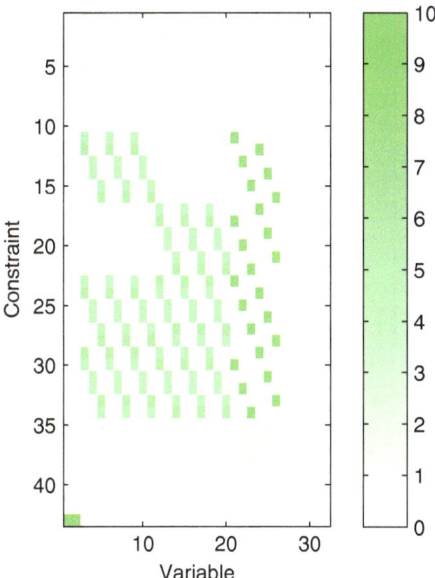

Fig. 10.1 The *A* matrix corresponding to the previously unsolved instance Pigeon-02 in the Chen/Padberg/Fasano formulation, with *colour* highlighting the absolute value of the coefficients

$$\sum_H (\lambda_{ij}^H + \lambda_{ji}^H) \leq \sum_H \delta_{1j}^H \tag{10.9}$$

$$\sum_H \delta_{1i}^H + \sum_H \delta_{1j}^H \leq 1 + \sum_H (\lambda_{ij}^H + \lambda_{ji}^H) \tag{10.10}$$

$$\sum_{i=1}^{n} \sum_H \left(\prod_\alpha L_{\alpha i} \right) \delta_{1i}^H \leq \prod_H D_H \tag{10.11}$$

$$\delta_{\alpha i}^H \in \{0, 1\}, \lambda_{ij}^H \in \{0, 1\}$$

$$L_{1i} \leq L_{2i} \leq L_{3i}, j(i) \text{ such that } L_{1j(i)} = \max\{L_{1j}\} \text{ for } 1 \leq i \neq j \leq n.$$

See Fig. 10.1 for the sparsity pattern of a small instance, known as Pigeon-02, where at most a single unit cube out of two can be packed into a single unit cube. The constraint matrix of Pigeon-02 is 43×32. In the figure, the column ordering is given by placing *D* first, δ second, λ third, and *x* at the end, with the highest-order-first indexing therein. *D* has dimension 2, δ has dimension 18, λ has dimension 6 and *x* has dimension 6. In the figure, the row ordering is given by the order of constraints (10.2)–(10.11) above. Notice that a similar ordering of columns and rows is naturally produced by a parser of an algebraic modelling language, such as AMPL, GAMS, MOSEL or OPL.

Table 10.2 The performance of various solvers on 3D Pigeon Hole Problem instances encoded in the Chen/Padberg/Fasano formulation

	Time (s)		
	Gurobi 4.0	CPLEX 12.4	SCIP 2.0.1 + CLP
Pigeon-01	< 1	< 1	< 1
Pigeon-02	< 1	< 1	< 1
Pigeon-03	< 1	< 1	< 1
Pigeon-04	< 1	< 1	< 1
Pigeon-05	< 1	< 1	3.3
Pigeon-06	< 1	< 1	37.9
Pigeon-07	1.5	< 1	779.3
Pigeon-08	7.4	< 1	–
Pigeon-09	88.6	66.4	–
Pigeon-10	1,381.4	686.3	–
Pigeon-11	–	–	–
Pigeon-12	–	–	–

"–" denotes that optimality of the incumbent solution has not been proven within an hour

This formulation has been studied a number of times. Notably, Fasano [11–13] suggested numerous improvements to the formulation. Padberg [29] has studied properties of the formulation and, in particular, identified the subsets of constraints with the integer property. Allen et al. [3] proposed further improvements, including symmetry-breaking constraints and means of exploitation of properties of the rotations. See [14–16] for further extensions to Tetris-like items and further references.

Nevertheless, whilst the addition of these constraints improves the performance somewhat, the formulation remains far from satisfactory. The formulations provide only weak lower bounds. As has been pointed out in Sect. 10.1 and can be confirmed in Table 10.2, Pigeon-k becomes very challenging as the number k of unit cubes to pack into ($k \times 1 \times 1$) box grows. See Fig. 10.2 for the sparsity patten of Pigeon-12, which is already a substantial challenge for any modern solver to date, although the constraint matrix is only 1333 × 552 and can be reduced to 738 rows and 288 columns in the presolve. Modern integer programming solvers fail to solve instances larger than this, even considering all the additional constraints described above.

Discretisations Discretised relaxations proved to be very strong in scheduling problems corresponding to one-dimensional packing [30, 33, 34] and can be shown to be asymptotically optimal for various geometric problems both in two dimensions [31] and in higher [35] dimensions. Beasley [6] has extended the formulation to 2D cutting applications, in the process of deriving a non-linear formulation, for which he proposed solvers. Allen et al. [3] have extended the formulation to the 3D problem of packing boxes into a larger box, with a considerable amount of work being done independently and subsequently [10, 22, 23]. In this formulation, the

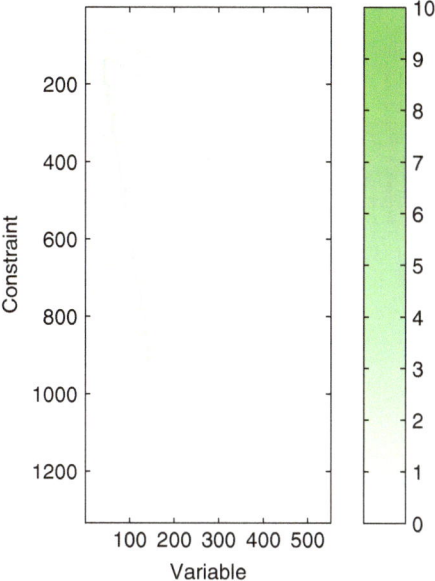

Fig. 10.2 The A matrix corresponding to instance Pigeon-12 in the Chen/Padberg/Fasano formulation, with *colour* highlighting the absolute value of the coefficients

small boxes are partitioned into types $t \in \{1, 2, \ldots, n\}$, where boxes of one type share the same triple of dimensions. A_t is the number of boxes of type t available. The large box is discretised into units of space, possibly non-uniformly, with the indices $(x, y, z) \in D \subset R^3$ used to index the space-indexed binary variable:

$$
\mu^t_{x,y,z} = \begin{cases} 1 & \text{if a box of type } t \text{ is placed such that its lower-back-left} \\ & \text{vertex is at coordinates } x, y, z \\ 0 & \text{otherwise} \end{cases} \tag{10.12}
$$

Without allowing for rotations, the formulation reads:

$$
\max \sum_{x,y,z,t} \mu^t_{xyz} w_t \tag{10.13}
$$

$$
\text{s.t.} \sum_t \mu^t_{xyz} \leq 1 \quad \forall x, y, z \tag{10.14}
$$

$$
\mu^t_{xyz} = 0 \quad \forall x, y, z, t \text{ where } x + L_{1t} > D_X \text{ or}
$$
$$
\text{or } y + L_{2t} > D_Y \text{ or } z + L_{3t} > D_Z \tag{10.15}
$$

$$
\sum_{x,y,z} \mu^t_{xyz} \leq A_t \quad \forall t \tag{10.16}
$$

Algorithm 1 $f(D, n, L, x, y, z, t)$

1: **Input:** Discretisation as indices $D \subset R^3$ of μ variables (10.12), number of boxes n, dimensions $L_{\alpha i} \in R \quad \forall \alpha = x, y, z, i = 1, \ldots, n$, indices $(x, y, z) \in D, t \in \{1, 2, \ldots, n\}$ of one scalar within μ (10.12)

2: **Output:** Set of indices of μ variables (10.12) to include in a set-packing inequality (10.17)

3: Set $S \leftarrow \{(x, y, z, t)\}$
4: **for** each other unit $(x', y', z') \in D, (x, y, z) \neq (x', y', z')$ **do**
5:     **for** each box type $t' \in \{1, 2, \ldots, n\}$ **do**
6:         **if** box of type t at (x, y, z) overlaps box of type t' at (x', y', z'), i.e., $(x \leq x' + L_{x't'} \leq x + L_{xt}) \wedge (y \leq y' + L_{y't'} \leq y + L_{yt})$
            $\wedge (z \leq z' + L_{z't'} \leq z + L_{zt})$ **then**
7:             $S \leftarrow S \cup \{(x', y', z', t')\}$
8:         **end if**
9:     **end for**
10: **end for**
11: **return** S

$$\sum_{x', y', z', t' \in f(D, n, L, x, y, z, t)} \mu^{t'}_{x'y'z'} \leq 1 \quad \forall x, y, z, t \tag{10.17}$$

$$\mu^{t}_{xyz} \in \{0, 1\} \quad \forall x, y, z, t \tag{10.18}$$

where one may use Algorithm 1 or similar to generate the index set f in Constraint (10.17). See Fig. 10.3 for an example.

The constraints are very natural: No region in space may be occupied by more than one box type (10.14), boxes must be fully contained within the container (10.15), there may not be more than A_t boxes of type t (10.16), and boxes cannot overlap (10.17). There is one non-overlapping constraint (10.17) for each discretised unit of space and type of box.

In order to support rotations, new box types need to be generated for each allowed rotation and linked via set packing constraints, which are similar to Constraint (10.16). In order to extend the formulation to the VLP, it suffices to add the payload capacity constraint $\sum_{x,y,z,t} \mu^{t}_{xyz} m_t \leq p$.

10.3 Finding the Precedence-Constrained Component

In the rest of the chapter, our goal is to extract the precedence-constrained component in Chen/Padberg/Fasano formulation from a general integer linear program, and reformulate it into the discretised formulation. Throughout, we distinguish between a submatrix of a larger matrix, whose columns and rows are selected arbitrarily, and a row-block, where the columns are selected arbitrarily, but the rows form a contiguous block in the larger matrix. First, let us state the problem of extracting the Chen/Padberg/Fasano relaxation formally:

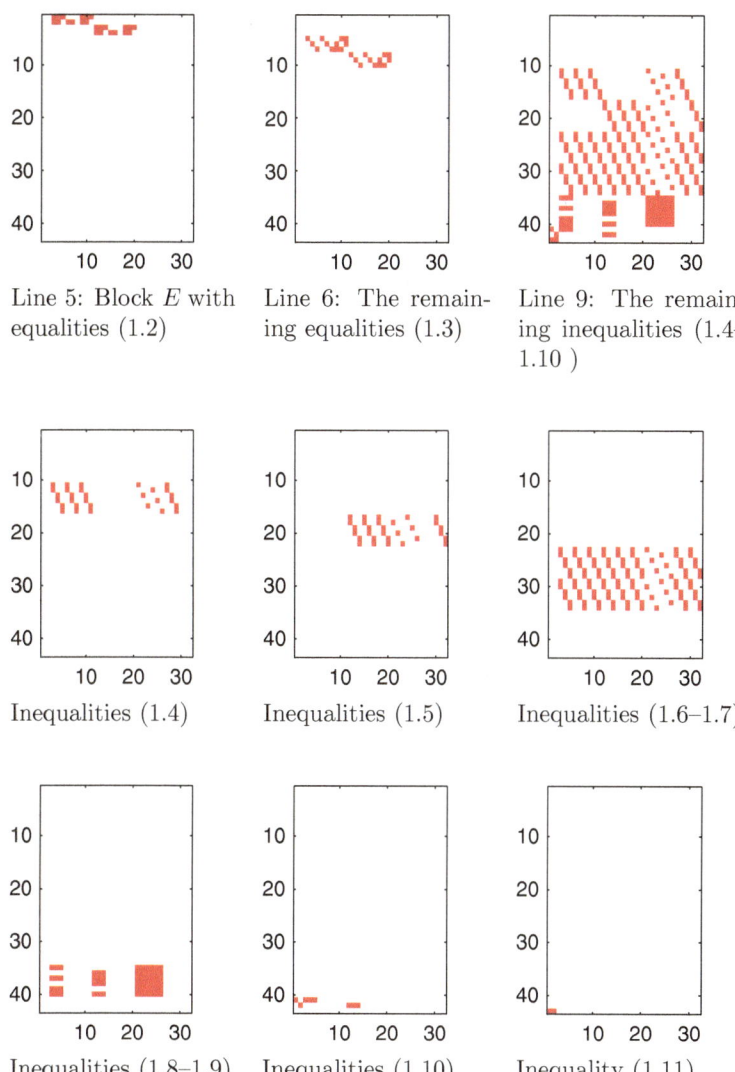

Fig. 10.3 The workings of Algorithm 1 illustrated on instance Pigeon-02 in the Chen/ Padberg/Fasano formulation, as introduced in Fig. 10.1

Problem 1.4. Precedence-Constrained Component/Row-Block Extraction: Given positive integers d, m_1, m_2, an $m_1 \times d$ integer matrix A_1, an $m_2 \times d$ integer matrix A_2, an m_1-vector b_1 of integers, and an m_2-vector b_2 of integers, corresponding to a mixed integer linear program with constraints $A_1 y = b_1$, $A_2 y \leq b_2$, find the largest integer n ("the maximum number of boxes"), such that:

- there exists a $4n \times 9n$ submatrix/row-block E of A_1, corresponding only to binary variables, which we denote δ, with zero coefficients elsewhere in the rows
- there exists a $9n(n-1)/2 + 6n + 1 \times 3n(n-1)/2 + 12n$ submatrix/row-block F of A_2, corresponding to $9n$ binary variables δ as before, $3n(n-1)/2$ binary variables denoted λ, and $3n$ continuous variables denoted x, with zero coefficients elsewhere in the rows
- E contains n rows with exactly two non-zero coefficients ± 1, corresponding to (10.2), and 0 in the right-hand side b_1
- E contains $3n$ rows with exactly four non-zero coefficients ± 1, corresponding to (10.3), and 0 in the right-hand side b_1
- F contains $6n$ rows with exactly five non-zero coefficients, some not necessarily ± 1, corresponding to (10.4)–(10.5), and 0 in the right-hand side b_2
- F contains $6n(n-1)$ rows with exactly nine non-zero coefficients, some not necessarily ± 1, corresponding to (10.6)–(10.7), and 0 in the right-hand side b_2
- F contains $2n(n-1)$ rows with exactly nine non-zero coefficients ± 1, corresponding to (10.8)–(10.9), and 0 in the right-hand side b_2
- F contains $n(n-1)$ rows with exactly 12 non-zero coefficients ± 1, corresponding to (10.10), and 1 in the right-hand side b_2
- F contains 1 rows with exactly $3n$ non-zero coefficients, not necessarily ± 1, corresponding to (10.11), and a positive number in the right-hand side b_2.

Notice that by maximising the number of rows involved, we also maximise the number of boxes, as number $r = 9n(n-1)/2 + 6n + 1$ of rows is determined by number n of boxes. The following can be seen easily:

Theorem. Precedence-Constrained Row-Block Extraction *is in* $\mathcal{P}$.

Proof sketch. A polynomial-time algorithm for extracting the precedence-constrained block can clearly rely on there being a polynomial number of blocks of the required size. See Algorithm 2. □

Algorithm 2 displays a very general algorithm schema for *Precedence-Constrained Row-Block Extraction*. Notably, the test of Line 10 requires elaboration. First, one needs to partition the block into the five families of rows (10.4)–(10.11). Some rows (10.8), (10.9), (10.4), (10.5) and (10.11) are clearly determined by the numbers of non-zeros (9, 5 and 3). One can distinguish between others (10.6), (10.7) and (10.10) by their right-hand sides. The test as to whether the rows represent the constraints (10.4)–(10.11) is based on identifying the variables. Continuous variables x are, however, identified easily and binary variables δ are determined in Line 6. What remains are variables λ. In certain situations, it may be possible to simplify the extraction by relying on further details provided by the parsers of algebraic modelling languages.

Algorithm 2 `PrecedenceConstrainedBlock`(A_1, b_1, A_2, b_2)

1: **Input:** $A_1 x = 1, A_2 x \leq 1$, that is $m_1 \times d$ matrix A_1 and $m_2 \times d$ matrix A_2, m_1-vector b_1, m_2-vector b_2
2: **Output:** Integer k and blocks E, F of A_1, A_2

3: Set $k_{\max}$ to the largest integer k such that there are k subsequent rows in A_1 with exactly 6 non-zero elements, all ± 1
4: **for** integer $k = 4n$ from $k_{\max}$ down to 4 **do**
5:      **for** $4n \times 9n$ block E in A_1 such that all rows have 6 non-zeros ± 1 **do**
6:          **if** there are no $6n$ other rows A_1 in corresponding to (10.3) **then**
7:              Continue
8:          **end if**
9:          **for** $7n + 9n(n-1)/2 \times 3n^2 + 9n$ block F in A_2 **do**
10:             **if** F cannot be partitioned into (10.4)–(10.11) **then**
11:                 Continue
12:             **end if**
13:             **return** n, E, F
14:         **end for**
15:     **end for**
16: **end for**

10.4 Exploiting the Packing Component

In order to reduce the number of regions of space, and thus the number of variables in the formulation, a sensible space-discretisation method should be employed. In many transport applications, for instance, there are only a small number of package types, with the ISO 269 standard giving the dimensions of the package. Using the discretisation of 1 mm, one could indeed introduce $162 \times 229 \times h$ variables to represent a package with an ISO 269 C5 base and the height of h millimeters, but if there are only packages with base-sizes specified by ISO 216 standard and larger than C5 to be packed in the batch, it would make sense to discretise to units of space representing 162×229 mm in two dimensions. The question is how to derive such a discretisation in a general-purpose system.

The greatest common divisor (GCD) reduction can be applied on a per-axis basis, finding the GCD between the length of the container for an axis and all the valid lengths of boxes that can be aligned along that axis and scaling by the inverse of the GCD. This is trivial to do and is useful when all lengths are multiples of a large number, which may be common in certain situations.

In other situations, this may not reduce the number of variables at all, and it may be worth tackling the optimisation variant of:

Problem 1.5. The DISCRETISATION DECISION: Given integers $k \geq n > 0$, dimensions of a large box ("container") $x, y, z > 0$, dimensions of n small boxes $D \in \mathbb{R}^{n \times 3}$ with associated values $w \in \mathbb{R}^n$ and specification $r = \{0, 1\}^{n \times 6}$ of what rotations are allowed, and k positions $S \in \mathbb{R}^{k \times 3}$, decide whether in any optimum solution of CONTAINER LOADING PROBLEM with x, y, z, D, w, r, the lower-bottom-left vertices of the n boxes can be positioned in n of the k positions in S.

Algorithm 3 DiscretisationDP(M, n, D, r)

1: **Input:** Dimensions $M \in \mathbb{R}^3$ of the container, dimensions of n small boxes $D \in \mathbb{R}^{n \times 3}$, and allowed rotations $r = \{0, 1\}^{n \times 6}$

2: **Output:** Integer k and k possible positions

3: $P = \emptyset$
4: **for** axis with limit $m \in M$ **do**
5:     $L = \{d \mid d \in D$ may appear along this axis, given allowed rotations $r\}$
6:     $P = $ closure of $P \cup \{p + l \mid p \in P, l \in L, p + l \leq m\}$,
        optionally pruning $p \in P$ that cannot occur due to the dependence of the axes and the fact
        each box can be packed at most once
7: **end for**
8: **return** $|P|, P$

Consider the following:

Theorem ([32]). *The decision whether the optimum of an instance of* Knapsack *is unique is Δ_2^p-Complete, where Δ_2^p is the class of problems that can be solved in polynomial time using oracles from $\mathcal{NP}$.*

Theorem. Discretisation Decision *is Δ_2^p-Hard.*

Proof sketch. One could check for the uniqueness of the optimum of an instance of KNAPSACK using any algorithm for DISCRETISATION DECISION (Problem 1.5) Consider $n = k$.   □

We can, however, use non-trivial non-linear space-discretisation heuristics. Early examples include [20]. We use the same values as before on a per-axis basis, i.e. the lengths of any box sides that can be aligned along the axis. We then use dynamic programming to generate all valid locations for a box to be placed. See Algorithm 3. For example, given an axis of length 10 and box lengths of 3, 4 and 6, we can place boxes at positions 0, 3, 4, 6, and 7. 8, 9 and 10 are also possible, but no length is small enough to still lie within the container if placed at these points. This has reduced the number of regions along that axis from 10 to 5. An improvement on this scale may not be particularly common in practice, but it is obvious that this approach can be no worse than the GCD method at discretising the container and that the approach can help when the GCD turns out to be 1. This also adds some implicit symmetry breaking into the model. Notice that the algorithm runs in time polynomial in the size of the output it produces, which may be exponential in the size of the input and considerably larger than the size of the best possible output.

10.5 Computational Experience

The approach was tested on two sets of instances, introduced in [3]:

- 3D Pigeon Hole Problem instances, Pigeon-n, where $n + 1$ unit cubes are to be packed into a container of dimensions $(1 + \epsilon) \times (1 + \epsilon) \times n$.

Table 10.3 The performance of Gurobi 4.0 on 3D Pigeon Hole Problem instances encoded in the Chen/Padberg/Fasano and the discretised formulations as reported in [3]

	Time (s)	
	Chen/Padberg/Fasano	Discretised
Pigeon-01	< 1	< 1
Pigeon-02	< 1	< 1
Pigeon-03	< 1	< 1
Pigeon-04	< 1	< 1
Pigeon-05	< 1	< 1
Pigeon-06	< 1	< 1
Pigeon-07	1.5	< 1
Pigeon-08	7.4	< 1
Pigeon-09	88.6	< 1
Pigeon-10	1,381.4	< 1
Pigeon-100	–	< 1
Pigeon-1000	–	1.0
Pigeon-10000	–	1.8
Pigeon-100000	–	4.2
Pigeon-1000000	–	45.1
Pigeon-10000000	–	664.0
Pigeon-100000000	–	–

"–" denotes that no integer solution has been found

- SA and SAX datasets, which are used to test the dependence of solvers' performance on parameters of the instances, notably the number of boxes, heterogeneity of the boxes, and physical dimensions of the container. There is 1 pseudo-randomly generated instance for every combination of container sizes ranging from 5–100 in steps of 5 units cubed and the number of boxes to pack ranging from 5–100 in steps of 5. The SA datasets are perfectly packable, i.e. are guaranteed to be possible to load the container with 100 % utilisation with all boxes packed. The SAX are similar but have no such guarantees; the total volume of the boxes is greater than the volume of the container.

All of the instances are available at http://discretisation.sf.net. Some of these instances have been included in MIPLIB 2010 by Koch [24] and have been widely utilised in benchmarking of integer programming solvers ever since.

For the 3D Pigeon Hole Problem, results obtained within 1 h using three leading solvers and the Chen/Padberg/Fasano formulation without any reformulation are shown in Table 10.2, while Table 10.3 compares the results on both formulations. These tests and further tests reported below were performed on a 64-bit computer running Linux, which was equipped with 2 quad-core processors (Intel Xeon E5472) and 16 GB memory. The solvers tested were IBM ILOG CPLEX 12.4, Gurobi Solver 4.0, and SCIP 2.0.1 of [1] with CLP as the linear programming

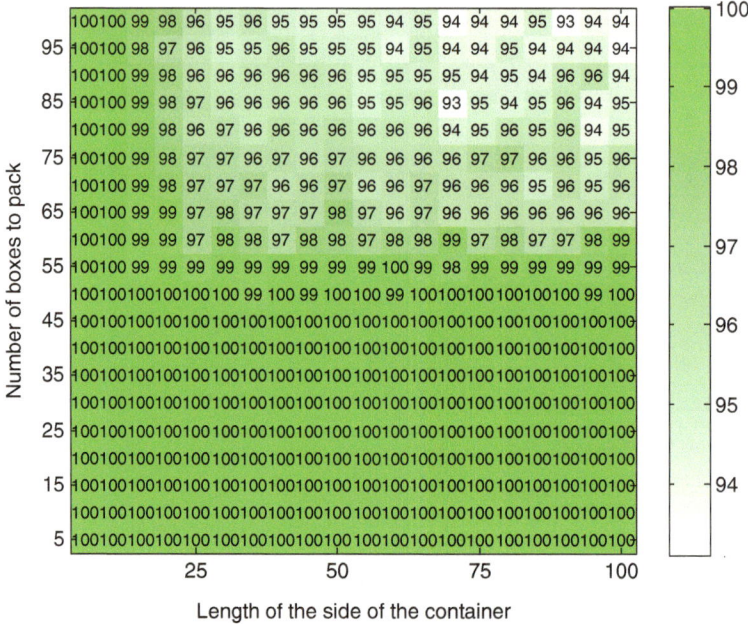

Fig. 10.4 The best solutions obtained within an hour per solver per instance from the SA dataset for varying number of boxes (*vertical axis*) and the length of the side of the container (*horizontal axis*). The *colours* highlight the volume utilisation in percent

solver. Pigeon-02 is easy to solve for any modern solver. IBM ILOG CPLEX 12.4 eliminates 17 rows and 26 columns in presolve and performs a number of further changes. The reduced instance has 23 rows and 22 columns and the reported runtime is 0.00 s. Pigeon-10 is the largest instance reliably solvable within an hour, but that should not be surprising, considering it has 525 rows, 220 columns, and 3,600 non-zeros in the constraint matrix after pre-solve of CPLEX 12.4. None of the solvers managed to prove optimality of the incumbent solution for Pigeon-12 within an hour using the Chen/Padberg/Fasano formulation, although the instance of linear programming had only 627 rows, 253 columns and 4,268 non-zeros after pre-solve. As of September 2014, instances up to pigeon-13 using the Chen/Padberg/Fasano have been solved in the process of testing integer programming solvers without the automatic reformulation, albeit at a great expense of computing time. In contrast, the reformulation and discretisation makes it possible to solve Pigeon-10000000 within an hour, where the instance of linear programming had 10,000,002 rows, 10,000,000 columns and 30,000,000 non-zeros. The time for the extraction and reformulation of the instance was under 1 s across of the instances.

For the SA and SAX datasets, Figs. 10.4 and 10.5 summarise solutions obtained within an hour using either the Chen/Padberg/Fasano formulation or the reformulation, whichever was faster. This shows that although it may be possible to solve certain instances with 10,000,000 boxes within an hour to optimality, real-life instances with hundreds of boxes may still be challenging, even considering

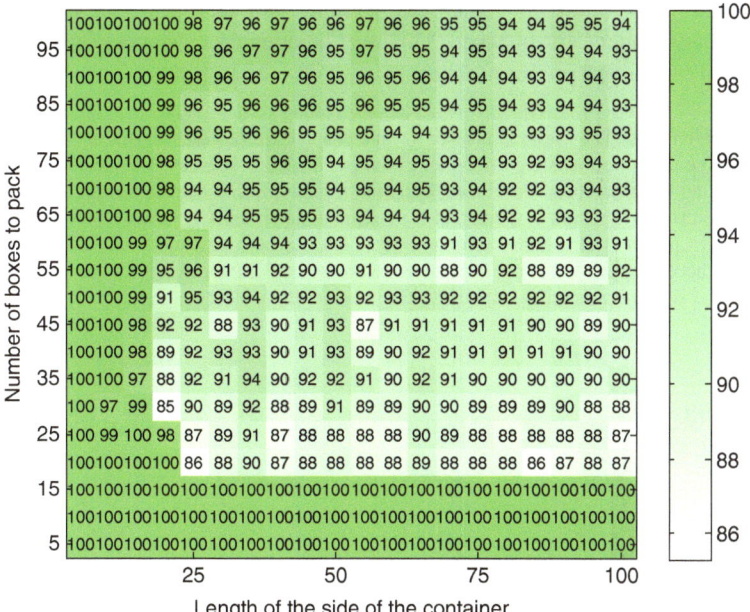

Fig. 10.5 As above for the SAX dataset. The *colours* highlight the quality in terms of $100(1 - s/b)$ for solution with value s and upper bound b after 1 h

the reformulation. Nevertheless, as has been pointed out by Allen et al. [3], the space-indexed relaxation provides a particularly strong upper bound. The mean integrality gap, or the ratio of the difference between root linear programming relaxation value and optimum to optimum has been 10.49 % and 0.37 % for the Chen/Padberg/Fasano and the space-indexed formulation, respectively, on the SA and SAX instances solved to optimality within the time limit of 1 h.

10.6   Conclusions

Overall, the discretisation formulation provides a particularly strong relaxation. It is easy to automate a more natural formulation to the discretisation formulation, programmatically, provided the constraints are a contiguous block of rows. This is the case, e.g., whenever one uses an algebraic modelling language.

Going forwards, it would be good to develop tests whether the reformulation is worthwhile, as the discretised relaxation may become prohibitively large and dense for instances with many distinct box-types, and box-types or containers of large sizes in terms of the units of discretisation. Alternatively, one may consider multi-level discretisations. Plausibly, one may also apply similar structure-exploiting approaches to "components" other than packing. Mareček [26] studied the graph colouring component, for instance. This may be open up new areas for research in computational integer programming.

Acknowledgements The views expressed in this chapter are personal views of the author and should not be construed as suggestions as to the product road map of IBM products. Some of the supporting code has been developed by Allen [2] for [3] and can be downloaded at http://discretisation.sf.net (September 30th, 2014). This material is loosely based upon an otherwise unpublished Chapter 8 of the dissertation of [26], but has been extended and improved greatly thanks to the comments of two anonymous referees, both in terms of the contents and the presentation. The author is most grateful for the referees' thoughtful suggestions.

References

1. Achterberg, T.: SCIP: solving constraint integer programs. Math. Program. Comput. **1**(1), 1–41 (2009)
2. Allen, S.D.: Algorithms and data structures for three-dimensional packing. Ph.D. thesis, University of Nottingham (2012)
3. Allen, S.D., Burke, E.K., Mareček, J.: A space-indexed formulation of packing boxes into a larger box. Oper. Res. Lett. **40**, 20–24 (2012)
4. Amor, H.B., de Carvalho, J.V.: Cutting stock problems. In: Column Generation, pp. 131–161. Springer, New York (2005)
5. Baldi, M.M., Perboli, G., Tadei, R.: The three-dimensional knapsack problem with balancing constraints. Appl. Math. Comput. **218**(19), 9802–9818 (2012)
6. Beasley, J.E.: An exact two-dimensional non-guillotine cutting tree search procedure. Oper. Res. **33**(1), 49–64 (1985)
7. Bienstock, D., Zuckerberg, M.: Solving lp relaxations of large-scale precedence constrained problems. In: Eisenbrand, F., Shepherd, F. (eds.) Integer Programming and Combinatorial Optimization. Lecture Notes in Computer Science, vol. 6080, pp. 1–14. Springer, Berlin/Heidelberg (2010)
8. Chen, C.S., Lee, S.M., Shen, Q.S.: An analytical model for the container loading problem. Eur. J. Oper. Res. **80**(1), 68–76 (1995)
9. Chlebík, M., Chlebíková, J.: Hardness of approximation for orthogonal rectangle packing and covering problems. J. Discrete Algorithms **7**(3), 291–305 (2009)
10. de Queiroz, T.A., Miyazawa, F.K., Wakabayashi, Y., Xavier, E.C.: Algorithms for 3d guillotine cutting problems: unbounded knapsack, cutting stock and strip packing. Comput. Oper. Res. **39**(2), 200–212 (2012)
11. Fasano, G.: Cargo analytical integration in space engineering: a three-dimensional packing model. In: Ciriani, T.A., Gliozzi, S., Johnson, E.L., Tadei, R. (eds.) Operational Research in Industry, pp. 232–246. Purdue University Press, West Lafayette, IN (1999)
12. Fasano, G.: A mip approach for some practical packing problems: balancing constraints and tetris-like items. Q. J. Belg. Fr. Ital. Oper. Res. Soc. **2**(2), 161–174 (2004)
13. Fasano, G.: Mip-based heuristic for non-standard 3d-packing problems. Q. J. Belg. Fr. Ital. Oper. Res. Soc. **6**(3), 291–310 (2008)
14. Fasano, G.: Erratum to: Chapter 3 model reformulations and tightening. In: Solving Non-standard Packing Problems by Global Optimization and Heuristics. Springer Briefs in Optimization, pp. E1–E2. Springer, New York (2014)
15. Fasano, G.: Model reformulations and tightening. In: Solving Non-standard Packing Problems by Global Optimization and Heuristics. Springer Briefs in Optimization, pp. 27–37. Springer, New York (2014)

16. Fasano, G.: Tetris-like items. In: Solving Non-standard Packing Problems by Global Optimization and Heuristics. Springer Briefs in Optimization, pp. 7–26. Springer, New York (2014)
17. Fasano, G., Pintér, J.: Modeling and Optimization in Space Engineering. Springer, New York (2013)
18. Gendreau, M., Iori, M., Laporte, G., Martello, S.: A tabu search algorithm for a routing and container loading problem. Transp. Sci. **40**(3), 342–350 (2006)
19. Gilmore, P.C., Gomory, R.E.: Multistage cutting stock problems of two and more dimensions. Oper. Res. **13**(1), 94–120 (1965)
20. Herz, J.C.: Recursive computational procedure for two-dimensional stock cutting. IBM J. Res. Dev. **16**(5), 462–469 (1972)
21. Iori, M., Salazar-González, J.J., Vigo, D.: An exact approach for the vehicle routing problem with two-dimensional loading constraints. Transp. Sci. **41**(2), 253–264 (2007)
22. Junqueira, L., Morabito, R., Yamashita, D.S.: Three-dimensional container loading models with cargo stability and load bearing constraints. Comput. Oper. Res. **39**(1), 74–85 (2012)
23. Junqueira, L., Morabito, R., Yamashita, D.S., Yanasse, H.H.: Optimization models for the three-dimensional container loading problem with practical constraints. In: Fasano, G., Pintér, János D. (eds.) Modeling and Optimization in Space Engineering. Springer Optimization and Its Applications, vol.73, pp. 271–293. Springer, New York (2013)
24. Koch, T., Achterberg, T., Andersen, E., Bastert, O., Berthold, T., Bixby, R.E., Danna, E., Gamrath, G., Gleixner, A.M., Heinz, S., Lodi, A., Mittelmann, H., Ralphs, T., Salvagnin, D., Steffy, D.E., Wolter, K.: MIPLIB 2010. Math. Program. Comput. **3**(2), 103–163 (2011)
25. Madsen, O.B.G.: Glass cutting in a small firm. Math. Program. **17**(1), 85–90 (1979)
26. Mareček, J.: Exploiting structure in integer programs. Ph.D. thesis, University of Nottingham (2012)
27. Martello, S., Pisinger, D., Vigo, D.: The three-dimensional bin packing problem. Oper. Res. **48**(2), 256–267 (2000)
28. Moreno, E., Espinoza, D., Goycoolea, M.: Large-scale multi-period precedence constrained knapsack problem: a mining application. Electron. Notes Discret. Math. **36**, 407–414 (2010) [ISCO 2010 - International Symposium on Combinatorial Optimization]
29. Padberg, M.: Packing small boxes into a big box. Math. Meth. Oper. Res. **52**(1), 1–21 (2000)
30. Pan, Y., Shi, L.: On the equivalence of the max-min transportation lower bound and the time-indexed lower bound for single-machine scheduling problems. Math. Program. **110**(3, Ser. A), 543–559 (2007)
31. Papadimitriou, C.H.: Worst-case and probabilistic analysis of a geometric location problem. SIAM J. Comput. **10**(3), 542–557 (1981)
32. Papadimitriou, C.H.: On the complexity of unique solutions. J. Assoc. Comput. Mach. **31**(2), 392–400 (1984)
33. Sousa, J.P., Wolsey, L.A.: A time indexed formulation of non-preemptive single machine scheduling problems. Math. Program. **54**, 353–367 (1992)
34. van den Akker, J.M.: LP-based solution methods for single-machine scheduling problems. Dissertation, Technische Universiteit Eindhoven, Eindhoven (1994)
35. Zemel, E.: Probabilistic analysis of geometric location problems. SIAM J. Algebraic Discret. Meth. **6**(2), 189–200 (1985)

Chapter 11
Robust Designs for Circle Coverings of a Square

Mihály Csaba Markót

Abstract In this chapter we investigate coverings of a square with uniform circles of minimal radius, with uncertainties in the actual locations of the circles. This setting is an example model of deploying sensors or other kind of observation units so that there are uncertainties in their deployments. Possible examples include scenarios when the deployment has to be made remotely (e.g., from the air) into a potentially dangerous place, deployments into a location with unknown terrain, or deployments influenced by the weather. Our goal is to produce coverings that are optimal in terms of a minimal radius, and are also robust in the following sense: wherever the circles are actually placed within a given uncertainty region, the result is still guaranteed to be a covering. We investigate three special uncertainty regions: first we prove that for uniform circular uncertainty regions the optimal robust covering can be created from the exact optimal covering without uncertainties, provided that the exact covering configuration is feasible for the robust scenario. For uncertainty regions given by line segments and by general convex polygons we design a bi-level optimization method combining a complete and rigorous global search and a derivative free black-box search, and show the efficiency of the method on some examples.

Keywords Circle covering • Uncertainty • Sensor network deployment • Robust design • Global optimization • Interval arithmetic • Complete search • Black box search

11.1 Introduction

In this study we are dealing with the problem of optimal coverings of a square with uniform circles, minimizing the required radius. In contrast to the classical and well-studied exact case, where the placement of the circles can be precisely carried out, we consider a case that is much more useful from the practical point of

M.C. Markót (✉)
Faculty of Mathematics, University of Vienna, Oskar-Morgenstern-Platz 1, 1090 Vienna, Austria
e-mail: mihaly.markot@univie.ac.at

© Springer International Publishing Switzerland 2015   225
G. Fasano, J.D. Pintér (eds.), *Optimized Packings with Applications*, Springer
Optimization and Its Applications 105, DOI 10.1007/978-3-319-18899-7_11

view: optimal coverings with uncertainties in the locations (i.e., the centers) of the circles. Throughout the work we often refer to a motivating example: the modeling of deployment scenarios, such as the placements of sensor networks or other kind of observation units in a real-life environment. Examples of these scenarios include remote deployment (e.g., from the air) into a potentially dangerous place, such as the location of a nuclear accident, deployments into a location with unknown terrain, or deployments influenced by the weather.

There are numerous earlier studies dealing with uncertainty modeling for various design and optimization problems: to name a few of the application areas, we mention facility location problems (see, e.g., [19] for a survey and [18] on the maximal covering location problem), scheduling and vehicle routing [12], structural optimization [6], space system design [2], etc. In any case, to the best of our knowledge the present work is the first one that employs interval numerical methods and complete global search for covering problems, in order to prove the covering property of the obtained designs with mathematical correctness.

11.2   Circle Coverings of a Square

The classical (exact) circle covering problem of a square is the following: given the unit square and the number N of uniform circles used for the covering, find those of the locations of the circle centers for which a covering of the square can be produced with minimal circle radius. In order to formalize the problem, we introduce the following notation: points in the plane are denoted by boldface, with their coordinates marked with x and y lower indices. That is, we write $\mathbf{a} \in \mathbb{R}^2$, $\mathbf{a} = (a_x, a_y)$.

Let $A = \{\mathbf{a}^i\}, \mathbf{a}^i \in [0, 1]^2, i = 1, \ldots, N$ be a set of N points in the unit square, called a *covering configuration*. The set of all covering configurations of N points is denoted by $\mathscr{A}_N$.

Proposition 1. *Let $A = \{\mathbf{a}^i\} \in \mathscr{A}_N$ be a covering configuration. The smallest possible radius for which N circles with centers A and with this uniform radius cover the square is given by*

$$r(A) = \max_{\mathbf{p} \in [0,1]^2} \min_{1 \le i \le N} ||\mathbf{p} - \mathbf{a}^i||_2.$$

Proof. First we prove that the circles with centers A and uniform radius $r(A)$ cover the square. Let $r(A)$ be attained at $\bar{\mathbf{p}} \in [0, 1]^2$, and let $\mathbf{p} \in [0, 1]^2$ be arbitrary. Then $r(A) = \min_i ||\bar{\mathbf{p}} - \mathbf{a}^i||_2 \ge \min_i ||\mathbf{p} - \mathbf{a}^i||_2$. That is, for $j = \arg\min_i ||\mathbf{p} - \mathbf{a}^i||_2$, the distance of p and $\mathbf{a}^j$ is at most $r(A)$, which means that $\mathbf{p}$ is covered by the circle centered at $\mathbf{a}^j$.

Next, assume that $t < r(A)$ is a radius that results in a covering with centers A. Then $\bar{\mathbf{p}}$ is also covered, so there is an index j such that $||\bar{\mathbf{p}} - \mathbf{a}^j||_2 \le t$. This implies $r(A) = \min_i ||\bar{\mathbf{p}} - \mathbf{a}^i||_2 \le ||\bar{\mathbf{p}} - \mathbf{a}^j||_2 \le t$, a contradiction.   □

The value $r(A)$ of the above proposition will be called the *optimal covering radius with respect to the configuration A*. That is, $r(A)$ is attained at that point **p** for which the smallest distance from the points $\mathbf{a}^i$ is maximal. For a fixed A, any $r > r(A)$ naturally also results in a covering; such an r radius will be called simply as a *covering radius with respect to A*. Then the problem of optimal circle covering is to find the covering configuration with the smallest optimal covering radius, that is,

$$\min_{A \in \mathscr{A}_N} r(A).$$

The minimum of this problem will be called the *optimal covering radius* of the problem. (It is important to note the difference between an optimal covering radius w.r.t. a configuration A, i.e., $r(A)$, and the overall optimal covering radius, $\min_{A \in \mathscr{A}_N} r(A)$.)

Note that finding the covering radius for a given covering configuration is itself a nontrivial problem, and thus, the problem with such a formulation can be attacked with some kind of bi-level programming methods only. Also note that there are other types of formalizations of the covering problem. For example, the current best known numerical approach [16]—that also employs bi-level optimization—uses the uncovered area as the inner level objective function to determine whether a radius is eligible to be a covering radius.

Similarly to other types of covering and packing problems, this problem is also treated in two typical ways: with purely mathematical tools (for finding good solutions, and proving their optimality when possible), or with computer methods. In particular, for $N \leq 10$ (the instances we investigate in the present study) the cases $N \leq 5$, $N = 7$ are proven to be optimal by mathematical methods [5], while the current best known, conjectured optimal solutions for $N = 6, 8$ [11], and for $N = 9, 10$ [20] are results of numerical methods.

In overall, the most extensive numerical study for the problem class (up to 30 circles) is [16]. This study found all optimal and previously best known solutions for the previously detailed cases. Table 11.1 contains the optimal radii up to $N = 10$, taken directly from [16]. The optimal structures are, however, only depicted there, showing the touching points between two circles and a circle and the square. In addition, the symmetry groups of the found configurations are also given in [16]. From this structural information we computed the approximate coordinates of the centers (by solving a set of polynomial system of equations). These will serve

Table 11.1 Optimal covering radii for $N = 1, \ldots, 10$ [16]

N	r	N	r
1	0.70710678118654752440	6	0.29872706223691915876
2	0.55901699437494742410	7	0.27429188517743176508
3	0.50389110926865935327	8	0.26030010588652494367
4	0.35355339059327376220	9	0.23063692781954790734
5	0.32616058400398728086	10	0.21823351279308384300

in the present study as reference values, and also, good starting points for our numerical procedures. Table 11.2 contains the computed coordinates. Assuming that the optimal radii are correct in all digits, the computed coordinates are correct up to 12–15 digits after the decimal dot. The optimal configurations are also depicted in the present work for comparison reasons, on the left sides of Figs. 11.1, 11.2, and 11.3 (in some cases rotated and/or reflected as compared to the coordinates of Table 11.1).

11.3   Circle Coverings with Uncertainties

The traditional way of designing robust methods that handle uncertainties (that is, methods which always lead to feasible, or even better, optimal solutions with a prescribed success probability) is based on establishing a joint probability distribution of the design variables. This is, however, often impossible in practice, since in many cases even the individual probability distributions are unknown: often there is not enough experimental data to create such functions, or the uncertainties are influenced by so many factors that it is not possible to build a model for it.

An alternative to such methods, introduced recently by Neumaier [15], and tried successfully for space system design [1, 2] is based on the concept of *clouds*. In this approach, one asks the experts to establish *confidence regions* of the design variables w.r.t. the desired success probability, instead of probability distributions. In practice this means that from the expert knowledge we can infer that *with the given success probability the realization (outcome) of all design variables will be in the specified regions*. Apart from that no further probability information on the design variables is needed. Given such specified regions we can then make a worst case analysis of the design outcome. In practice this needs numerical methods the are able to analyze the whole search space (given by the Cartesian product of the confidence regions) and return results with mathematical rigor. Such tools are available in the forms of numerical methods based on *interval analysis* and *complete and rigorous global search* [4, 7, 13, 14]. The appropriate solution methods can naturally be bi-level programming techniques: in the inner level a worst case analysis is done (with interval global search), to determine feasible design solutions, while in the outer level another (not necessarily complete) method, e.g., black-box optimization is used to select the optimal or satisfactory designs among the feasible ones.

For our covering problem, the uncertainty modeling can be carried out in the following way: the design variables are naturally the centers of the circles, and a confidence region (w.r.t. a given success probability) of each center needs to be specified by experts. For example, in our motivating example, when sensors are dropped into a dangerous environment, experts of the carrier vehicle and the payload experts will determine that all payload items will fall into a given neighborhood of the desired target positions with a prescribed probability. Our job then, knowing the confidence regions, is to determine the target positions (the centers) and the required sensor range (the common radius) in such a way that the outcome configuration will

Table 11.2 Coordinates of the optimal coverings, $N = 1, \ldots, 10$

X	Y	X	Y
$N = 1$		$N = 8$	
0.5000000000000000	0.5000000000000000	0.1556943961115090	0.2086034518985931
$N = 2$		0.5000000000000000	0.1793933037363553
0.2500000000000000	0.5000000000000000	0.8443056038884910	0.2086034518985931
0.7500000000000000	0.5000000000000000	0.2467821880640653	0.5000000000000000
$N = 3$		0.7532178119359346	0.5000000000000000
0.2500000000000000	0.4374999999999999	0.1556943961115090	0.7913965481014069
0.7500000000000000	0.4374999999999999	0.5000000000000000	0.8206066962636447
0.5000000000000000	0.9375000000000000	0.8443056038884910	0.7913965481014069
$N = 4$		$N = 9$	
0.2500000000000000	0.2500000000000000	0.1387697944017800	0.1842181767245759
0.7500000000000000	0.2500000000000000	0.4581546916026700	0.1434279509541508
0.2500000000000000	0.7500000000000000	0.8193848972008900	0.1434279509541508
0.7500000000000000	0.7500000000000000	0.1894317802806140	0.5000000000000000
$N = 5$		0.5506619858788360	0.5000000000000000
0.2500000000000000	0.7905227302120298	0.9118921914770558	0.5000000000000000
0.7500000000000000	0.7905227302120298	0.1387697944017800	0.8157818232754241
0.1482473264108672	0.2905227302120298	0.4581546916026700	0.8565720490458493
0.5000000000000000	0.2548848764200725	0.8193848972008900	0.8565720490458493
0.8517526735891328	0.2905227302120298	$N = 10$	
$N = 6$		0.1666666666666667	0.1408832436034581
0.1416449464353010	0.2630105831749758	0.5000000000000000	0.1408832436034581
0.4597827072935948	0.2410148214731849	0.8333333333333334	0.1408832436034581
0.8181377608582938	0.2369894168250242	0.0000000000000000	0.5000000000000000
0.1818622391417062	0.7630105831749758	0.3333333333333333	0.5000000000000000
0.5402172927064052	0.7589851785268151	0.6666666666666666	0.5000000000000000
0.8583550535646990	0.7369894168250242	0.1666666666666667	0.8591167563965419
$N = 7$		0.5000000000000000	0.8591167563965419
0.2500000000000000	0.1128540574112841	0.8333333333333334	0.8591167563965419
0.7500000000000000	0.1128540574112841		
0.0000000000000000	0.5000000000000000		
0.5000000000000000	0.5000000000000000		
1.0000000000000000	0.5000000000000000		
0.2500000000000000	0.8871459425887159		
0.7500000000000000	0.8871459425887159		

result in a covering *regardless of the actual locations within the confidence regions*, and the radius is the smallest possible. The respective model is the following:

Given a configuration $A = \{\mathbf{a}^i\}$ in the unit square and given a fixed success probability, we model the uncertainty by a function U that maps each $\mathbf{a}^i$ into a confidence (or uncertainty) region around $\mathbf{a}^i$, denoted by $U(\mathbf{a}^i) \subseteq \mathbb{R}^2$. In the present

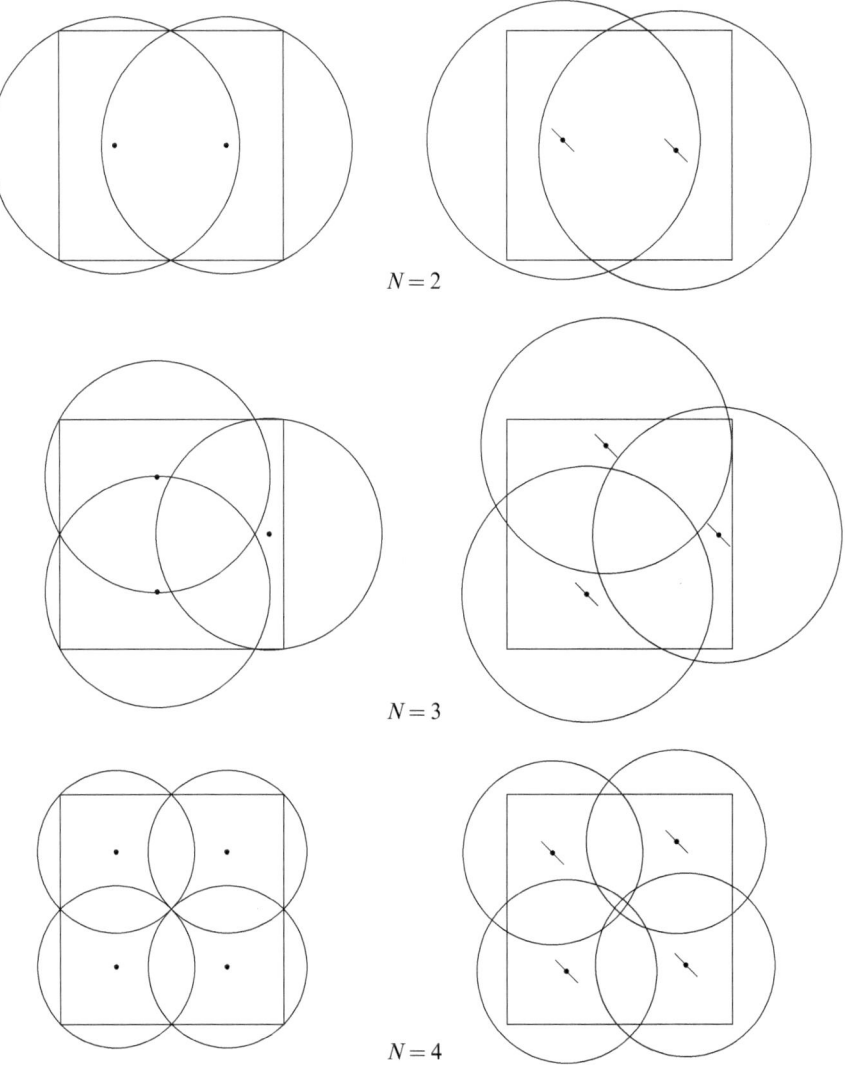

$N = 2$

$N = 3$

$N = 4$

Fig. 11.1 Exact (*left*) and uncertain (*right*) optimal covering with line segment shaped uncertainty regions; $N = 2, 3, 4$

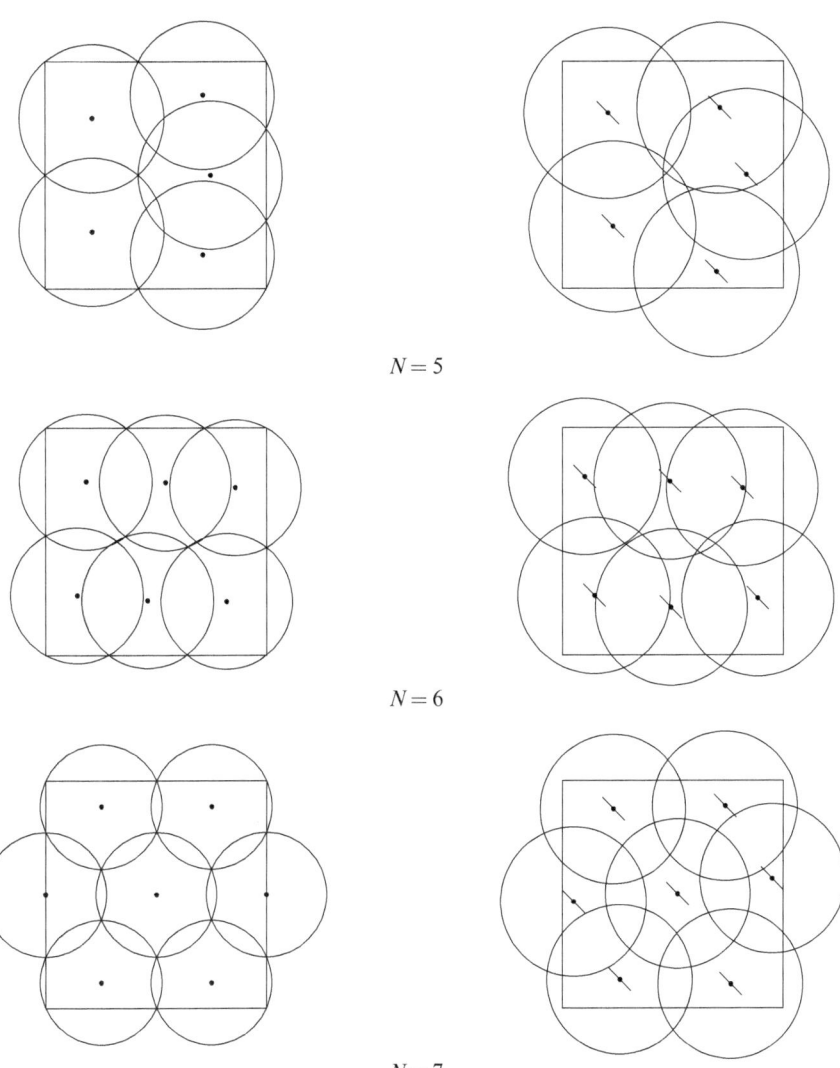

$N = 5$

$N = 6$

$N = 7$

Fig. 11.2 Exact (*left*) and uncertain (*right*) optimal covering with line segment shaped uncertainty regions; $N = 5, 6, 7$

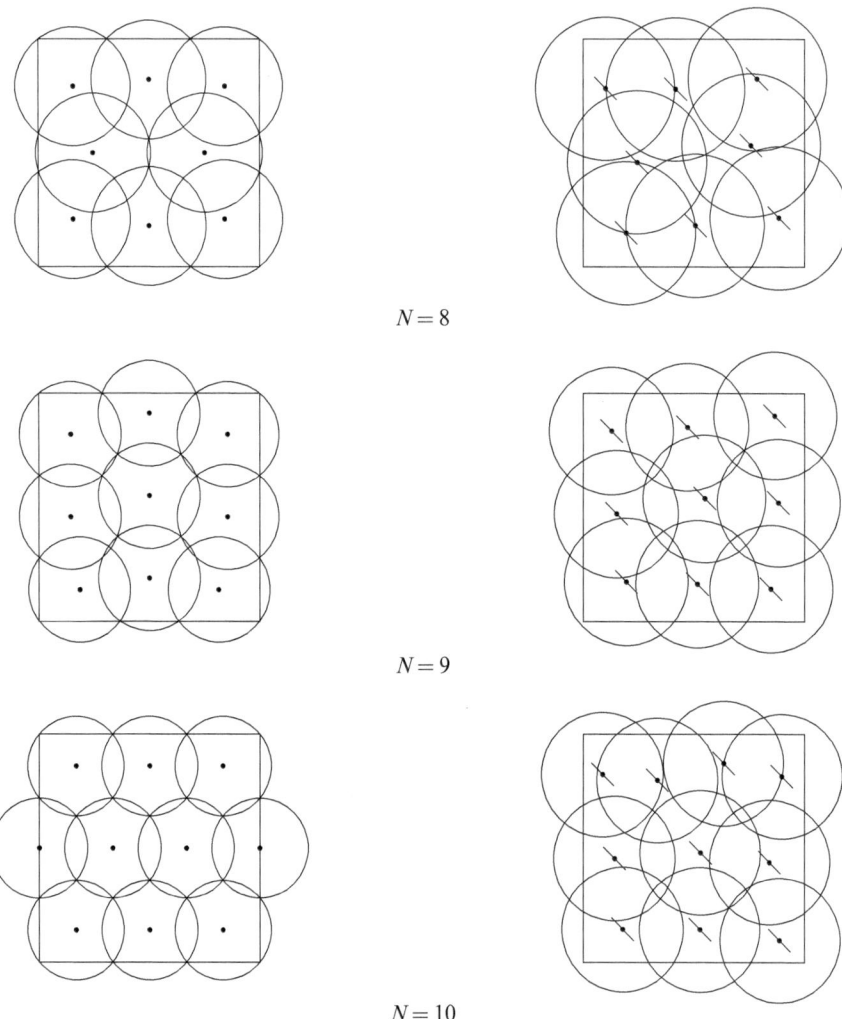

Fig. 11.3 Exact (*left*) and uncertain (*right*) optimal covering with line segment shaped uncertainty regions; $N = 8, 9, 10$

study we assume that $U(\mathbf{a}^i) \subseteq [0, 1]^2$ for all i, that is, with the given probability, a point will never be placed outside the square. (In practice this corresponds to the natural assumption that with the given probability the payload will always be placed inside the target region, avoiding, e.g., its loss or destruction.) This implies that for each given U, we need to restrict our search to the configurations

$$\mathscr{A}_N(U) := \{A \subseteq \mathscr{A}_N : U(\mathbf{a}^i) \subseteq [0, 1]^2, \ i = 1, \dots, N\}.$$

The elements of $\mathscr{A}_N(U)$ will be called *feasible configurations w.r.t. U*. Furthermore, *we assume that $U(\mathbf{a}^i)$ is closed for all $\mathbf{a}^i$*. The set $\{U(\mathbf{a}^i),\ i = 1,\ldots,N\}$ will be denoted by $U(A)$. For a set of points $B = \{\mathbf{b}^i \mid \mathbf{b}^i \in U(\mathbf{a}^i),\ i = 1,\ldots,N\}$ we use the shorthand containment notation $B \in U(A)$. That is, A denotes the target position of the centers, and $B \in U(A)$ denotes the realization of the placement.

Proposition 2. *Let $A = \{\mathbf{a}^i\} \in \mathscr{A}_N(U)$ be a covering configuration with uncertainty regions $U(A)$. The smallest possible radius that results in a covering for all possible sets of circle centers $B = \{\mathbf{b}^i\} \in U(A)$ is given by*

$$r(U(A)) = \max_{B \in U(A)} \max_{\mathbf{p} \in [0,1]^2} \min_{1 \leq i \leq N} ||\mathbf{p} - \mathbf{b}^i||_2 = \max_{B \in U(A)} r(B). \qquad (11.1)$$

Proof. For all realizations $B \in U(A)$, $r(B)$ is a respective covering radius, so any radius larger than or equal to this also results in a covering for B. Thus, the maximum of these radii over $U(A)$ results in coverings for all B, and this will obviously be the smallest appropriate radius. $\qquad \square$

The value $r(U(A))$ above will be called the *uncertain optimal covering radius with respect to A and $U(A)$*, and an $r > r(U(A))$ will be called simply as an *uncertain covering radius with respect to A and $U(A)$*. Then our problem of optimal circle covering with uncertainties is to find the covering target configuration with the smallest possible uncertain optimal covering radius, that is,

$$\min_{A \in \mathscr{A}_N(U)} r(U(A)).$$

Similarly to the exact case, this minimum will be called the *uncertain optimal covering radius* of the given problem.

Note that the two outer maximization levels in (11.1) can be merged or even swapped, so that

$$r(U(A)) = \max_{\mathbf{p} \in [0,1]^2, B \in U(A)} \min_{1 \leq i \leq N} ||\mathbf{p} - \mathbf{b}^i||_2 = \qquad (11.2)$$

$$= \max_{\mathbf{p} \in [0,1]^2} \max_{B \in U(A)} \min_{1 \leq i \leq N} ||\mathbf{p} - \mathbf{b}^i||_2. \qquad (11.3)$$

Furthermore, as the next lemma shows, in the last expression the inner maximization and the minimization levels can also be swapped, resulting in a convenient formula for calculating with given uncertainty regions:

Lemma 1. *Let $A = \{\mathbf{a}^i\} \in \mathscr{A}_N(U)$ be a covering configuration with uncertainty regions $U(A)$, and let $\mathbf{p} \in [0,1]^2$. Then*

$$\max_{B \in U(A)} \min_{1 \leq i \leq N} ||\mathbf{p} - \mathbf{b}^i||_2 = \min_{1 \leq i \leq N} \max_{\mathbf{b}^i \in U(\mathbf{a}^i)} ||\mathbf{p} - \mathbf{b}^i||_2. \qquad (11.4)$$

Proof. Construct a configuration $\bar{B} = \{\bar{\mathbf{b}}^i\} \in U(A)$ such that for each i index $||\mathbf{p} - \bar{\mathbf{b}}^i||_2$ is maximal. (The maximum is attained since all $U(\mathbf{a}^i)$ are closed.) This means that for each i and for all $\mathbf{b}^i \in U(\mathbf{a}^i)$

$$||\mathbf{p} - \bar{\mathbf{b}}^i||_2 \geq ||\mathbf{p} - \mathbf{b}^i||_2. \tag{11.5}$$

Setting $j = \arg\min_i ||\mathbf{p} - \bar{\mathbf{b}}^i||_2$, we conclude that $d := ||\mathbf{p} - \bar{\mathbf{b}}^j||_2$ is equal to the right-hand side of (11.4).

Now we claim that $d \geq \min_i ||\mathbf{p} - \mathbf{b}^i||_2$ for all $B \in U(A)$, that is, that d is also equal to the left-hand side of (11.4). (Obviously, d is attained in the left-hand side maximization, namely, at the configuration $\bar{B}$.) Assume the contrary of the claim, i.e., the existence of a configuration $C = \{\mathbf{c}^i\} \in U(A)$, such that $\min_i ||\mathbf{p} - \mathbf{c}^i||_2 > ||\mathbf{p} - \bar{\mathbf{b}}^j||_2$. But then $||\mathbf{p} - \mathbf{c}^j||_2 > ||\mathbf{p} - \bar{\mathbf{b}}^j||_2$, and since $\mathbf{c}^j \in U(\mathbf{a}^j)$, this contradicts (11.5). That completes the proof.   □

The essence of the above lemma is that the uncertainty regions are independent of each other, thus, the maximal distances between them and a given point (i.e., the values $\max_{\mathbf{b}^i \in U(\mathbf{a}^i)} ||\mathbf{p} - \mathbf{b}^i||_2$) can be calculated one by one during the computation of $r(U(A))$. This calculation can be easily done for the most common shapes used for modeling uncertainty regions (circles, ellipses, convex polygons, etc.)

In the next sections we analyze three covering problem classes defined for various simple shapes of uncertainty regions. From now on we will assume that within each problem class the uncertainty regions will be uniformly shaped, i.e., their size and orientation does not depend on the target location.

11.4   Circular Uncertainty Regions

The simplest model of an uncertainty region for covering problems is a circular region. This corresponds, for instance, to the situation when the probability of the actual placement follows a bivariate symmetric normal distribution; then the uncertainty regions corresponding to the various success probabilities will be concentric circles centered at the desired target positions.

Somewhat surprisingly, it turns out that the centers of the optimal uncertain coverings will be identical to the centers of the exact coverings (provided that the exact covering is a feasible configuration w.r.t. the uncertainty modeling function), as the next theorem shows:

Theorem 1. *Let N be given and let the uncertainty regions U be uniform circles of radius s around each target point. Let us denote r^* and $\tilde{r}^*$ the optimal covering radius of the exact and the uncertain covering problems, respectively. Then, if the exact covering configuration is a feasible configuration w.r.t. U, then $\tilde{r}^* = r^* + s$, and the optimal uncertain covering configuration is identical to that of the exact problem. Otherwise, if the exact covering configuration is not a feasible configuration w.r.t. U, then $\tilde{r}^* \geq r^* + s$.*

Proof. Consider a target point $\mathbf{a} \in [0, 1]^2$, a circle of radius s around $\mathbf{a}$ as its uncertainty region $U(\mathbf{a})$, and a point $\mathbf{p} \in [0, 1]^2$. A simple geometrical observation shows that $\max_{\mathbf{b} \in U(\mathbf{a})} ||\mathbf{p} - \mathbf{b}||_2 = ||\mathbf{p} - \mathbf{a}||_2 + s$. (The maximal distance is obtained by drawing a ray from $\mathbf{p}$ passing through $\mathbf{a}$, when $\mathbf{p} \neq \mathbf{a}$, and taking any point on the boundary of $U(\mathbf{a})$ when $\mathbf{p} = \mathbf{a}$.) Then, using (11.3) and Lemma 1, we obtain that for all $A \in \mathscr{A}_N(U)$

$$
\begin{aligned}
r(U(A)) &= \max_{\mathbf{p} \in [0,1]^2} \max_{B \in U(A)} \min_{1 \le i \le N} ||\mathbf{p} - \mathbf{b}^i||_2 = \\
&= \max_{\mathbf{p} \in [0,1]^2} \min_{1 \le i \le N} \max_{\mathbf{b}^i \in U(\mathbf{a}^i)} ||\mathbf{p} - \mathbf{b}^i||_2 = \\
&= \max_{\mathbf{p} \in [0,1]^2} \min_{1 \le i \le N} (||\mathbf{p} - \mathbf{a}^i||_2 + s) = \\
&= (\max_{\mathbf{p} \in [0,1]^2} \min_{1 \le i \le N} ||\mathbf{p} - \mathbf{a}^i||_2) + s = \\
&= r(A) + s.
\end{aligned}
$$

Now assume that the exact covering configuration is feasible w.r.t. U, i.e., it is in $\mathscr{A}_N(U)$. This means that $\min_{A \in \mathscr{A}_N(U)} r(A) = \min_{A \in \mathscr{A}_N} r(A)$. Then

$$
\tilde{r}^* = \min_{A \in \mathscr{A}_N(U)} r(U(A)) = \min_{A \in \mathscr{A}_N(U)} r(A) + s = \min_{A \in \mathscr{A}_N} r(A) + s = r^* + s,
$$

and the two optima $\tilde{r}^*$ and r^* are obviously attained at the same A.

On the other hand, if the exact covering configuration is not feasible w.r.t. U, then we have $\min_{A \in \mathscr{A}_N(U)} r(A) \ge \min_{A \in \mathscr{A}_N} r(A)$, thus, from the above chain of equations we obtain $\tilde{r}^* \ge r^* + s$. $\qquad\square$

11.5 Line Segment Shaped Uncertainty Regions

Another possible type of uncertainty regions is a line segment that contains the target points and has uniform lengths and directions. Different success probabilities would then correspond to line segments of different lengths. This setting can be appropriate to model the real-life sensor placement scenario of, e.g., a target terrain with a certain slope, or the effect of wind during payload drops.

A mathematical formalization of such uncertainty regions is as follows. For simplicity, we assume that the uncertainty regions are parallel to the line $y = -x$, that is, they are in the NW–SE direction, and the target point is the midpoint of the line segment. Furthermore, we assume that the common length of the line segments is $2\sqrt{2}t$ for a fixed $t > 0$ value. The respective uncertainty region for a point $\mathbf{a}$ is then

$$U(\mathbf{a}) = \{(a_x + l, a_y - l) \mid l \in [-t, t]\}.$$

Remark 1. In practice, the effect of a slope or a wind direction would be naturally modeled by a line segment with one of its endpoint as the target point. Nevertheless, it is easy to see that once the shape of the uncertainty region is fixed, it actually does not matter which point of it is chosen as a target point. For our numerical computation the model with the target point as the midpoint is more convenient to formulate and the results will be comparable to those approximated from the circular case, see Sect. 11.5.3.

In contrast to the circular uncertainty regions, $\max_{\mathbf{b} \in U(\mathbf{a})} ||\mathbf{p} - \mathbf{b}||_2 - ||\mathbf{p} - \mathbf{a}||_2$ will not be a constant for all $\mathbf{p}$, like in the proof of Theorem 1, so the method there for finding the optimal solution in a theoretical way will fail. Instead, for such uncertainty regions we opt for a numerical solution.

The numerical optimization for this case could go as follows: the inner optimization problem is to maximize the expression

$$\min_{1 \le i \le N} \max_{\mathbf{b}^i \in U(\mathbf{a}^i)} ||\mathbf{p} - \mathbf{b}^i||_2 \tag{11.6}$$

over $\mathbf{p} \in [0, 1]^2$, for a fixed $A = \{\mathbf{a}^i\}$. It is easy to see that given a line segment $\overline{\mathbf{uv}}$ and a point $\mathbf{p}$ in the plane, the maximal distance from $\mathbf{p}$ to a point of $\overline{\mathbf{uv}}$ is always attained at either $\mathbf{u}$ or $\mathbf{v}$. (To prove this, let us denote $\overrightarrow{\mathbf{v}}$, $\overrightarrow{\mathbf{v}}$, and $\overrightarrow{\mathbf{b}}$ the vectors from $\mathbf{p}$ to $\mathbf{u}$, $\mathbf{v}$, and to an arbitrary point $\mathbf{b} \in \overline{\mathbf{uv}}$, respectively. Then for some $0 \le \lambda \le 1$, $\overrightarrow{\mathbf{b}} = \lambda \overrightarrow{\mathbf{u}} + (1 - \lambda) \overrightarrow{\mathbf{v}}$, thus, $||\overrightarrow{\mathbf{b}}|| = ||\lambda \overrightarrow{\mathbf{u}} + (1 - \lambda) \overrightarrow{\mathbf{v}}|| \le \lambda ||\overrightarrow{\mathbf{u}}|| + (1 - \lambda) ||\overrightarrow{\mathbf{v}}|| \le \max\{||\overrightarrow{\mathbf{u}}||, ||\overrightarrow{\mathbf{v}}||\}$.) This implies that the objective function value (11.6) can be easily computed for all $\mathbf{p} \in [0, 1]^2$. The maximum of the inner problem is the optimal covering radius $r(U(A))$ for A. The outer optimization problem is then to minimize $r(U(A))$ over $A \in \mathscr{A}_N(U)$.

It is essential to note that the inner optimization problem *must be solved to global optimality with a rigorous search method* (considering all $\mathbf{p} \in [0, 1]^2$), so that the obtained maximum is guaranteed to be a covering radius for the configuration A. More precisely, it must be a guaranteed upper bound of the exact maximum. Without this it may happen that the design fails, i.e., it will not result in a covering (with the given radius) for some realizations.

As a demonstrative case study of such uncertain regions, we solve the uncertain covering problems for $N = 2, \ldots, 10$ and $t = 0.05$. In the next subsections the details of the proposed optimization method are discussed.

11.5.1 Solving the Inner Problem

Since this level needs a complete search method, we applied the coco_gop_ex interval branch-and-bound solver [10] of the COCONUT Environment [21]. Note that the objective function (11.6) is nonsmooth, but it is always only two-dimensional, regardless of N. For this particular problem we enabled only those tools in coco_gop_ex that require no derivative information. The only addition to the standard tools was a kind of special *constraint propagation* method that targeted the elimination of those points of a rectangular region of the unit square that are in distance closer to both endpoints of the current line segments than a given distance value. (This tool is very similar to one introduced in [9] for circle *packing* problems with interval methods.) The outputs of coco_gop_ex are the mathematically rigorous interval enclosures of all global maximizers and the global maximum value, respectively. According to the previous notes, the upper bound of the latter enclosure can be used as a guaranteed uncertain covering radius w.r.t. the input configuration. We used 10^{-6} as the output precision of the algorithm (the width of the enclosure of the maximum is approximately equal to this value).

As we found, coco_gop_ex was extremely efficient in solving the inner problem: most of the total of 800,000 problem instances were completed within a few hundreds of a second only (on a PC with an Intel Mobile CPU of 1.73 GHz), and even the hardest cases—some instances for $N = 9, 10$—required less than 0.4 s.

11.5.2   Solving the Outer Problem

The problem of minimizing (the upper estimates of) $r(U(A))$ over $A \in \mathscr{A}_N(U)$ was solved by the sequential version of the HOPSPACK derivative-free solver [17]. HOPSPACK employs a generating set search (GSS) method based on pattern search ideas [3, 8]. For each $N = 2, \ldots, 10$, HOPSPACK was started from $15N$ randomly generated starting points, which were obtained the following way. $5N$ points were generated in the neighborhood of the exact optimal covering configuration such that for each coordinate c of the exact covering the respective component of the starting point was taken from the interval $(c + [-0.1, 0.1]) \cap [0, 1]$. The other $10N$ starting points were generated from the whole search space $[0, 1]^{2N}$ so that the points were 'spread out' in the square in order to result in a reasonable starting configuration. To achieve this, for all N the square was split into either $k \times k$ or $k \times (k + 1)$ uniform pieces with the smallest k for which $k^2 \geq N$ or $k(k + 1) \geq N$ holds. Each starting configuration was then generated in such a way that each piece contained at most one point of the configuration.

HOPSPACK converged from each starting points and stopped in around 500–2,000 iterations, however, it is important to note that it ended up in different (presumably locally optimal) solutions in many cases. It was already observed in [16] for the exact case that the covering problem possesses many local optimizers with often only very small differences in the optimal values. According to the present experiences, for uncertain covering problems this phenomenon also appears, but in a much pronounced form. This makes the present problem class very hard for global optimization methods, even for smaller dimensions and simple uncertainty regions.

Table 11.3 Best found (r_L) and easy-to-obtain covering radii (r_C) for the line segment shaped problem class, and the respective improvement ratios (r_L/r_C)

N	r_L	r_C	r_L/r_C (%)
2	0.606646	0.629728	96.3
3	0.556570	0.574602	96.9
4	0.399951	0.424265	94.3
5	0.376100	0.396872	94.8
6	0.343817	0.369438	93.1
7	0.328612	–	–
8	0.313715	0.331011	94.8
9	0.279036	0.301348	92.6
10	0.273233	–	–

11.5.3 The Results

The best found optimal uncertain covering configurations are shown in the right columns of Figs. 11.1, 11.2, and 11.3. For reference, on the left side the exact covering configurations are also depicted, so that the differences in the pairs of configurations can be observed. It is worth to mention that due to the shape of the uncertainty regions, the uncertain coverings show less symmetries than the exact ones (and thus the ones with circular uncertainty regions). One can observe that for most N values the found uncertain configurations somewhat resemble the exact ones, but as N grows, there are less and less similarities. For $N = 10$, the uncertain configuration shows no relation at all to the exact one.

The found covering radii (named r_L) are presented in column 1 of Table 11.3, upward rounded to six decimal digits after the decimal dot.

To show the usefulness and efficiency of the numerical method above, we can compare the obtained covering radii with some (not necessarily optimal) covering radii obtained without numerical optimization: we enclose the line segments into circles of radius $v = \sqrt{2}t = \sqrt{2}/20$, i.e., we transform the problem to the circular uncertain covering. Obviously, the covering radius of each circular uncertain configuration will naturally be a covering radius for the respective line segment shaped uncertain configuration as well. The exact configurations are feasible for circular uncertain problems if and only if all circles of radius v drawn around the points of the exact configuration are fully within the square (i.e., all of its coordinates are in $[v, 1 - v]$). In the present case this holds for all problem instances except $N = 7, 10$. Thus, for $N = 2, \ldots, 6, 8, 9$, the exact configuration is a feasible one for the line segment shaped problems, and so, from Theorem 1, an uncertain covering radius for the latter problem is $r_C = r^* + v$, where r^* is the optimal exact covering radius (column 2 of Table 11.3).

Column 3 of Table 11.3 shows the improvement ratios r_L/r_C. The results show that the numerical procedure improved the approximate solutions by 3–7 % (with larger improvements for larger N values), which is fairly significant for covering problems.

11.6 Convex Polygon Shaped Uncertainty Regions

A natural generalization of the line segment shaped case is to consider convex polygons as the uncertainty regions. Using the fact that all points of a convex polygon can be written as a convex linear combination of its vertices, one can see that the maximal distance from a given point to the polygon is always attained at one of the vertices. Thus the respective bi-level optimization model and the suggested optimization procedure are essentially the same as described in Sect. 11.5. As a demonstrative example, we solved an uncertain covering problem with $N = 5$ for uncertainty regions given as identical quadrilaterals, defined by the following formula. Assuming that the target point is $\mathbf{a} = (a_x, a_y)$, the quadrilateral is the convex hull of the points $(a_x + d_{1j}, a_y + d_{2j})$, $j = 1, \ldots, 4$, where the matrix $D = (d_{ij})$ is specified as

$$\begin{pmatrix} 0.04 & -0.11 & -0.01 & 0.07 \\ 0.05 & -0.01 & -0.08 & 0.00 \end{pmatrix}.$$

The optimization was carried out with exactly the same methodology as in Sect. 11.5. HOPSPACK was started from 100 starting points generated in the whole search space and from 50 starting points generated in the neighborhood of the exact configuration. The obtained best solution is depicted in Fig. 11.4; in particular, in part (a) the covering circles are centered at the target points, while parts (b)–(d) show three possible realizations of the placements, to demonstrate the covering in extremal placement situations. In Fig. 11.4b the upper left and middle and the lower left target points were moved to the extremes, which shows a tight covering of the upper side and the lower left corner. (Observe that the latter point has the tightest covering when the lower left target point is placed on a side, but not on a vertex, of the uncertainty region.) Figure 11.4c shows another situation when all but the lower right target points are changed, which result in a tight covering on the upper and left sides of the square. Finally, Fig. 11.4d shows what happens when the upper middle and the two lower target points are moved to the extremes: the inner part of the square still remains covered.

The best found optimal uncertain covering radius was 0.397945 (rounded upward). As before, a theoretical uncertain covering radius can also be created from the circular case: the smallest circle that encloses the polygon is of radius ≈ 0.090139; and if we take the exact optimal configuration as the midpoint of these circles, the resulting configuration will be feasible for the circular uncertainty regions. Thus the approximated covering radius (upward rounded) is 0.326161 +

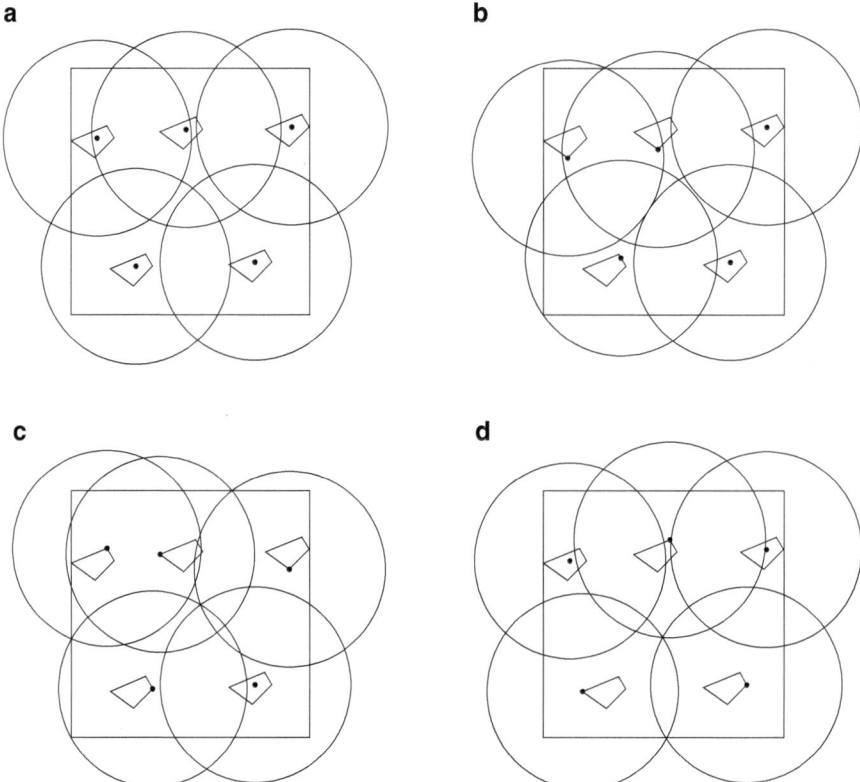

Fig. 11.4 Optimal uncertain covering with convex polygon shaped uncertainty regions; the found optimal solution (**a**), and some possible extremal realizations (**b**)–(**d**)

$0.090139 = 0.416300$. The numerical optimization thus produced a 4.4 % smaller optimum, i.e., again significantly improved the approximate solution obtained from theoretical considerations.

11.7 Summary

We investigated the problem of finding the optimal covering of a square with congruent circles, with uncertainties in the placement of the circles—as a simple model to, e.g., robust deployment designs in unknown environments. We treated the uncertainties by confidence regions around the planned placement and created the respective optimization problems. We studied three types of (uniform) uncertainty regions. For circular regions we showed that the uncertain optimal covering problem has the same optimal configurations as the exact ones without uncertainties (provided that the exact configurations are feasible for the uncertain case). For other types of regions with less internal symmetry we proposed a bi-level numerical optimization technique: a complete global search in the inner level with interval based global optimization, and an extensive black-box search in the outer level. We showed the efficiency and usefulness of the proposed technique on example covering problems with line segment and convex polygon shaped uncertainty regions. The study shows the applicability of the presented approaches in modeling relatively simple uncertainty scenarios, so that they can serve as a basis for designing more complex methods for real-life applications.

Acknowledgements The research was supported by the Hungarian National Development Agency (NFÜ) Grant TÁMOP-4.2.2/08/1/2008-0008 and by the Austrian Science Fund (FWF) Grant Nr. P25648-N25. The author is grateful to Hermann Schichl (Faculty of Mathematics, University of Vienna) for his valuable suggestions.

References

1. Fuchs, M., Neumaier, A.: Autonomous robust design optimization with potential clouds. Int. J. Reliab. Safety **3**, 23–34 (2009)
2. Fuchs, M., Neumaier, A., Girimonte, D.: Uncertainty modeling in autonomous robust space-craft system design. Proc. Appl. Math. Mech. **7**, 2060041–2060042 (2007)
3. Griffin, J.D., Kolda, T.G.: Nonlinearly-constrained optimization using heuristic penalty methods and asynchronous parallel generating set search. Appl. Math. Res. Express **25**, 36–62 (2010)
4. Hansen, E.R.: Global Optimization Using Interval Analysis. Marcel Dekker, New York (1992)
5. Heppes, A., Melissen, H.: Covering a rectangle with equal circles. Period. Math. Hungar. **34**, 65–81 (1997)
6. Kang, Z.: Robust design optimization of structures under uncertainties. Thesis work, University of Stuttgart, 161 p (2005)
7. Kearfott, R.B.: Rigorous Global Search: Continuous Problems. Nonconvex Optimization and Its Applications. Springer, New York (2010)
8. Kolda, T.G., Lewis, R.M., Torczon, V.: Optimization by direct search: new perspectives on some classical and modern methods. SIAM Rev. **45**, 385–482 (2003)
9. Markót, M.C.: An interval method to validate optimal solutions of the "packing circles in a unit square" problems. Cent. Eur. J. Oper. Res. **8**, 63–78 (2000)
10. Markót, M.C., Schichl, H.: Bound constrained interval global optimization in the COCONUT environment. J. Global Optim. **60**, 751–776 (2014)
11. Melissen, J.B.M., Schuur, P.C.: Improved coverings of a square with six and eight equal circles. Electron. J. Combin. 3R32, 10 (1996)
12. Mousavi, S.M., Vahdani, B., Tavakkoli-Moghaddam, R., Hashemi, H.: Location of cross-docking centers and vehicle routing scheduling under uncertainty: a fuzzy possibilistic–stochastic programming model. Appl. Math. Model. **38**, 2249–2264 (2014)
13. Neumaier, A.: Interval Methods for Systems of Equations. Cambridge University Press, Cambridge (1990)
14. Neumaier, A.: Complete search in continuous global optimization and constraint satisfaction. In: Iserles, A. (ed.) Acta Numerica, pp. 271–369. Cambridge University Press, Cambridge (2004)
15. Neumaier, A.: Clouds, fuzzy sets and probability intervals. Reliab. Comput. **10**, 249–272 (2004)

16. Nurmela, K.J., Östergaard, P.R.J.: Covering a square with up to 30 equal circles. Technical Report HUT-TCS-A62, Helsinki University of Technology (2000)
17. Plantenga, T.D.: HOPSPACK 2.0 user manual. Technical Report SAND2009-6265. Sandia National Laboratories, Albuquerque and Livermore (2009)
18. Shahanaghi, K., Ghezavati, V.R.: Efficient solution procedure to develop maximal covering location problem under uncertainty (Using GA and Simulation). Int. J. Ind. Eng. Prod. Res. **19**, 21–29 (2008)
19. Snyder, L.V.: Facility location under uncertainty: a review. IIE Trans. **38**, 537–554 (2005)
20. Tarnai, T. Gáspár, Z.: Covering a square by equal circles. Elem. Math. **50**, 167–170 (1995)
21. The COCONUT Environment.: Software, University of Vienna. http://www.mat.univie.ac.at/coconut-environment (2000-2015)

Chapter 12
Batching-Based Approaches for Optimized Packing of Jobs in the Spatial Scheduling Problem

Sudharshana Srinivasan, J. Paul Brooks, and Jill Hardin Wilson

Abstract Spatial resources are often an important consideration in shipbuilding and large-scale manufacturing industries. Spatial scheduling problems (SSP) involve the non-overlapping arrangement of jobs within a limited physical workspace such that some scheduling objective is optimized. The jobs are typically heavy and occupy large areas, requiring that the same contiguous units of space be assigned throughout the duration of their processing time. This adds an additional level of complexity to the general scheduling problem. Since solving large instances using exact methods becomes computationally intractable, there is a need to develop alternate solution methodologies to provide near optimal solutions for these problems. Much of the literature focuses on minimizing the makespan of the schedule. We propose two heuristic methods for the minimum sum of completion times objective. Our approach is to group jobs into a batch and then apply a scheduling heuristic to the batches. We show that grouping jobs earlier in the schedule, although intuitive, can result in poor performance when jobs have sufficiently large differences in processing times. We provide bounds on the performance of the algorithms and also present computational results comparing the solutions to the optimal objective obtained from the integer programming formulation for SSP. With a smaller number of jobs, both algorithms produce comparable solutions. For instances with a larger number of jobs and a higher variability in spatial dimensions, we observe that the efficient area model outperforms the iterative model both in terms of solution quality and run time.

Keywords Spatial scheduling • Integer programs • Approximation algorithms • Optimal packings

S. Srinivasan (✉) • J.P. Brooks
Virginia Commonwealth University, Richmond, VA 23284, USA
e-mail: srinivasans3@vcu.edu; jpbrooks@vcu.edu

J.H. Wilson
Northwestern University, Evanston, IL 60208, USA
e-mail: jill.wilson@northwestern.edu

© Springer International Publishing Switzerland 2015   243
G. Fasano, J.D. Pintér (eds.), *Optimized Packings with Applications*, Springer
Optimization and Its Applications 105, DOI 10.1007/978-3-319-18899-7_12

12.1   Introduction

In large-scale production and manufacturing industries, assembly units are often heavy and occupy large areas. Since physical processing space is limited at such facilities, the assembly line scheduling needs to assign non-overlapping locations (spatial characteristic) and starting times (temporal characteristic) for each job. Further, the schedule should ensure that the locations assigned are the same contiguous units of space for the entire duration of processing as jobs cannot be moved once set up. Mathematically, the spatial scheduling problem (SSP) can be described as follows: Given a set J of jobs with processing times p_j, heights h_j, and widths w_j, and a workspace of height H and width W, does there exist a schedule of the jobs that effectively utilizes the workspace such that some scheduling objective is minimized? Figure 12.1 shows the layout of jobs before and after applying spatial scheduling solution procedures. We can see that initially the space is not utilized effectively and some jobs are waiting to be processed. On applying some spatial scheduling method, we get a better utilization of the space and no delays in processing of jobs.

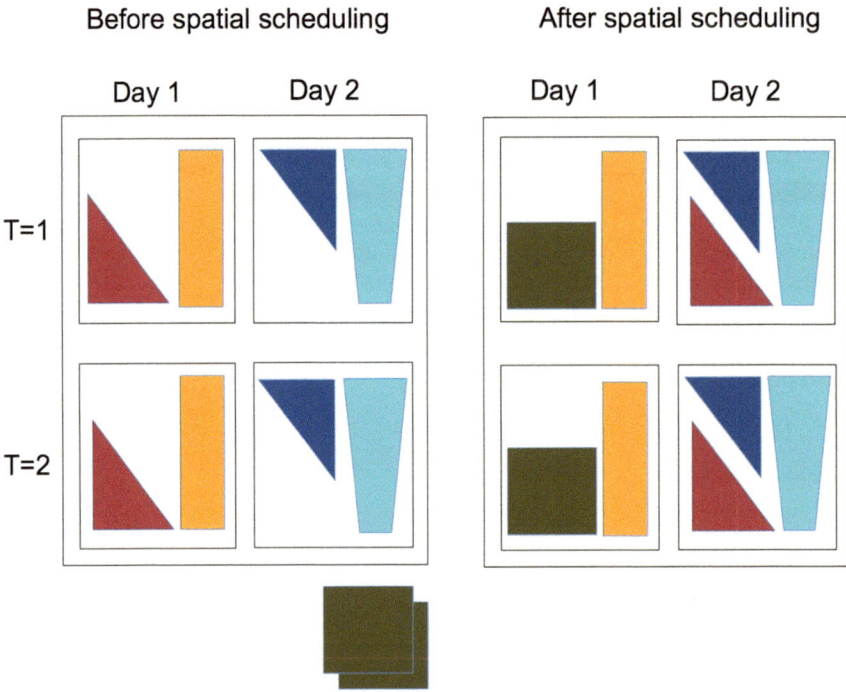

Fig. 12.1 Depicting the motivation for spatial scheduling with jobs each requiring two time units scheduled over a 2-day horizon

When the space required by all jobs are identical to the dimensions of the workspace, the problem reduces to single machine scheduling (SMS) and is polynomially solvable. However, when the dimensions of the jobs are allowed to vary, the spatial constraints add an additional level of complexity to the traditional scheduling problem [5]. Also spatial resources are not divisible and distributable like normal renewable resources. Due to the computational intractability in solving large instances of the problem, there is a need to develop alternate solution methodologies that provide near optimal solutions. Previous work in spatial scheduling has mostly been in the context of shipbuilding applications [4, 11, 15, 18]. Much of the literature focusses on approaches with the objective of minimizing the makespan or $\max_{j \in J} C_j$, where C_j denotes the completion time of job j in a given schedule [2, 9, 16, 22, 23]. Garcia and Rabadi [7] provides a meta heuristic algorithm to minimize the total tardiness for instances with release dates and multiple processing areas. To the best of our knowledge, this is the first study to consider the minimum sum of completion times ($\sum_{j \in J} C_j$) objective for this problem. When jobs are independent and competing for the same resource, the cost associated with individual completion times becomes more relevant and natural [12]. Evaluating completion times for individual jobs also becomes important while measuring the time in the system for each job. The motivation, here, is to examine the most simple form of the problem by considering a workspace area of fixed dimensions. We disallow precedence constraints, due-dates, rotation of jobs, and set-up times. By doing so, we are able to focus on the relationship between the spatial and temporal components in the problem and gain a better understanding of the problem characteristics.

Our approach in developing solution procedures for the problem is to take ideas from two-dimensional bin packing (2DBP) and group jobs similar in processing times to form a batch. Once the batches are determined we can schedule them using some heuristic rule. This approach lets us relax the temporal constraints in the original problem. We identify scenarios where batching can be effective or disadvantageous. For the minimum sum of completion times objective, it seems intuitive to schedule as many jobs as we can ahead in the schedule to produce a lower objective [19]. However, we show that grouping jobs with different processing times earlier in the schedule actually results in an objective value larger than if each job were to be assigned its own batch. We also determine that the sequence in which the batches are scheduled is another factor affecting the objective.

After introducing the problem in Sect. 12.2, we propose two methods (iterative and efficient area) to determine the batches in Sect. 12.3. Section 12.4 analyzes the performance of the batching methods. Computational results comparing the two methods to the optimal objective obtained from the integer programming formulation for SSP are presented in Sect. 12.5.

12.2   The Spatial Scheduling Problem

In this section, we formally introduce the SSP, which involves determining spatial layouts and starting times for a set J of N jobs to be scheduled on a single $W \times H$ work space. Each job $j \in J$ requires a processing time (p_j), width (w_j), and height (h_j). Throughout the reminder of this chapter we consider minimizing the sum of completion times (denoted by Z) as the objective for SSP. We assume that the jobs cannot be rotated or preempted. Also we do not consider due dates, release dates, or precedence relationships for the jobs. Further, without loss of generality, we assume all problem data to be integer. The problem can then be denoted using the following mixed-integer programming (MIP) formulation adapted from [8].

$$\min \sum_{j \in J} z_j \qquad (12.1)$$

subject to:

$$-x_i + x_j - W\alpha_{ij} \geq -W + w_i \ \forall i,j \in J, i \neq j \qquad (12.2)$$

$$-y_i + y_j - H\beta_{ij} \geq -H + h_i \ \forall i,j \in J, i \neq j \qquad (12.3)$$

$$-z_i + z_j - T\gamma_{ij} \geq -T + p_i \ \forall i,j \in J, i \neq j \qquad (12.4)$$

$$\alpha_{ij} + \alpha_{ji} + \beta_{ij} + \beta_{ji} + \gamma_{ij} + \gamma_{ji} \geq 1 \ \forall i,j \in J, i \neq j \qquad (12.5)$$

$$-x_i - w_i \geq -W \qquad \forall i \in J \qquad (12.6)$$

$$-y_i - h_i \geq -H \qquad \forall i \in J \qquad (12.7)$$

$$x_i, y_i, z_i \geq 0 \qquad \forall i \in J \qquad (12.8)$$

$$\alpha_{ij}, \beta_{ij}, \gamma_{ij} \in \{0, 1\} \qquad \forall i,j \in J \qquad (12.9)$$

where
J is the set of all jobs
x_j is the x-coordinate of job $j \in J$
y_j is the y-coordinate of job $j \in J$
z_j is the z-coordinate (start time) for job $j \in J$

$$\alpha_{ij} = \begin{cases} 1 \text{ if no overlap occurs between jobs i and j in the x direction} \\ 0 \text{ otherwise} \end{cases}$$

$$\beta_{ij} = \begin{cases} 1 \text{ if no overlap occurs between jobs i and j in y direction} \\ 0 \text{ otherwise} \end{cases}$$

$$\gamma_{ij} = \begin{cases} 1 \text{ if no overlap occurs between jobs i and j in z direction} \\ 0 \text{ otherwise} \end{cases}$$

For $i, j \in J$

Here Z is obtained by adding $\sum_{j \in J} p_j$ to the objective in (12.1). Constraints (12.2)–(12.5) prevent overlap from occurring in the x (width), y (height), and z (time) dimensions. We use constraints (12.6) and (12.7) to ensure that the jobs are confined to the physical dimensions of the workspace. Duin and Sluis [5] shows that scheduling problems with varying spatial resource requirements are NP Hard. Hence obtaining optimal solutions to large problem instances is computationally intractable. Therefore, the motivation here is to develop methods that provide provably good solutions to minimize Z for large instances of SSP, quickly and efficiently. Approximation algorithms deliver solutions with provable quality that are bounded in runtime. The following definition of an approximation algorithm can be found in [20, 21]. Suppose we wish to solve an NP-hard minimization problem consisting of instances in $\mathscr{I}$. Let $z(I) = \min\{c_I x : x \in S_I\} \ \forall I \in \mathscr{I}$. Let $\mathscr{A}$ be an algorithm that operates on instances in $\mathscr{I}$, and let $\mathscr{A}(I)$ be the objective value resulting from the application of $\mathscr{A}$ to I. Let $\rho \geq 1$.

Definition 1. $\mathscr{A}$ is a ρ-approximation algorithm for $\mathscr{I}$ if for each $I \in \mathscr{I}$, $\mathscr{A}$ runs in time polynomial in the size of I, and $\mathscr{A}(I) \leq \rho z(I)$. $\mathscr{A}$ is said to have a factor ρ, also referred to as the performance guarantee of $\mathscr{A}$.

Observe that we compare the objective value obtained by the application of the algorithm to instance I with the optimal objective value $z(I)$ for that instance. In practice, however, this is not possible, because if $z(I)$ is known then there would be no need to approximate it. To overcome this issue and calculate ρ, we compare $\mathscr{A}(I)$ with a lower bound for $z(I)$, say $L(I)$. Lower bounds can be obtained using LP or combinatorial relaxations. Since $L(I) \leq z(I)$ we have

$$\mathscr{A}(I) \leq \rho L(I) \implies \mathscr{A}(I) \leq \rho z(I).$$

In the following sections, we describe the development of an approximation algorithm (with two variants) based on existing packing algorithms and discuss its performance.

12.3 Batch-Scheduling

12.3.1 Introduction

SSP requires that jobs be arranged without overlap in a two-dimensional space while minimizing some scheduling objective. The spatial component of SSP can be attributed to optimized multi-dimensional packing problems. Lodi et al. [13] provides a survey of the models and algorithms used to solve the 2DBP problem. Castillo et al. [3] presents applications and approaches to solve circle packing problems encountered in container loading. Batch-scheduling ideas originated from the problem of scheduling "burn-in" operations at large-scale integrated circuit manufacturing [1, 10]. Mathirajan and Sivakumar [14] surveys the literature for scheduling of batching processors in the semi-conductor industry. The central idea in batch-scheduling is grouping similar jobs together to form a "batch." All jobs in a batch start at the same time and the next batch starts upon completion of the longest job in the previous batch. The processing time of a batch is equal to the largest processing time of any job in the batch. Our goal is to utilize ideas from 2DBP to design batch-scheduling strategies that identify the batches consisting of jobs that can simultaneously fit the space to minimize the sum of completion times.This approach lets us to relax the temporal constraints in the original problem.

Assume we have a set J of N jobs such that $p_1 \leq p_2 \leq \cdots \leq p_N$. When all the jobs fit in the space simultaneously, irrespective of the difference in their processing times p_j they are placed in the same batch. So $Z = \sum_{j \in J} p_j$. If no pair of jobs simultaneously fits the space, SSP reduces to SMS. Then each job is its own batch and $Z = \sum_{j=1}^{N} \sum_{i=1}^{j} p_i$. Smith [19] proved that ordering jobs in the nondecreasing sequence of their processing times is optimal for SMS. In general, while minimizing the sum of completion times, the more jobs we can fit earlier in our schedule the lower the objective. Therefore, it seems intuitive to always group jobs together rather than assign them to individual batches. Consider an instance of SSP with $W=H=3$ and job data as given in Table 12.1. Jobs 1 and 3 are the only jobs that fit the space simultaneously.

Let the processing times $[p_1, p_2, p_3] = [2, 3, 7]$ and let us assume we schedule the batch with the lowest processing time first. We define batch processing time as the maximum processing time of jobs in a batch. Therefore, the batch sequence is $\{2\}$ and $\{1, 3\}$ as seen in Fig. 12.2a. Then the objective value for batched jobs is calculated as $Z = p_1 + 3p_2 + p_3 = 2 + 9 + 7 = 18$. Alternately, if we schedule the batches in their own batch, the sequence is $\{1\}, \{2\}, \{3\}$ as seen in Fig. 12.2b and $Z = 3p_1 + 2p_2 + p_3 = 6 + 6 + 7 = 19$. This shows that grouping jobs can result in a lower sum of completion times objective.

Now suppose, $[p_1, p_2, p_3] = [2, 24, 25]$. When jobs 1 and 3 are batched, $Z = p_1 + 3p_2 + p_3 = 2 + 72 + 25 = 99$. Without batching, $Z = 3p_1 + 2p_2 + p_3 = 6 + 49 + 25 = 79$. Thus in scenarios where jobs with large differences in processing times are grouped together, the batching approach does not necessarily lead to improvement in the objective.

Table 12.1 Example instance with three jobs such that only jobs 1 and 3 simultaneously fit the space

Job	Width	Height
1	3	2
2	2	3
3	3	1

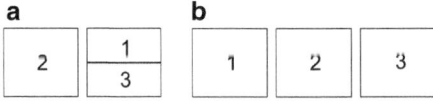

Fig. 12.2 Batching sequence for example with three jobs. (**a**) Batching. (**b**) No batching

Proposition 1. *For N jobs, assume that $p_1 < p_2 < \cdots < p_{N-1} < p_N$. When jobs with both the largest and smallest processing times are assigned to the same batch, that is $\{1, N\}$ form a batch, and $(N-1)p_1 - p_2 - \cdots - p_{N-1} < 0$, the sum of completion times obtained by batching is greater than the objective value obtained without batching.*

Proof. Let $\sum_{j \in J} C_j^b$ be the sum of completion times obtained when batching and $\sum_{j \in J} C_j^n$ be the sum of completion times obtained without batching. Jobs $\{1, N\}$ form a batch, while the other jobs are each assigned individual batches. Since $p_1 < p_2 < \cdots < p_{N-1} < p_N$, batch $\{1, N\}$ is processed at the end of the schedule (see Fig. 12.3). So $\sum_{j \in J} C_j^b = p_1 + Np_2 + \cdots + 3p_{N-1} + p_N$. If each job is assigned its own batch, then $\sum_{j \in J} C_j^n = Np_1 + (N-1)p_2 + \cdots + 2p_{N-1} + p_N$. Therefore,

$$\sum_{j \in J} C_j^b - \sum_{j \in J} C_j^n$$

$$= [p_1 + Np_2 + \cdots + 3p_{N-1} + p_N] - [Np_1 + (N-1)p_2 + \cdots + 2p_{N-1} + p_N]$$

$$= (N-1)p_1 - p_2 - \cdots - p_{N-1}$$

Hence, when $(N-1)p_1 - p_2 - \cdots - p_{N-1} < 0$, the result follows.

This contradicts the notion of scheduling as many jobs earlier in the schedule to minimize our objective. So, our intuitions about general scheduling problems do not always apply directly to problems with spatial resources. Batching seems to be beneficial only when processing times are similar.

When looking at the instance with $[p_1, p_2, p_3] = [2, 24, 25]$, we observed that scheduling batches in the sequence $\{2\}$ then $\{1, 3\}$ as seen in Fig. 12.4a results in an objective $Z = p_1 + 3p_2 + p_3 = 2 + 72 + 25 = 99$. Instead, if we were to schedule the batches in the sequence $\{1, 3\}$ then $\{2\}$, as seen in Fig. 12.4b, the objective value is calculated as $Z = p_1 + p_2 + 2p_3 = 2 + 24 + 50 = 76$. This suggests that scheduling the jobs in the increasing order of batch processing times is not always effective.

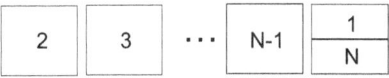

Fig. 12.3 Sequence in which batches are scheduled for Proposition 1

Fig. 12.4 Comparing strategies for sequencing batches. (**a**) Batch sequence. (**b**) Alt. sequence

Proposition 2. *Consider a set J of N jobs such that $p_1 < p_2 < \cdots < p_{N-1} < p_N$. If m of those jobs are in a batch B, including Job 1 and Job N, and $p_N > mp_i, \forall i \in J \setminus B$, then placing batch B at the end of a schedule provides a better objective than placing it at the beginning of the schedule.*

Proof. Let $\sum_{j \in J} C_j^b$ be the sum of completion times obtained when processing batch B at the end of the schedule and $\sum_{j \in J} C_j^a$ be the sum of completion times obtained using an alternate sequencing of batches (batch B is the first batch to be scheduled). Jobs $\{1, N\}$ along with $(m - 2)$ other jobs form a batch B, while the remaining jobs are each assigned individual batches. Let $\{1, u_1, u_2, \cdots, u_{m-2}, N\}$ be the m jobs in batch B such that

$$p_1 < p_2 < \cdots < p_{u_1 - 1} < p_{u_1} < \cdots < p_{u_{m-2}} < p_{u_{m-1}} < \cdots < p_N.$$

The sequence of batches scheduled in increasing order of batch processing times is $\{2\}, \{3\}, \cdots, \{u_1 - 1\}, \cdots, \{u_{m-1}\}, \cdots, \{N - 1\}, \{B\}$.

So $\sum_{j \in J} C_j^b = (p_1 + p_{u_1} + \cdots + p_{u_{m-2}} + p_N) + (Np_2 + \cdots + (m + 1)p_{N-1})$.

Alternately, if we place batch B at the beginning of the schedule, $\sum_{j \in J} C_j^a$ is given by

$$p_1 + p_{u_1} + \cdots + p_{u_{m-2}} + (N - m + 1)p_N + \cdots + p_{N-1}.$$

Therefore, $\sum_{j \in J} C_j^a - \sum_{j \in J} C_j^b$

$$= \begin{aligned}&[(p_1 + p_{u_1} + \cdots + p_{u_{m-2}} + (N - m + 1)p_N + (N - m)p_2 + \cdots + p_{N-1})] - \\ &[(p_1 + p_{u_1} + \cdots + p_{u_{m-2}} + p_N) + (Np_2 + \cdots + (m + 1)p_{N-1})]\end{aligned}$$

$$= -mp_2 - mp_3 - \cdots - mp_{N-1} + (N - m)p_N$$

Hence, when $p_N > mp_2, p_N > mp_3, \cdots, p_N > mp_{N-1}$, the result follows.

In summary, placing jobs with the smallest and largest processing times in the same batch or scheduling jobs in the increasing order of batch processing times does not necessarily result in a good batching scheme.

12.3.2 Forming the Batches

Using the insights gained from our previous analysis, we group jobs similar in processing time that also efficiently utilize the space to form a batch. We present two MIP models, iterative and efficient area, that identify the assignment of jobs to batches. The objective for the iterative model is to minimize the maximum difference in processing times among jobs for each batch. The efficient area model extends this idea by also minimizing the total unused area in each batch. Both MIP formulations have been adapted from the 2DBP model found in [17]. Let J denote the set of jobs and B the set of batches. Since at most each job can be its own batch, the number of batches equals the number of jobs (N).

12.3.2.1 Iterative Model

In the iterative model (M1), we add a constraint to limit the number of batches (S) being used by the model. We do not chose S as part of the model, because the objective here is not to reduce the number of batches used, but to find the best assignment of jobs (to batches) that minimizes the sum of completion times. Therefore, the strategy is to iterate through possible values for S, starting at $S = N - 1$ and decreasing by 1 in each iteration. From the set of all solutions, we can then chose the batching that results in the lowest sum of completion times objective value. The formulation for the iterative model is given by

$$\min \sum_{b \in B} (Zmax_b - Zmin_b) \tag{12.10}$$

$$\sum_{b \in B} r_{jb} = 1 \qquad \forall j \in J \tag{12.11}$$

$$x_j + w_j \leq W \qquad \forall j \in J \tag{12.12}$$

$$y_j + h_j \leq H \qquad \forall j \in J \tag{12.13}$$

$$x_i + w_i - x_j \leq W(1 - l_{ij}) \quad \forall i, j \in J, i < j, b \in B \tag{12.14}$$

$$y_i + h_i - y_j \leq H(1 - b_{ij}) \quad \forall i, j \in J, i < j, b \in B \tag{12.15}$$

$$l_{ij} + l_{ji} + b_{ij} + b_{ji} + (1 - r_{ib}) + (1 - r_{jb}) \geq 1 \qquad \forall i, j \in J, b \in B \tag{12.16}$$

$$Zmin_b \leq (p_j - M)r_{jb} + Mq_b \qquad \forall j \in J, b \in B \tag{12.17}$$

$$Zmax_b \geq p_j r_{jb} \qquad \forall j \in J, b \in B \tag{12.18}$$

$$r_{jb} \leq q_b \qquad \forall j \in J, b \in B \tag{12.19}$$

$$\sum_{j \in J} r_{jb} - \epsilon q_b \geq 0 \qquad \forall b \in B \tag{12.20}$$

$$\sum_{b \in B} q_b = S \tag{12.21}$$

$$x_j, y_j \geq 0 \qquad \forall j \in J \tag{12.22}$$

$$Zmin_b, Zmax_b \geq 0 \qquad \forall b \in B \tag{12.23}$$

$$l_{ij}, b_{ij} \in \{0, 1\} \qquad \forall i, j \in J \tag{12.24}$$

$$r_{jb} \in \{0, 1\} \qquad \forall j \in J, b \in B \tag{12.25}$$

$$q_b \in \{0, 1\} \qquad \forall b \in B \tag{12.26}$$

where
    J is the set of all jobs
    B is the set of all batches

x_j is the x-coordinate of job $j \in J$

y_j is the y-coordinate of job $j \in J$

$Zmax_b$ is the maximum processing time of jobs in batch $b \in B$

$Zmin_b$ is the minimum processing time of jobs in batch $b \in B$

$r_{jb} = \begin{cases} 1 \text{ if job j is in batch b} \\ 0 \text{ otherwise} \end{cases}$

$l_{ij} = \begin{cases} 1 \text{ if job i is to the left of job j} \\ 0 \text{ otherwise} \end{cases}$

$b_{ij} = \begin{cases} 1 \text{ if job i is below job j} \\ 0 \text{ otherwise} \end{cases}$

$q_b = \begin{cases} 1 \text{ if batch b is nonempty} \\ 0 \text{ otherwise} \end{cases}$

For $i, j \in J$ and $b \in B$.

Here, constraint (12.11) ensures that each job is assigned to only one batch. Constraints (12.12) and (12.13) ensure that jobs do not exceed the width and height of the space. We use constraints (12.14)–(12.16) to prevent overlap of jobs within the space. Constraint (12.17) determines the minimum processing time within a batch, while (12.18) identifies the maximum processing time for each batch. If job j is in batch b ($r_{jb} = 1$), then constraint (12.19) makes sure batch b is non-empty ($q_b = 1$). When no jobs are present in a batch, constraint (12.20) ensures that the batch is empty or $q_b = 0$. Constraint (12.21) sets the number of batches to be used by the model to some value S. We set $\epsilon = 0.5$ and define $M = 1 + \max_{j \in J} p_j$.

12.3.2.2   Efficient Area Model

While solving $N - 1$ instances of M1 for different values of S finds the best possible batch assignment, the second approach or efficient area model (M2) proposes to solve just one MIP to decide when and where to place jobs. The efficient area model includes an area utilization component to the existing objective. So, model 2 minimizes the maximum difference in processing times and the amount of workspace area that remains unused for each batch. The formulation for the efficient area model is given by

$$\min \sum_{b \in B} (Zmax_b - Zmin_b + UA_b) \qquad (12.27)$$

$$\sum_{b \in B} r_{jb} = 1 \qquad \forall j \in J \qquad (12.28)$$

$$x_j + w_j < W \qquad \forall j \in J \qquad (12.29)$$

$$y_j + h_j \leq H \qquad \forall j \in J \qquad (12.30)$$

$$x_i + w_i - x_j \leq W(1 - l_{ij}) \ \forall i, j \in J, i < j, b \in B \ (12.31)$$

$$y_i + h_i - y_j \leq H(1 - b_{ij}) \ \forall i, j \in J, i < j, b \in B \ (12.32)$$

$$l_{ij} + l_{ji} + b_{ij} + b_{ji} + (1 - r_{ib}) + (1 - r_{jb}) \geq 1 \qquad \forall i, j \in J, b \in B \qquad (12.33)$$

$$Zmin_b \leq (p_j - M)r_{jb} + Mq_b \qquad \forall j \in J, b \in B \qquad (12.34)$$

$$Zmax_b \geq p_j r_{jb} \qquad \forall j \in J, b \in B \qquad (12.35)$$

$$r_{jb} \leq q_b \qquad \forall j \in J, b \in B \qquad (12.36)$$

$$\sum_{j \in J} r_{jb} - \epsilon q_b \geq 0 \qquad \forall b \in B \qquad (12.37)$$

$$WHq_b - \sum_{j \in J} w_j h_j r_{jb} = UA_b \qquad \forall b \in B \qquad (12.38)$$

$$x_j, y_j \geq 0 \qquad \forall j \in J \qquad (12.39)$$

$$Zmin_b, Zmax_b, UA_b \geq 0 \qquad \forall b \in B \qquad (12.40)$$

$$l_{ij}, b_{ij} \in \{0, 1\} \qquad \forall i, j \in J \qquad (12.41)$$

$$r_{jb} \in \{0, 1\} \qquad \forall j \in J, b \in B \qquad (12.42)$$

$$q_b \in \{0, 1\} \qquad \forall b \in B \qquad (12.43)$$

where

J is the set of all jobs

B is the set of all batches

x_j is the x-coordinate of job $j \in J$

y_j is the y-coordinate of job $j \in J$

$Zmax_b$ is the maximum processing time of jobs in batch $b \in B$

$Zmin_b$ is the minimum processing time of jobs in batch $b \in B$

UA_b is the unused area in batch $b \in B$

$r_{jb} = \begin{cases} 1 \text{ if job } j \text{ is in batch } b \\ 0 \text{ otherwise} \end{cases}$

$l_{ij} = \begin{cases} 1 \text{ if job } i \text{ is to the left of job } j \\ 0 \text{ otherwise} \end{cases}$

$b_{ij} = \begin{cases} 1 \text{ if job } i \text{ is below job } j \\ 0 \text{ otherwise} \end{cases}$

$q_b = \begin{cases} 1 \text{ if batch } b \text{ is nonempty} \\ 0 \text{ otherwise} \end{cases}$

For $i, j \in J$ and $b \in B$.

Here, constraint (12.28) ensures that each job is assigned to only one batch. Constraints (12.29) and (12.30) ensure that jobs do not exceed the width and height of the space. We use constraints (12.31)–(12.33) to prevent overlap of jobs within the space. Constraint (12.34) determines the minimum processing time within a batch, while (12.35) identifies the maximum processing time for each batch. If job

j is in batch b ($r_{jb} = 1$), then constraint (12.36) makes sure batch b is non-empty ($q_b = 1$). When no jobs are present in a batch, constraint (12.37) ensures that the batch is empty or $q_b = 0$. Constraint (12.38) calculates the unused area for each batch b. We set $\epsilon = 0.5$ and define $M = 1 + \max_{j \in J} p_j$.

12.3.3   Scheduling the Batches

Once the batches are identified using either M1 or M2, it is also important to decide the sequence in which to schedule the batches. Smith [19] proved that the shortest processing time (SPT) rule, ordering jobs in the nondecreasing sequence of their job processing times, is optimal for the SMS problem. The idea is that by scheduling shorter jobs earlier in the schedule, more jobs can finish early resulting in a smaller sum. For SSP, the rule translates to scheduling the batches in the nondecreasing sequence of their batch processing times. For example, if P_1 is the maximum processing time of all jobs in batch 1 and P_2 is the maximum processing time of all jobs in batch 2, then batch 1 is scheduled before batch 2 if and only if $P_1 \leq P_2$. However, as noted before, there are instances for which this rule does not necessarily provide a better objective value. Therefore, we also consider scheduling jobs in the non-decreasing order of the average batch processing times, or the average processing time of all the jobs in a batch. We indicate the two scheduling rules as MAX and AVG, respectively.

12.3.4   Post Processing Algorithm

By solving each instance of SSP using the iterative and efficient area models, we determine the assignments of jobs to batches that minimize the maximum difference in processing times while efficiently utilizing the workspace. With this information, we then schedule the batches by applying either the MAX or AVG rules. Once a schedule is created, we calculate the sum of completion times for the jobs as $Z_H = \sum_{j \in J} C_j^H$, where C_j^H is the completion time for job j. With this batching algorithm, each job must wait until the previous batch has completed before it can start processing. In reality there may be jobs in the current batch that finish processing before the final job in the batch. This means that jobs in later batches may be able to start earlier in the schedule. Since neither MIP model takes into account the temporal dimension, we use a post-processing algorithm to incorporate this observation and improve Z_H. For each batch the algorithm determines if jobs can start processing earlier in the schedule. If job j can be moved ahead in time by say t_j units, then the completion time is updated as, $\hat{C}_j = C_j^H - t_j$ and $\hat{Z} = \sum_{j \in J} \hat{C}_j$ is the new objective value.

Proposition 3. *For instances defined by N=nk jobs, $n, k \in \mathbb{Z}_+^*$, where $w_j = \frac{W}{k}$, $h_j = H \; \forall \, j \in J$, and $p_1 \leq p_2 \leq \cdots \leq p_N$, the solution obtained after the post-processing routine is optimal.*

Proof. Consider the instances with $N = nk$ jobs, such that k jobs can simultaneously fit the space. Let C_j^H, $\hat{C}_j$, and C_j^{OPT} denote the completion time for job j and Z_H, $\hat{Z}$, and Z_{OPT} denote the objective value for the batch-scheduling algorithm, the post-processing routine, and the optimal solution, respectively. First we observe that if k jobs can simultaneously fit within the workspace that there are n batches. So for all jobs $j \leq k$, $\hat{C}_j = C_j^{OPT}$.

Let $U = \{u_1, u_2, \cdots, u_k\}$ denote the k jobs in the next batch waiting to be scheduled, such that $p_{u_1} \leq p_{u_2} \leq \cdots \leq p_{u_k}$. Then by definition, if job j can be moved ahead in time by say t_j units, the new completion time is given by, $\hat{C}_j = C_j^H - t_j$. Since job u_i can be processed as soon as u_{i-k} completes and space becomes available, we get the following recursive improvement on job completion times:

$$\hat{C}_{u_i} = C_{u_i}^H - [(p_{u_i-1} - p_{u_i-k}) + \cdots + (p_k - p_1)] \; \forall i \in \{1, \cdots, k-1\} \text{ and}$$
$$\hat{C}_{u_k} = C_{u_k}^H$$
$$\text{So, } \hat{Z} = Z_H - \sum_{j=1}^{n} \sum_{i=1}^{j} (p_{ik} - p_{(ik-k+1)}) = Z_{OPT}$$

12.4 Performance Analysis

In this section, we present solution guarantees on the objective values Z_H generated by both the batch-scheduling algorithms. We refer to Z_{OPT} as the optimal objective for the SSP formulation. We begin by analyzing special instances of SSP with a set J of N jobs such that at any given time k jobs can simultaneously fit the space $(W \times H)$ and $p_1 \leq p_2 \leq \cdots \leq p_N$.

Theorem 1. *Suppose there are N=nk jobs for any $n, k \in \mathbb{Z}_+^*$, $w_j \leq W$ and $h_j \leq H$ $\forall \, j \in J$, and $p_1 \leq p_2 \leq \cdots \leq p_N$, where k jobs can simultaneously fit the space, then batch-scheduling is a k-approximation algorithm.*

Proof. Let J denote the set of nk jobs and B the set of batches. If the first k jobs are scheduled in a batch at the beginning of the schedule, job $k + 1$ does not start until any of the jobs finish processing. The first job to finish processing would be job 1. So completion time, $C_{k+1} = p_{k+1} + p_1$. Applying this reasoning we note that a lower bound on the optimal objective for these instances is given by, $Z_{OPT} \geq \sum_{j=1}^{n} (n - j + 1)(p_{jk} + p_{jk-1} + \cdots + p_{jk-k+1})$, since $p_1 \leq p_2 \leq \cdots \leq p_N$ and we are trying to minimize the sum of completion times. In the following discussion p_{jk} is defined as the processing time of the job in the j times k position in the sequence $p_1 \leq p_2 \leq \cdots \leq p_N$.

Since only k jobs can occupy the space at any given time, the number of batches is $\frac{nk}{k} = n$. If we use the MAX rule, $Z_b = max_{j \in b} p_j$ for each batch $b \in B$ and $Z_1 \leq Z_2 \leq \cdots \leq Z_n$. Let us order the jobs in the sequence of the batches they are assigned and in the increasing order of their processing times within each batch, so that $p_{|jk|}$ refers to the processing time of the jkth job in the scheduling sequence and not p_{jk}. The completion time of job j, C_j^H, based on this new ordering is then calculated as the sum of its processing time and the completion times of the batches scheduled ahead of it. For example, if jobs j is in batch b, the completion time is calculated as: $C_j^H = p_j + Z_{b-1} + \cdots + Z_1$.

$$Z_H = \sum_{j \in J} C_j^H \tag{12.44}$$

$$= \sum_{j \in J} p_j + kZ_1 + k(Z_1 + Z_2) + k(Z_1 + Z_2 + Z_3) + \cdots +$$

$$+ k(Z_1 + Z_2 + \ldots + Z_{n-1}) \tag{12.45}$$

$$= \sum_{j \in J} p_j + k[(n-1)Z_1 + (n-2)Z_2 + \cdots + 2Z_{n-2} + Z_{n-1}] \tag{12.46}$$

$$= \sum_{j \in J} p_j + k[(n-1)p_{|k|} + (n-2)p_{|2k|} + \cdots +$$

$$+ 2p_{|(n-2)k|} + p_{|(n-1)k|}] \tag{12.47}$$

$$= \sum_{j=1}^{n} (p_{|jk-1|} + \cdots + p_{|jk-k+1|}) + \sum_{j=1}^{n} ((nk - jk + 1)p_{|jk|}) \tag{12.48}$$

$$= \sum_{j=1}^{n} (p_{|jk-1|} + \cdots + p_{|jk-k+1|}) + k \sum_{j=1}^{n} ((n - j + \frac{1}{k})p_{|jk|}) \tag{12.49}$$

$$= \sum_{j=1}^{n} (p_{|jk-1|} + \cdots + p_{|jk-k+1|}) + \sum_{j=1}^{n} ((n - j + \frac{1}{k})p_{|jk|})$$

$$+ (k-1) \sum_{j=1}^{n} ((n - j + \frac{1}{k})p_{|jk|}) \tag{12.50}$$

$$\leq \sum_{j=1}^{n} (p_{|jk-1|} + \cdots + p_{|jk-k+1|}) + \sum_{j=1}^{n} ((n - j + \frac{1}{k})p_{|jk|})$$

$$+ (k-1) \sum_{j=1}^{n} ((n - j + 1)p_{|jk|}) \tag{12.51}$$

$$\leq \sum_{j-1}^{n} (p_{|jk-1|} + \cdots + p_{|jk-k+1|}) + \sum_{j=1}^{n} ((n - j + \frac{1}{k})p_{|jk|})$$

$$+ (k-1)Z_{OPT} \tag{12.52}$$

$$\leq Z_{OPT} + (k-1)Z_{OPT} \tag{12.53}$$

$$= kZ_{OPT} \tag{12.54}$$

Equation (12.45) is obtained from the definition of completion times, C_j^H and we get Eq. (12.46) per the definition of Z_b. In each batch b of k jobs, the batch processing time is the processing time of the kth job in the batch, $p_{|bk|}$. This is the only processing time included in the calculation of completion times for the batches scheduled later. The processing times of the remaining $(k-1)$ jobs are not repeated in this objective as seen in Eq. (12.47). Equations (12.52) and (12.53) follow from the lower bound on the optimal objective, $Z_{OPT} \geq \sum_{j=1}^{n}(n-j+1)(p_{jk} + p_{jk-1} + \cdots + p_{jk-k+1})$.

The bound shown helps us understand what makes instances of SSP hard. The real difficulty in solving instances of SSP lies in the spatial constraints as reflected by the bound, which is dependent on k, the number of jobs that can simultaneously fit within the given workspace. Also, recall that when minimizing the sum of completion times, we want to schedule more jobs earlier in the schedule. This is because the completion time of a job includes the completion times of the jobs earlier in the schedule. When $k = 1$, SSP reduces to SMS and our batching heuristic becomes SPT, which we know is optimal [19]. Our bound depicts that as k increases, the spatial component plays a larger role in the objective obtained from the batch-scheduling algorithm.

Consider the instance data with six jobs shown in Table 12.2 and a 10×10 workspace. We can fit three (k) jobs within the space, so the batches formed are $\{1,2,3\}$ and $\{4,5,6\}$ as shown in Fig. 12.5a. The sum of completion times before post-processing, $Z_H = p_1 + p_2 + 4p_3 + p_4 + p_5 + p_6 = 192$. Using the lower bound we know that $Z_{OPT} \geq 2(p_1 + p_2 + p_3) + (p_4 + p_5 + p_6) = 153$. So, $Z_H \leq 3Z_{OPT}$.

Now, if we were to schedule the batches as seen in Fig. 12.5b in the sequence $\{1,2\}, \{3,4\}$, and $\{5,6\}$ such that $k=2$, then $Z_H = p_1 + 5p_2 + p_3 + 3p_4 + p_5 + p_6 = 181$. Therefore, packing more jobs (larger k) that are sufficiently different in processing times because they efficiently utilize the space does not result in a lower sum of completion times objective.

Table 12.2 SSP instance with N=6 jobs to depict that grouping more jobs in a batch does not guarantee lower objective value

Job	Processing time	Width	Height
1	1	2	H
2	2	4	H
3	21	2	H
4	22	4	H
5	41	2	H
6	42	4	H

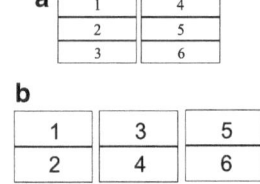

Fig. 12.5 Example schedule with two and three jobs in a batch. (**a**) three job batch (**b**) two job batch

12.5 Computational Analysis

In this section we provide the computational results obtained by evaluating the two proposed procedures for solving the SSP and comparing it to the optimal solution or the best solution obtained after a certain time limit for the integer programming formulation of SSP.

12.5.1 Instance Generation

We tested both the iterative model (M1) and the efficient area model (M2) on generated instances of SSP. The instance class denoted as $NnPpRr < ABC > i$ has $n = 5$ or 10 jobs, processing times generated in the discrete uniform interval of $(1, p)$ with workspace area dimension $W = H = r$. The value for r is 10 or 20 units and i is an instance indicator. A, B, C classifiers are used to indicate the distributions from which the width and height of jobs are sampled.

Class A $w_j \in$ Uniform Discrete $[1, \frac{W}{2}]$ and $h_j \in$ Uniform Discrete $[1, \frac{H}{2}]$
Class B $w_j \in$ Uniform Discrete $[1, \frac{W}{2}]$ and $h_j \in$ Uniform Discrete $[\frac{H}{2}, H]$
Class C $w_j \in$ Uniform Discrete $[\frac{W}{2}, W]$ and $h_j \in$ Uniform Discrete $[\frac{H}{2}, H]$

Five instances of each class-type were generated, resulting in a total of 60 instances. All of the instances had jobs sorted in the increasing order of processing times. Instances in Class C have jobs that occupy more than half the area. This results in each job getting its individual batch and SSP reduces to SMS which can be solved to optimality. So for the computational analysis we only consider instances in classes A and B. By design, instances in Class B should be relatively harder to solve than instances in class A. This is because all of the jobs in class A are small compared to the dimensions of the workspace, so we can fit more jobs together. Difficult instances of the problem occur, when some jobs are small and some are large (Class B).

Larger instances were modified from [6]. The instances have 100, 500, and 1,000 jobs with a 10×7 workspace. For each job:

$w_j \in UniformDiscrete[1, 10]$
$h_j \in UniformDiscrete[1, 7]$
$p_j \in UniformDiscrete[5, 25]$

Since we did not permit rotation of jobs, we had to interchange the widths and heights in certain cases to ensure that the jobs would fit within the space.

12.5.2 Initial Feasible Solution Heuristic

The motivation behind creating the batching models (M1 and M2) was to reduce the complexity of the original SSP by looking only at the packing component of the problem. Nevertheless, we need to understand that M1 and M2 are still MIPs and as the instances grow larger, these models could take longer to solve to optimality. Further, an optimal solution to the batching model does not necessarily guarantee an optimal solution to SSP. In order to improve the solution time for these MIP formulations, we provide the solver with an initial feasible solution obtained from a greedy packing heuristic. Basically, we start with an instance of SSP sorted in the increasing order of job processing times, i.e. $p_1 \leq p_2 \leq \cdots \leq p_N$. We sequentially begin grouping jobs into a batch until they fit the space. Once the job can no longer fit the space, we create a new batch. This process is repeated until all jobs are assigned a batch.

12.5.3 Computational Results

In this section, we compare the solutions generated by the batch-scheduling approaches (iterative and efficient area models) to the optimal solution (OPT) obtained by solving the mixed-integer program for SSP. The batching MIPs, M1 and M2, and the SSP MIP formulation were all implemented using the C programming language and solved using Gurobi 5.0 with a thread count of 1 and cuts parameter set to default on a RedHat Enterprise 6.5 x86_64 server. The following tables compare the objective values and runtimes for the small instances with 5 jobs or 10 jobs and the large instances with 25 jobs or 100 jobs (defined at the beginning of Sect. 12.5).

Table 12.3 lists the objective values obtained from solving instances with five and ten jobs for M1 and M2 using the MAX rule and the optimal solution (OPT) for the original MIP formulation of SSP. Note that the objective reported for M1 is the best possible value among the $N - 1$ potential solutions it obtains and the run time is the total time taken to iteratively solve all of the models. We observe that M2 seems to perform at least as well as M1, and both models return values close to the optimal

Table 12.3 Comparison of objectives obtained from M1, M2, and OPT for small instances of batch-scheduling

Instance	M1 (Best)	M2	OPT	Factors	
				M1/OPT	M2/OPT
N5P10R10A	27	27	27	1.00	1.00
N5P19R10B	33	33	29	1.14	1.14
N5P10R20A	26	26	26	1.00	1.00
N5P10R20B	26	25	23	1.13	1.12
N10P10R10A	49	49	49	1.00	1.00
N10P10R10B	80	72	66	1.22	1.10
N10P10R20A	54	54	51	1.05	1.05
N10P10R20B	101	88	77	1.31	1.14

Table 12.4 Comparison of M1, M2, and OPT runtimes for small instances of batch-scheduling

Instance	Runtime (s)		
	M1 (Total)	M2	OPT
N5P10R10A	0.19	0.01	0.01
N5P19R10B	0.14	0.04	0.02
N5P10R20A	0.17	0.01	0.01
N5P10R20B	0.13	0.07	0.01
N10P10R10A	43.98	0.11	0.03
N10P10R10B	110.21	82.79	287.69
N10P10R20A	300.81	0.23	0.30
N10P10R20B	223.26	114.17	244.33

solution. For these set of instances, the objective values returned by both models for the MAX and AVG rules were identical for instances with five jobs and ten jobs.

Table 12.4 presents the runtimes for solving the instances with five and ten jobs using M1, M2, and the original MIP formulation. We observe that with smaller number of jobs, all three methods produce results quickly. The runtimes for M1 are larger because it iteratively solves $N - 1$ models for each instance with N jobs.

Table 12.5 lists the objective values obtained from solving larger instances (25 and 100 jobs) for M1 and M2 using the MAX rule and the objective Z_{IP} for the original MIP formulation of SSP. Note that the objective reported for M1 is the best possible value among the $N - 1$ potential solutions it obtains, with each iteration of M1 allowed 2 min of execution time. M2 and Z_{IP} report the best objective obtained after 20 min of execution. To improve upon the solution, M2 is given an initial feasible solution. The resulting solution is then updated using the post-processing algorithm. Although both models produce objectives close to optimal, it is observed that with a larger number of jobs, M2 outperforms M1, and on some occasions, after 20 min, M2 is able to produce better solutions than the original SSP formulation.

Table 12.5 Comparison of objectives obtained from M1, M2, and Z_{IP} for large instances of batch-scheduling

Instance	Objective			Factor	
	M1 (Best)	M2 (Updated)	Z_{IP}	M1/Z_{IP}	M2/Z_{IP}
N25P25E11	1,697	1,421	1,215	1.40	1.17
N25P25E12	1,518	1,409	1,022	1.49	1.38
N25P25E13	2,204	2,046	1,540	1.43	1.33
N25P25E14	1,555	1,292	995	1.56	1.30
N25P25H11	1,819	1,762	1,353	1.34	1.30
N25P25H12	1,587	1,332	965	1.64	1.38
N25P25H13	1,929	1,712	1,169	1.65	1.46
N25P25H14	1,625	1,525	1,066	1.52	1.43
N100P25E1	34,372	24,495	28,205	1.22	0.87
N100P25H1	45,571	24,919	27,672	1.65	0.90

In conclusion, the efficient area model seems to be more effective for larger instances both in terms of runtime and solution quality. Further investigations on the weights in the multi-objective function in the efficient area model (M2) could result in potential improvements in objective value.

12.6 Conclusions

The study aims to develop solution methods for SSP with good approximations for the minimum sum of completion times objective. We conclude by summarizing the main contributions and key results presented and by suggesting possible directions for future research. We explored the relationship between the spatial and temporal components of the problem. We considered just the spatial restrictions and utilized bin-packing strategies to identify batches of jobs that will efficiently utilize the space. We then scheduled the jobs using rules to minimize the sum of completion times objective.

When minimizing the sum of completion times objective sometimes counterintuitive policies are better. Here we proved an approximation factor under certain conditions and also identified scenarios when grouping jobs did not necessarily result in a better objective. We also gave a post-processing algorithm to improve the objective value of the batching models, which resulted in optimal solutions for certain instances. Based on the instances we tested for both the iterative and efficient-area approaches, our assessment is that scheduling jobs similar in processing times within the same space yields good solutions. If processing times are sufficiently different, then grouping jobs together because they effectively utilize the space does not necessarily result in a lower sum of completion times. The efficient area model outperforms the iterative model both in terms of solution quality and run time.

Directions for future research are plentiful. We provide two MIP formulations to decide the assignment of jobs to batches, the iterative and efficient area model. Currently, we solve at most $N - 1$ instances for the iterative procedure and weigh the two objectives in the efficient area model equally. Possible enhancements could be to implement a binary search procedure that improves runtimes for the iterative model or tweak the weights in the multi-objective efficient area model. This study assumes that a single spatial resource of fixed dimension is available. An interesting extension would be to look at multiple workspace problems with varying area. We may be able to use ideas from variable size bin packing to design algorithms for this problem. Another area that merits investigation is to consider weights on the completion times of the jobs. If l_j is the weight on completion time for job $j \in J$, and we assign the number of jobs in the batch containing job j as a weight on its completion time, can we get similar results for our procedures? Lastly, although the results and analyses presented in this study pertain to the sum of completion times objective, the solution methods developed here can easily be applied to other objective functions of the problem.

References

1. Brucker, P., Kovalyov, M.Y., Shafransky, Y.M., Werner, F.: Batch scheduling with deadlines on parallel machines. Ann. Oper. Res. **83**, 23–40 (1998)
2. Caprace, J.D., Petcu, C., Velarde, M., Rigo, P.: Optimization of shipyard space allocation and scheduling using a heuristic algorithm. J. Mar. Sci. Technol. **18**(3), 404–417 (2013)
3. Castillo, I., Kampas, F.J., Pintér, J.D.: Solving circle packing problems by global optimization: numerical results and industrial applications. Eur. J. Oper. Res. **191**(3), 786–802 (2008)
4. Cho, K., Chung, K., Park, C., Park, J., Kim, H.: A spatial scheduling system for block painting process in shipbuilding. CIRP Ann. Manuf. Technol. **50**(1), 339–342 (2001)
5. Duin, C., Sluis, E.: On the complexity of adjacent resource scheduling. J. Sched. **9**(1), 49–62 (2006)
6. Garcia, C.J.: Optimization models and algorithms for spatial scheduling. Ph.D. thesis, Old Dominion University, Norfolk (2010)
7. Garcia, C., Rabadi, G.: A meta-raps algorithm for spatial scheduling with release times. Int. J. Plann. Sched. **1**, 19–31 (2011)
8. Garcia, C., Rabadi, G.: Exact and approximate methods for parallel multiple-area spatial scheduling with release times. OR Spectr. **35**(3), 639–657 (2013)
9. Koh, S., Logendran, R., Choi, D., Woo, S.: Spatial scheduling for shape-changing mega-blocks in a shipbuilding company. Int. J. Prod. Res. **49**(23), 7135–7149 (2011)
10. Lee, C.Y., Uzsoy, R., Martin-Vega, L.A.: Efficient algorithms for scheduling semiconductor burn-in operations. Oper. Res. **40**(4), 764–775 (1992)
11. Lee, K., Jun, K.L., Park, H.K., Hong, J.S., Lee, J.S.: Developing scheduling systems for Daewoo shipbuilding: {DAS} project. Eur. J. Oper. Res. **97**(2), 380–395 (1997)
12. Leung, J., Kelly, L., Anderson, J.H.: Handbook of Scheduling: Algorithms, Models, and Performance Analysis. CRC Press, Boca Raton (2004)
13. Lodi, A., Martello, S., Monaci, M.: Two-dimensional packing problems: a survey. Eur. J. Oper. Res. **141**(2), 241–252 (2002)
14. Mathirajan, M., Sivakumar, A.: A literature review, classification and simple meta-analysis on scheduling of batch processors in semiconductor. Int. J. Adv. Manuf. Technol. **29**(9–10), 990–1001 (2006)

15. Park, K., Lee, K., Park, S., Kim, S.: Modeling and solving the spatial block scheduling problem in a shipbuilding company. Comput. Ind. Eng. **30**(3), 357–364 (1996)
16. Perng, C., Lai, Y.C., Ho, Z.P.: A space allocation algorithm for minimal early and tardy costs in space scheduling. In: International Conference on New Trends in Information and Service Science, 2009 (NISS '09), pp. 33–36 (2009)
17. Pisinger, D., Sigurd, M.: The two-dimensional bin packing problem with variable bin sizes and costs. Discret. Optim. **2**(2), 154–167 (2005)
18. Raj, P., Srivastava, R.K.: Analytical and heuristic approaches for solving the spatial scheduling problem. In: 2007 IEEE International Conference on Industrial Engineering and Engineering Management, pp. 1093–1097 (2007)
19. Smith, W.E.: Various optimizers for single-stage production. Nav. Res. Logist. Q. **3**(1–2), 59–66 (1956)
20. Vazirani, V.V.: Approximation Algorithms. Springer, New York (2001)
21. Williamson, D.P., Shmoys, D.B.: The Design of Approximation Algorithms, 1st edn. Cambridge University Press, New York (2011)
22. Zhang, Z., Chen, J.: Solving the spatial scheduling problem: a two-stage approach. Int. J. Prod. Res. **50**(10), 2732–2743 (2012)
23. Zheng, J., Jiang, Z., Chen, Q., Liu, Q.: Spatial scheduling algorithm minimising makespan at block assembly shop in shipbuilding. Int. J. Prod. Res. **49**(8), 2351–2371 (2011)

Chapter 13
Optimized Object Packings Using Quasi-Phi-Functions

Yuriy Stoyan, Tatiana Romanova, Alexander Pankratov, and Andrey Chugay

Abstract In this chapter we further develop the main tool of our studies, phi-functions. We define new functions, called *quasi-phi-functions*, that we use for analytic description of relations of geometric objects placed in a container taking into account their continuous rotations, translations, and distance constraints. The new functions are substantially simpler than phi-functions for some types of objects. They also are simple enough for some types of objects for which phi-functions could not be constructed. In particular, we derive quasi-phi-functions for certain 2D&3D-objects. We formulate a basic optimal packing problem and introduce its exact mathematical model in the form of a nonlinear continuous programming problem, using our quasi-phi-functions. We propose a general solution strategy, involving: a construction of feasible starting points, a generation of nonlinear subproblems of a smaller dimension and decreased number of inequalities; a search for local extrema of our problem using subproblems. To show the advantages of our quasi-phi-functions we apply them to two packing problems, which have a wide spectrum of industrial applications: packing of a given collection of ellipses into a rectangular container of minimal area taking into account distance constraints; packing of a given collection of 3D-objects, including cuboids, spheres, spherocylinders and spherocones, into a cuboid container of minimal height. Our efficient optimization algorithms allow us to get local optimal object packings and reduce considerably computational cost. We applied our algorithms to several inspiring instances: our new benchmark instances and known test cases.

Keywords Packing 2D- and 3D-objects • Continuous rotations • Mathematical model development • Quasi-phi-functions • Nonlinear optimization

Y. Stoyan (✉) • T. Romanova • A. Pankratov • A. Chugay
Department of Mathematical Modeling and Optimal Design, Institute for Mechanical Engineering Problems, National Academy of Sciences of Ukraine, Pozharskyi Str., 2/10, Kharkiv 61046, Ukraine
e-mail: tarom27@yahoo.com

© Springer International Publishing Switzerland 2015   265
G. Fasano, J.D. Pintér (eds.), *Optimized Packings with Applications*, Springer Optimization and Its Applications 105, DOI 10.1007/978-3-319-18899-7_13

13.1 Introduction

Optimal packing problem is a part of operational research and computational geometry. It has multiple applications in modern biology, mineralogy, medicine, materials science, nanotechnology, robotics, coding, pattern recognition systems, control systems, space apparatus control systems, as well as in the chemical industry, power engineering, mechanical engineering, shipbuilding, aircraft construction, civil engineering, logistics, etc. At present, the interest in finding effective solutions for packing problems is growing rapidly. This is due to a large and growing number of applications and an extreme complexity of methods used to handle many of them. We refer the reader to [1] for typology of the class of problems.

These problems are NP-hard [2], and, as a result, solution methodologies generally employ heuristics, e.g. [3–16]. Some researchers develop approaches based on mathematical modeling and general optimization procedures; e.g. [17–25].

Our approach is based on mathematical modeling of relations between geometric objects and thus reducing the Optimal Packing Problem to a nonlinear programming problem. We use the phi-function technique [26, 27] for an analytic description of relations of objects to be packed in a container taking into account their continuous rotations, translations, and distance constraints. In [28] we review our *phi*-functions. One may also find there a clear definition of a phi-function. There we construct a mathematical model of a basic placement (cutting and packing) problem using phi-functions as a constrained optimization problem. We propose a solution strategy for placement problems. The paper also considers a layout problem encountered in space engineering and provides a number of computational results for 2D- and 3D-applications. The complete class of phi-functions for basic 2D-objects are derived in [29]. The functions allow us to cover a wide spectrum of irregular packing problems involving arbitrary shaped 2D-objects, bounded by circular arcs and line segments; see, e.g., [30]. Phi-functions for the simplest 3D-objects under continuous rotations, such as parallelepipeds, convex polytopes, and spheres, are considered in [31, 32]. But some of these phi-functions (especially for 3D-objects) happen to be rather complicated, analytically, and difficult in practical use. Our attempts to construct convenient phi-functions for more general types of objects have been futile.

In this chapter we further develop the concept of phi-functions, introducing a new class of functions, called *quasi-phi-functions*. The functions can be described by analytical formulas that are substantially simpler than those used for phi-functions, for pairs of some types of 2D- and 3D-objects (convex polygons, circles, circular segments, cuboids, spheres, cylinders, disks, and convex polytopes). They also are simple enough for some types of rotating objects for which phi-functions could not be constructed. In particular, we find convenient quasi-phi-functions for ellipses, and for certain 3D-objects including, so-called, spherocylinders, spherocones. The use of quasi-phi-functions allows us to handle new types of objects, but there is a price to pay. now the optimization has to be performed over a larger set of parameters, including the extra variables used by our new functions. To demonstrate high efficiency of our quasi-phi-functions we consider two practical problems of packing a collection of ellipses into a rectangular container of minimal area as well as packing a collection of given 3D-objects (cuboids, spheres, spherocylinders, spherocones) into a cuboid container of minimal height. We derive here quasi-phi-functions to describe non-overlapping and containment constraints for appropriate pairs of rotating objects and develop efficient optimization algorithms. In this chapter the reader will find theoretical results presented in our works [33, 34].

The chapter is organized as follows: in Sect. 13.2 we define our new quasi-phi-functions for an analytical description of non-overlapping, containment, and distance constraints; we also discuss their general properties. In Sect. 13.3 we define quasi-phi-functions for certain types of convex 2D- and 3D-objects needed in applications. In Sect. 13.4 we formulate a basic optimal packing problem, construct its mathematical model, using our quasi-phi-functions, in the form of a nonlinear programming problem with nonsmooth functions, and develop a general solution strategy. In Sect. 13.5 we formulate the optimal packing problem of ellipses taking into account continuous ellipse rotations and distance constraints as a continuous nonlinear programming problem with smooth functions; describe the algorithm to search for "good" local optimal solutions for the problem which involves a fast starting point and efficient local optimization procedures. In Sect. 13.6 we formulate the optimal packing problem of 3D-objects, including spherocylinders and spherocones, and based on characteristics of its mathematical model, describe an efficient solution algorithm, using local and global optimization methods. We provide some computational results of several instances for 2D- and 3D-optimal packing problems, illustrated with pictures, in Sect. 13.7, and finish with some concluding remarks in Sect. 13.8.

13.2   Quasi-Phi-Functions and Their Properties

Let $A \subset R^d$ and $B \subset R^d$ be closed phi-objects, $d = 2, 3$; one can find a precise definition of phi-objects, e.g., in [26, 27]. We assume that at least one of these objects is bounded. Position of the object A is defined by a vector of placement parameters (v_A, θ_A), where v_A is a translation vector and θ_A is a vector of rotation parameters: for 2D object $v_A = (x_A, y_A)$ and θ_A is a rotation angle; for 3D-object $v_A = (x_A, y_A, z_A)$ and $\theta_A = (\theta_z, \theta_x, \theta_y)$, where $\theta_z, \theta_x, \theta_y$ are rotation angles, respectively: from axis OX to OY, from axis OY to OZ and from axis OX to OZ. We denote the vector of variables for the object A by $u_A = (v_A, \theta_A)$ and the vector of variables for the object B by $u_B = (v_B, \theta_B)$. The object A, rotated by angles $\theta_z, \theta_x, \theta_y$ (in this order), translated by vector v_A, will be denoted by $A(u_A)$.

Definition 1 A continuous and everywhere defined function $\Phi'^{AB}(u_A, u_B, u')$ is called a *quasi-phi-function* for two phi-objects $A(u_A)$ and $B(u_B)$ if $\max_{u' \in U} \Phi'^{AB}(u_A, u_B, u')$ is a phi-function $\Phi^{AB}(u_A, u_B)$ for the objects. Here u' is a vector of auxiliary variables, that takes values in some domain $U \subset R^n$ (which may depend on the shapes of objects A and B).

The concept of quasi-phi-functions and basic characteristics of quasi-phi-functions formulated in the form of theorems are introduced in [33].

We emphasize that according to the definition, a quasi-phi-function Φ'^{AB} for a pair of objects A and B can be constructed by many different formulas, and we can choose the most convenient ones for our optimization algorithms.

Next we discuss general properties of quasi-phi-functions. Let $\Phi'^{AB}(u_A, u_B, u')$ be a quasi-phi-function for two phi-objects $A(u_A)$ and $B(u_B)$.

Property 1 If $\Phi'^{AB}(u_A, u_B, u') \geq 0$ for some u', then $\text{int} A(u_A) \cap \text{int} B(u_B) = \varnothing$. Here $\text{int} A$ denotes the topological interior of object A.

Property 2 Let $P(u_P) = \{(x, y, z) : \psi_P = \alpha \cdot x + \beta \cdot y + \gamma \cdot z + \mu_P \leq 0\}$ be a half-space (for $d = 2$ it will be a half-plane; see below); here, $u_P = (\theta_{xP}, \theta_{yP}, \mu_P)$, $\alpha = \sin \theta_{yP}$, $\beta = \sin \theta_{xP} \cdot \cos \theta_{yP}$, $\gamma = \cos \theta_{xP} \cdot \cos \theta_{yP}$ (note $\alpha^2 + \beta^2 + \gamma^2 = 1$). If $A, B \subset R^2$, then $P(u_P) = \{(x, y) : \psi_P = \alpha \cdot x + \beta \cdot y + \mu_P \leq 0\}$, where $u_P = (\theta_P, \mu_P)$ $\alpha = cos\theta_P$, $\beta = sin\theta_P$. Suppose $\Phi^{AP}(u_A, u_P)$ is a phi-function for $A(u_A)$ and $P(u_P)$ and $\Phi^{BP^*}(u_B, u_P)$ is a phi-function for $B(u_B)$ and $P^*(u_P) = R^d \setminus \text{int} P(u_P)$, $d = 2, 3$.

Then a function defined by

$$\Phi'^{AB}(u_A, u_B, u_P) = \min \left\{ \Phi^{AP}(u_A, u_P), \Phi^{BP^*}(u_B, u_P) \right\}, \tag{13.1}$$

is a quasi-phi-function for the pair of bounded objects $A(u_A)$ and $B(u_B)$. Here $u' = u_P$.

Property 3 If $\Phi'^{AP}(u_A, u_P, u'_1)$ is a quasi-phi-function for $A(u_A)$ and $P(u_P)$, $\Phi'^{BP^*}(u_B, u_P, u'_2)$ is a quasi-phi-function for $B(u_B)$ and $P^*(u_P)$, then function

$$\Phi'^{AB}(u_A, u_B, u') = \min \left\{ \Phi'^{AP}(u_A, u_P, u'_1), \Phi'^{BP^*}(u_B, u_P, u'_2) \right\}, \tag{13.2}$$

is a quasi-phi-function for the pair of bounded objects $A(u_A)$ and $B(u_B)$. Here $u' = (u_P, u'_1, u'_2)$.

We adapt the concept of quasi-phi-functions to model distance constraints. To this end we define normalized and adjusted quasi-phi-functions [33], based on similar terms for phi-functions [27].

Let $\text{dist}(A, B) = \min_{a \in A, b \in B} d(a, b)$, where $d(a, b)$ stands for the Euclidean distance between points $a, b \in R^d$, $d = 2, 3$, and let $\rho^- > 0$ denote minimal allowable distances between objects $A(u_A)$ and $B(u_B)$.

We remind the reader that by definition (see for instance [27]) a phi-function $\tilde{\Phi}^{AB}(u_A, u_B)$ for objects $A(u_A)$ and $B(u_B)$ is said to be a *normalized* phi-function if $\tilde{\Phi}^{AB}(u_A, u_B) = \text{dist}(A(u_A), B(u_B))$ whenever $\text{int}\, A(u_A) \cap \text{int}\, B(u_B) = \varnothing$.

Definition 2 A quasi-phi-function $\tilde{\Phi}'^{AB}(u_A, u_B, u')$ is called a *normalized* quasi-phi-function for objects $A(u_A)$ and $B(u_B)$, if function $\max\limits_{u' \in U} \tilde{\Phi}'^{AB}(u_A, u_B, u')$ is a normalized phi-function.

Thus, $\max\limits_{u' \in U} \tilde{\Phi}'^{AB} \geq \rho^- \iff \text{dist}(A, B) \geq \rho^-$.

Definition 3 Function $\widehat{\Phi}'^{AB}(u_A, u_B, u')$ is called an *adjusted* quasi-phi-function for objects $A(u_A)$ and $B(u_B)$, if function $\max\limits_{u' \in U} \widehat{\Phi}'^{AB}(u_A, u_B, u')$ is an adjusted phi-function.

Thus, $\max\limits_{u' \in U} \widehat{\Phi}'^{AB} \geq 0 \iff \text{dist}(A, B) \geq \rho^-$.

Let $\tilde{\Phi}^{AP}(u_A, u_P)$, $\tilde{\Phi}^{BP*}(u_B, u_P)$ be normalized phi-functions. Assume

$$\Phi'^{AB}(u_A, u_B, u_P) = \min\left\{\tilde{\Phi}^{AP}(u_A, u_P), \tilde{\Phi}^{BP*}(u_B, u_P)\right\}.$$

Then a quasi-phi-function

$$\tilde{\Phi}'^{AB}(u_A, u_B, u_P) = 2\Phi'^{AB}(u_A, u_B, u_P), \tag{13.3}$$

is a normalized quasi-phi-function, and a quasi-phi-function

$$\widehat{\Phi}'^{AB}(u_A, u_B, u_P) = \Phi'^{AB}(u_A, u_B, u_P) - 0.5\rho^-, \tag{13.4}$$

is an adjusted quasi-phi-function.

13.3 Construction of Quasi-Phi-Functions

Here we derive quasi-phi-functions for certain 2*D*- and 3*D*-objects, based on our general formulas (13.1)–(13.3).

A *quasi-phi-function for convex polygons.* Let $K_1(u_1)$ and $K_2(u_2)$ be convex *polygons*, given by their vertices $p_i^1, i = 1, ..., m_1$, and $p_i^2, i = 1, ..., m_2$, respectively. Then $\Phi^{K_1P}(u_1, u_P) = \min\limits_{1 \leq i \leq m_1} \psi_P(p_i^1)$ and $\Phi^{K_2P}(u_2, u_P) = \min\limits_{1 \leq i \leq m_2} (-\psi_P(p_i^2))$ are phi-functions for $K_1(K_2)$ and $P(P*)$, respectively.

Now the function

$$\Phi'^{K_1K_2}(u_1, u_2, u_P) = \min\left\{\Phi^{K_1P}(u_1, u_P), \Phi^{K_2P*}(u_2, u_P)\right\}, \tag{13.5}$$

is a quasi-phi-function for $K_1(u_1)$ and $K_2(u_2)$.

Note that function $2\Phi'^{K_1K_2}(u_1, u_2, u_P)$ is a normalized quasi-phi-function.
An adjusted quasi-phi-function for $K_1(u_1)$ and $K_2(u_2)$ is defined by

$$\widehat{\Phi}^{\prime K_1 K_2}(u_1, u_2, u_P) = \min\left\{\Phi^{K_1 P}(u_1, u_P), \Phi^{K_2 P^*}(u_2, u_P)\right\} - 0.5\rho^-. \quad (13.6)$$

A quasi-phi-function for a convex polygon $K(u_1)$ and a circle $C(u_2)$. Let $K(u_1)$ be a convex polygon given by its vertices $p_i, i = 1, ..., m$. Let p_C and r_C be the center and radius of circle $C(u_2)$. Then $\Phi^{KP}(u_1, u_P) = \min\limits_{1 \le i \le m} \psi_P(p_i)$ and $\Phi^{CP^*}(u_2, u_P) = -\psi_P(p_C) - r_C$ are phi-functions.

Now a quasi-phi-function for $K(u_1)$ and $C(u_2)$ may be defined as:

$$\Phi'^{CK}(u_1, u_2, u_P) = \min\left\{\Phi^{KP}(u_1, u_P), \Phi^{CP^*}(u_2, u_P)\right\}. \quad (13.7)$$

It should be noted that function $2\Phi'^{CK}(u_1, u_2, u_P)$ is a normalized quasi-phi-function.

Quasi-phi-functions defined by (13.5)–(13.7) can be applied to *convex polytopes* and *spheres*.

A quasi-phi-function for circular segments $D_1(u_1)$ and $D_2(u_2)$. Let $D_1(u_1) = T_1(u_1) \cap C_1(u_1)$, $D_2(u_2) = T_2(u_2) \cap C_2(u_2)$ be two circular segments, where $T_1(u_1)$ $(T_2(u_2))$ denotes a triangle given by its vertices $p_i^1(p_i^2), i = 1, 2, 3$ (we note that two sides of T have to be tangents to C and one side is a chord of C) and $p_C^1 = (x_1, y_1)$ $(p_C^2 = (x_2, y_2))$ and r_C^1 (r_C^2) denote the center and radius of $C_1(u_1)$ (resp., $C_2(u_2)$). Then, following (13.1), a quasi-phi-function for $D_1(u_1)$ and $D_2(u_2)$ may be defined by

$$\Phi'^{D_1 D_2}(u_1, u_2, u_P) = \min\left\{\Phi^{D_1 P}(u_1, u_P), \Phi^{D_2 P^*}(u_2, u_P)\right\}, \quad (13.8)$$

where $\Phi^{D_1 P}(u_1, u_P) = \max\left\{\Phi^{T_1 P}, \Phi^{C_1 P}\right\}$, $\Phi^{D_2 P^*}(u_2, u_P) = \max\left\{\Phi^{T_2 P^*}, \Phi^{C_2 P^*}\right\}$, are phi-functions, and $\Phi^{T_1 P}(u_1, u_P) = \min\limits_{i=1,2,3} \psi_P(p_i^1)$, $\Phi^{C_1 P}(u_1, u_P) = \psi_P(p_C^1) - r_C^1$, $\Phi^{T_2 P^*}(u_2, u_P) = \min\limits_{i=1,2,3}(-\psi_P(p_i^2))$, $\Phi^{C_2 P^*}(u_2, u_P) = -\psi_P(p_C^2) - r_C^2$.

We can define a quasi-phi-function for $D_1(u_1)$ and $D_2(u_2)$ using formula (13.2)

$$\Phi'^{D_1 D_2}(u_1, u_2, u') = \min\left\{\Phi'^{D_1 P}\left(u_1, u_P, u_1'\right), \Phi'^{D_2 P^*}\left(u_2, u_P, u_2'\right)\right\},$$

where $u' = \left(u_P, u_1', u_2'\right)$, $u_1' \in [0, 1] \subset R^1$, $u_2' \in [0, 1] \subset R^1$.

To this end, first, we construct quasi-phi-functions $\Phi'^{D_1 P}\left(u_1, u_P, u_1'\right)$ and $\Phi'^{D_2 P^*}\left(u_2, u_P, u_2'\right)$. Let $\Phi^{C_1 P}(u_1, u_P)$ be a phi-function for $C_1(u_1)$ and $P(u_p)$. We introduce function

$$\Phi'^{D_1 P} \left(u_1, u_P, u_1' \right) = \min \left\{ \psi_P \left(p_1^1 \right), \psi_P \left(p_2^1 \right), \chi_1 \left(u_1, u_P, u_1' \right) \right\},$$

$$\chi_1 \left(u_1, u_P, u_1' \right) = \psi_P \left(p_3^1 \right) - u_1' \psi_P \left(p_3^1 \right) + u_1' \Phi^{C_1 P} \left(u_1, u_P \right),$$

where $u_1' \in [0, 1] \subset R^1$, p_i^1, $i = 1, 2$, are the endpoints of the chord of $D_1(u_1)$.

By analogy we have

$$\Phi'^{D_2 P} \left(u_2, u_P, u_2' \right) = \min \left\{ -\psi_P \left(p_1^2 \right), -\psi_P \left(p_2^2 \right), \chi_2 \left(u_2, u_P, u_2' \right) \right\},$$

$$\chi_2 \left(u_2, u_P, u_2' \right) = -\psi_P \left(p_3^2 \right) - u_2' \left(-\psi_P \left(p_3^2 \right) \right) + u_2' \Phi^{C_2 P^*} \left(u_2, u_P \right),$$

where $u_2' \in [0, 1] \subset R^1$, p_i^2, $i = 1, 2$, are the endpoints of the chord of $D_2(u_2)$.

The a quasi-phi-function defined by (13.8) may be adapted to a pair of spherical segments defined as intersections of right circular cones with solid spheres.

A quasi-phi-function for ellipses. Let $E_1(u_1)$ and $E_2(u_2)$ be two ellipses with semi-axes a_i and b_i, $a_i > b_i$ $i = 1, 2$.

Then, a quasi-phi-function for $E_1(u_1)$ and $E_2(u_2)$ may be defined as follows:

$$\Phi'^{E_1 E_2} \left(u_1, u_2, u' \right) = \min \left\{ \chi \left(\theta_1, \theta_2, u' \right), \chi^+ \left(u_1, u_2, u' \right), \chi^- \left(u_1, u_2, u' \right) \right\}, \quad (13.9)$$

where θ_1 and θ_2 are rotation angles and $u' = (t_1, t_2)$ is a vector of auxiliary parameters, $0 \leq t_i \leq 2\pi$, $i = 1, 2$; functions χ, χ^+, χ^- are defined below.

The parameter t_i specifies a point on ellipse E_i. In the local coordinate system of ellipse E_i that point is $\left(x_i^t, y_i^t \right) = (a_i \cos t_i, b_i \sin t_i)$, and after rotation and translation its coordinates are $\left(x_i', y_i' \right) = v_i + M(\theta_i) \cdot \left(x_i^t, y_i^t \right)$, where $M(\theta)$ denotes the standard rotation matrix, $v_i = (x_i, y_i)$ is a translation vector of E_i.

Now we define the three functions mentioned in (13.9): $\chi = -\left\langle N_1', N_2' \right\rangle$, where $N_i' = \left(\alpha_i', \beta_i' \right) = M(\theta_i)(\alpha_i, \beta_i)$, $\alpha_i = \frac{\cos t_i}{a_i}$, $\beta_i = \frac{\sin t_i}{b_i}$; $\chi^\pm = \psi_1 \left(x_2^\pm - x_1, y_2^\pm - y_1 \right) = \alpha_1' \left(x_2^\pm - x_1 \right) + \beta_1' \left(y_2^\pm - y_1 \right) - 1$, where $\left(x_2^\pm, y_2^\pm \right)$ are coordinates of two points $q_2^\pm$ on the second tangent line, $\left(x_2^\pm, y_2^\pm \right) = \left(x_2', y_2' \right) \pm \eta \left(-\beta_2', \alpha_2' \right)$, $\eta = (a_2)^2$, $\left\langle N_1', N_2' \right\rangle$ is a scalar product of vectors N_1' and N_2'.

Alternatively, a quasi-phi-function for $E_1(u_1)$ and $E_2(u_2)$ may be defined according to (13.2):

$$\Phi'^{E_1 E_2} \left(u_1, u_2, u' \right) = \min \left\{ \Phi'^{E_1 P} \left(u_1, u_P, u_1' \right), \Phi'^{E_2 P^*} \left(u_2, u_P, u_2' \right) \right\}.$$

It remains to define a quasi-phi-function for an ellipse $E(u_E)$ and a half-plane $P(u_P)$. This can be done as follows:

$$\Phi'^{EP} (u_E, u_P, t) = \min \left\{ \chi (\theta_E, \theta_P, t), \psi_P^+ (u_E, u_P, t), \psi_P^- (u_E, u_P, t) \right\}, \quad (13.10)$$

where $u_P = (\theta_P, \mu_P), 0 \leq t \leq 2\pi$ is auxiliary parameter.

Here the half-plane is defined by $\psi_P (x, y) = \alpha_P x + \beta_P y + \mu_P \leq 0$, where $\alpha = \cos\theta_P, \beta = \sin\theta_P$.

Note that $N_P = (\alpha_P, \beta_P)$ is the corresponding outer normal vector for the half-plane. For ellipse $E(u_E)$ we adopt our previous formulas introduced for $E_2(u_2)$, i.e. $N_2' = \left(\alpha_2', \beta_2'\right)$ and $(x_2^\pm, y_2^\pm)$, we just replace the subscript 2 with E in those formulas. Thus $\psi_P^\pm (x_E^\pm, y_E^\pm) = \alpha_P x_E^\pm + \beta_P y_E^\pm + \mu_P \leq 0$. Lastly we define $\chi = -\left\langle N_P, N_E' \right\rangle$, which completes our construction of (13.10), here $\langle N_P, N_E' \rangle$ is a scalar product of vectors N_P and N_E'.

Now let a minimal allowable distance between two ellipses E_1 and E_2 be given, we denote it by ρ^-. Assume that $\widehat{\Phi}'^{E_1 P} (u_1, u_P)$, $\widehat{\Phi}'^{E_2 P^*} (u_2, u_P)$ are adjusted quasi-phi-functions provided that $\max\limits_{u_P \in U} \widehat{\Phi}'^{E_1 P} (u_1, u_P) \geq 0$ if $\operatorname{dist}(E_1, P) \geq 0.5\rho^-$ and $\max\limits_{u_P \in U} \widehat{\Phi}'^{E_2 P^*} (u_2, u_P) \geq 0$ if $\operatorname{dist}(E_2, P^*) \geq 0.5\rho^-$. Then

$$\widehat{\Phi}'^{E_1 E_2} (u_1, u_2, u_P) = \min \left\{ \widehat{\Phi}'^{E_1 P} (u_1, u_P), \widehat{\Phi}'^{E_2 P^*} (u_2, u_P) \right\}, \quad (13.11)$$

is an adjusted quasi-phi-function for distance constraint $\operatorname{dist}(E_1, E_2) \geq \rho^-$.

A quasi-phi-function for ellipse $E(u_1)$ and the complement of the interior of Ω

Let $E(u_1)$ be an ellipse with variable parameters $u_1 = (x_1, y_1, \theta_1)$, and let Ω be a rectangular container with vertices $p_1 = (0, 0), p_2 = (l, 0), p_3 = (l, w), p_4 = (0, w)$. We denote $\Omega^* = R^2 \setminus \operatorname{int} \Omega$.

Then a quasi-phi-function for E and Ω^* may be defined as

$$\Phi'^{E\Omega^*} \left(u_1, t_1', t_2'\right) = \min \left\{ \varphi_{11} (p_1), \varphi_{11} (p_2), \varphi_{12} (p_3), \varphi_{12} (p_4), \varphi_{21} (p_2), \right.$$

$$\left. \varphi_{21} (p_3), \varphi_{22} (p_1), \varphi_{22} (p_4) \right\}, \quad (13.12)$$

where $t_2' \neq t_1' \in [0, 2\pi]$, $\varphi_{11} = A_1 x + B_1 y + C_1 - 1$, $\varphi_{12} = -A_1 x - B_1 y - C_1 - 1$, $A_1 = \alpha_1 \cdot \cos \theta_1 + \beta_1 \cdot \sin \theta_1$, $B_1 = -\alpha_1 \cdot \sin \theta_1 + \beta_1 \cdot \cos \theta_1$, $\alpha_1 = \frac{\cos t_1'}{a}$, $\beta_1 = \frac{\sin t_1'}{b}$, $C_1 = -A_1 x_1 - B_1 y_1$, $\varphi_{21} = A_2 x + B_2 y + C_2 - 1$, $\varphi_{22} = -A_2 x - B_2 y - C_2 - 1$, $A_2 = \alpha_2 \cdot \cos \theta_2 + \beta_2 \cdot \sin \theta_2$, $B_2 = -\alpha_2 \cdot \sin \theta_2 + \beta_2 \cdot \cos \theta_2$, $C_2 = -A_2 x_2 - B_2 y_2$, $\alpha_2 = \frac{\cos t_2'}{u}$, $\beta_2 = \frac{\sin t_2'}{b}$

Let a minimal allowable distance ρ^- between an ellipse $E(u_1)$ and the frontier of the rectangle Ω be given. Then function

$$\overset{\frown}{\Phi}^{E\Omega^*} \left(u_1, t_1', t_2'\right) = min \left\{ \varphi_{11}\left(p_2^-\right), \varphi_{11}\left(p_2^-\right), \varphi_{12}\left(p_3^-\right), \varphi_{12}\left(p_4^-\right), \right. \tag{13.13}$$
$$\left. \varphi_{21}\left(p_2^-\right), \varphi_{21}\left(p_3^-\right), \varphi_{22}\left(p_1^-\right), \varphi_{22}\left(p_4^-\right) \right\},$$

is an adjusted quasi-phi-function enforcing the distance constraint dist $(E_1, \Omega^*) \geq \rho^-$, where $p_i^-, i = 1, 2, 3, 4$ are vertices of region $\Omega^* \oplus C(\rho^-)$, $C(\rho^-)$ is circle of radius ρ^-, i.e. $p_1^- = (\rho^-, \rho^-)$, $p_2^- = (l - \rho^-, \rho^-)$, $p_3^- = (l - \rho^-, w - \rho^-)$, $p_4^- = (\rho^-, w - \rho^-)$, $\oplus$ is a symbol of Minkovski sum [35].

A quasi-phi-function for two spherocones $\overset{\frown}{\mathbb{T}}_1$ *and* $\overset{\frown}{\mathbb{T}}_2$. Further a convex object $\overset{\frown}{\mathbb{T}}$ we call a spherocone, if $\overset{\frown}{\mathbb{T}} = \mathbb{D}_1 \cup \overset{\frown}{\mathbb{T}} \cup \mathbb{D}_2$, where: $\overset{\frown}{\mathbb{T}}$ is a truncated cone of height $2e$, with radius r_1 of the upper base and radius r_2 of the lower base, $r_1 \geq r_2$; $\mathbb{D}_k$ is a spherical segment of sphere $\mathbb{S}_k$ of radius $R_k = \frac{r_k^2 + \varpi_k^2}{2\varpi_k}$, $k = 1, 2$, $\mathbb{D}_1$ is an upper spherical segment of height ϖ_1 and the base radius r_1; $\mathbb{D}_2$ is a lower spherical segment of height ϖ_2 and the base radius r_2.

A quasi-phi-function for spherocones $\overset{\frown}{\mathbb{T}}_1(u_1)$ and $\overset{\frown}{\mathbb{T}}_2(u_2)$ can be derived as

$$\Phi'^{\overset{\frown}{\mathbb{T}}_1 \overset{\frown}{\mathbb{T}}_2}\left(u_1, u_2, u_p\right) = min \left\{ \Phi^{\overset{\frown}{\mathbb{T}}_1 P}\left(u_1, u_p\right), \Phi^{\overset{\frown}{\mathbb{T}}_2 P*}\left(u_2, u_p\right) \right\}, \tag{13.14}$$

where $\Phi^{\overset{\frown}{\mathbb{T}}_1 P}$ and $\Phi^{\overset{\frown}{\mathbb{T}}_2 P*}$ are phi-functions for objects $\overset{\frown}{\mathbb{T}}_1$ and P, and objects $\overset{\frown}{\mathbb{T}}_2$ and $P*$ respectively. Now we define

$$\Phi^{\overset{\frown}{\mathbb{T}}_1 P}\left(u_1, u_p\right) = min \left\{ \Phi^{\mathbb{D}_{11} P}\left(u_1, u_p\right), \Phi^{\mathbb{D}_{12} P}\left(u_1, u_p\right) \right\}, \tag{13.15}$$

$$\Phi^{\overset{\frown}{\mathbb{T}}_2 P*}\left(u_2, u_p\right) = min \left\{ \Phi^{\mathbb{D}_{21} P*}\left(u_2, u_p\right), \Phi^{\mathbb{D}_{22} P*}\left(u_2, u_p\right) \right\}, \tag{13.16}$$

where $\Phi^{\mathbb{D}_{11} P}\left(u_1, u_p\right)$, $\Phi^{\mathbb{D}_{12} P}\left(u_1, u_p\right)$, $\Phi^{\mathbb{D}_{21} P*}\left(u_2, u_p\right)$, $\Phi^{\mathbb{D}_{22} P*}\left(u_2, u_p\right)$ are phi-functions for $\mathbb{D}_{11}$ (or $\mathbb{D}_{12}$) and P, and for $\mathbb{D}_{21}$ (or $\mathbb{D}_{22}$) and $P*$, respectively.

It remains to define a phi-function for a spherical segment $\mathbb{D}(u_1)$ and a half-space $P(u_p)$. This can be done as follows:

$$\Phi^{\mathbb{D}P}\left(u_1, u_p\right) = max \left\{ min \left\{ \rho_1\left(u_1, u_p\right), \rho_3\left(u_1, u_p\right) \right\}, \rho_2\left(u_1, u_p\right) \right\}, \tag{13.17}$$

where $\rho_1\left(u_1, u_p\right) = \psi_p + e\zeta_p - r\sqrt{1 - \zeta_p^2}$, $\rho_2\left(u_1, u_p\right) = \psi_p - R + q\zeta_p$, $\zeta_p = \alpha sin\theta_{y_p} - \beta sin\theta_{x_p} cos\theta_{y_p} + \gamma cos\theta_{x_p} cos\theta_{y_p}$, $q = e + \varpi - R$, $\rho_3\left(u_1, u_p\right) = \psi_p + \left(e + \frac{r^2}{e - q}\right)\zeta_p$.

By analogy, replacing ψ_p by $-\psi_p$, we can derive a phi-function for a spherical segment $\mathbb{D}(u_2)$ and a half-space $P*(u_p)$.

Remark By altering the values of the sizes of $\overset{\frown}{\mathbb{T}}$ we can obtain the following shapes of 3D-objects: spherocylinder $\mathbb{C}$ if $r_1 = r_2$; truncated cone $\mathbb{T}$ if $\varpi_1 = \varpi_2 = 0$;

circular cylinder $\mathbb{C}$ if $r_1 = r_2$ and $\varpi_{1i}^0 = \varpi_{2i}^0 = 0$; cone $\widehat{\mathbb{T}}$ if $r_1 = \varpi_1 = \varpi_2 = 0$; spherical segment $\mathbb{D}$ if $r_2 = \varpi_2 = e = 0$ or $r_1 = \varpi_1 = e = 0$; spherical disk $\mathbb{E}$ if $e = 0$ and $r_1 = r_2$.

Based on the quasi-phi-function for spherocones defined by relations (13.14)–(13.17) we can derive the following quasi-phi-functions:
for spherical segments $\mathbb{D}_1$ and $\mathbb{D}_2$

$$\Phi'^{\,\mathbb{D}_1\mathbb{D}_2}\left(u_1, u_2, u_p\right) = min\left\{\Phi^{\mathbb{D}_1 P}\left(u_1, u_p\right), \Phi^{\mathbb{D}_2 P*}\left(u_2, u_p\right)\right\}; \qquad (13.18)$$

for truncated cones $\overline{\mathbb{T}}_1$ and $\overline{\mathbb{T}}_2$

$$\Phi'^{\,\overline{\mathbb{T}}_1\overline{\mathbb{T}}_2}\left(u_1, u_2, u_p\right) = min\left(\rho_{11}^1\left(u_1, u_p\right), \rho_{21}^1\left(u_1, u_p\right),\right.$$
$$\left.\rho_{11}^2\left(u_2, u_p\right), \rho_{21}^2\left(u_2, u_p\right)\right), \qquad (13.19)$$

where ρ_{1j}^i and ρ_{2j}^i, $i, j = 1, 2$, are defined as ρ_1 and ρ_2 in (13.17);
for cones $\mathbb{T}_1$ and $\mathbb{T}_2$

$$\Phi'^{\,\mathbb{T}_1\mathbb{T}_2}\left(u_1, u_2, u_p\right) = min\left\{\rho_{11}^1\left(u_1, u_p\right), \psi_p\left(\widehat{p}_1\right), \rho_{11}^2\left(u_2, u_p\right), \psi_p\left(\widehat{p}_2\right)\right\}, \quad (13.20)$$

where $\widehat{p}_i = \left(-e_i \cos\theta_{xi} sin\theta_{yi}, e_i \sin\theta_{xi}, e_i cos\theta_{xi} \cos\theta_{yi}\right)$, $i = 1, 2$.

A *quasi phi-function* $\Phi'^{\,\mathbb{C}_1\mathbb{C}_2}$ *for cylinders* $\mathbb{C}_1$ and $\mathbb{C}_2$ may be defined by formula (13.19).

Using (13.5) and (13.16), we define *a quasi phi-function for cuboid* $\mathbb{K}_1$ *and spherocone* $\widehat{\mathbb{T}}_2$ in the form

$$\Phi'^{\,\mathbb{K}_1\mathbb{T}_2}\left(u_1, u_2, u_p\right) = min\left\{\Phi^{\mathbb{K}_1 P}\left(u_1, u_p\right), \Phi^{\mathbb{T}_2 P*}\left(u_2, u_p\right)\right\}.$$

We refer the reader to papers [33] and [34] for details of construction of the quasi-phi-functions mentioned above.

13.4   A Mathematical Model and a General Solution Strategy

We consider here a packing problem in the following setting. Let a collection of objects $O_i \subset R^d$, $i \in \{1, 2, \ldots, n\} = I_n$, $d = 2, 3$, be given. And let Ω denote a rectangle of length l and width w in two-dimensional case, and a cuboid of length l, width w and height η in three-dimensional case. Each of the sizes of Ω may be variable. We denote an objective function by F.

We assemble a complete set of variables for our optimization problem. We denote: the vector of variable sizes of container Ω by u_Ω; the vector of placement parameters of object O_i by u_i, $i \in I_n$; the vector of all additional variables, taken from quasi-phi-functions (13.5)–(13.20), by τ.

Thus, a vector of all our variables can be described as follows: $u = (u_\Omega, u_1, u_2, \ldots, u_n, \tau) \in R^\sigma$, where R^σ denotes the σ-dimensional Euclidean space.

Optimal packing problem. Pack the set of objects O_i, $i \in I_n$, into a given container Ω taking into account distance constraints, such that objective function F will reach its minimal value.

A mathematical model of the *optimal packing problem* may now be stated in the form:

$$\min F(u), \quad \text{s.t.} \quad u \in W \subset R^\sigma \tag{13.21}$$

$$W = \left\{ u \in R^\sigma : \widehat{\Phi}'_{ij} \geq 0, i < j \in I_n, \widehat{\Phi}'_i \geq 0, i \in I_n \right\}, \tag{13.22}$$

where $\widehat{\Phi}'_{ij}$ is an adjusted quasi-phi-function derived for the pair of objects O_i and O_j, taking into account minimal allowable distance ρ^-_{ij}, $\widehat{\Phi}'_i$ is an adjusted (or normalized) phi-function derived for objects O_i and $\Omega^* = R^d \backslash \operatorname{int} \Omega$ (to hold the *containment* constraint), also taking into account minimal allowable distance ρ^-_i. If $\rho^-_{ij} = 0$, then we replace an adjusted quasi-phi-function $\widehat{\Phi}'_{ij}$ by a quasi-phi-function Φ'_{ij} for objects O_i and O_j; as well as an adjusted quasi-phi-function $\widehat{\Phi}'_i$ – by a quasi-phi-function Φ'_i (or a phi-function Φ_i) for objects O_i and Ω^*.

Our problem (13.21)–(13.22) is NP-hard, in general, nonlinear programming problem with nonsmooth functions. The feasible region W defined by (13.22) has a complicated structure: it is, in general, a disconnected set, each connected component of W is multiconnected, the frontier of W is usually made of nonlinear surfaces containing valleys, ravines. A matrix of the inequality system which specifies W is strongly sparse and has a block structure. The feasible region W is specified by a system of nonlinear inequalities with piecewise continuously differentiable functions (quasi-phi-functions or phi-functions), which involve operations of maximum and minimum of smooth functions. This means that the feasible region W can be represented as a finite union of subregions W_s, $s = 1, \ldots, \eta$. Each subregion W_s is described by a system of inequalities with smooth functions. Now we may reduce the problem (13.21)–(13.22) to the following optimization problem:

$$F(u^*) = \min \left\{ F(u^{s*}), s = 1, \ldots, \eta \right\},$$

where $F(u^{s*}) = \min\limits_{u \in W_s} F(u)$ is a nonlinear programming problem with smooth functions.

To solve the problem (13.21)–(13.22) we use the strategy, which employs the following optimization procedures:

1. Generation of a starting point from the feasible region of the problem (13.21)–(13.22). To this aim we use the starting point algorithm (SPA), based on homothetic transformations of geometric objects.
2. Search for a local minimum of the objective function $F(u)$ of problem (13.21)–(13.22) by means of the Local Optimization with Feasible Region Transformation (LOFRT) procedure. The LOFRT procedure considerably reduces the dimension of the optimal packing problem, the number of inequalities in (13.22), as well as, the computational time.
3. Non-exhaustive search of local minima to get "good" local optimal solution of the problem (13.21)–(13.22).

Now we consider two practical problems: (1) packing of a set of ellipses into a rectangular container of minimal area; (2) packing of a set of certain 3D-objects into a cuboid container of minimal height. We use quasi-phi-functions defined in Sect. 13.3 for appropriate pairs of rotating objects in model (13.21)–(13.22) and, following the general solution strategy given above, we describe efficient optimization algorithms based on characteristics of our problems.

13.5   Application of Quasi-Phi-Functions for Optimal Packing of Ellipses

In the subsection we follow work [33]. Suppose a set of ellipses E_i, $i \in I_n$, is given to be placed in a rectangular container $\Omega = \{(x, y) \in R^2 : 0 \leq x \leq l, 0 \leq y \leq w\}$. Each ellipse E_i is defined by its semi-axes a_i and b_i, whose values are fixed. With each ellipse E_i we associate its eigen coordinate system whose origin coincides with the center of the ellipse and the coordinate axes are aligned with the ellipse's axes. In that system the ellipse is described by parametric equations $x = a \cos t$, $y = b \sin t$, $0 \leq t \leq 2\pi$. Continuous ellipse rotations and translation are allowed. In addition, minimal allowable distance ρ_{ij}^- between two ellipses E_i and E_j, as well as between ellipse E_i and the frontier of container Ω may be given.

Optimal ellipse packing problem. Pack the set of ellipses E_i, $i \in I_n$, into a rectangular container Ω of minimal area taking into account distance constraints.

It should be noted that one of the dimensions (l or w) may be fixed.

Our approach, which is based on quasi-phi-functions, is capable of handling precise ellipses (without approximations) and thus finding an exact local optimal solution. The only other method of that sort was developed in [23]. The paper is entirely devoted to the problem of cutting ellipses from a rectangular plate of minimal area. It offers a good overview of related publications. The key idea of [23], just like ours, is to use separating lines to ensure that the ellipses do not overlap with each other. But their implementation of this idea is technically different. For a small number of ellipses they are able to compute a globally optimal solution subject to the finite arithmetic of global solvers at hand. However, for more than 14 ellipses none of the nonlinear programming (NLP) solvers available in GAMS can even compute a locally optimal solution. The authors of [23] develop a heuristic approach, in which the ellipses are added sequentially in a strip of a given width and variable length. The algorithm allows the authors to compute good solutions for up to 100 ellipses.

In order to compare the performance of the two methods, we applied our algorithm to some instances of the ellipse packing problem as used in [23] (see Sect. 13.7.1).

The vector $u = (u_\Omega, u_1, u_2, \ldots, u_n, \tau)$ of all variables in the ellipse packing problem is defined as follows: $u_\Omega = (l, w)$ contains the variable length and width of rectangular container Ω; $u_i = (x_i, y_i, \theta_i)$ contains placement parameters of ellipse E_i, $i \in I_n$; vector of additional variables τ now is defined as follows: $\tau = (t, u_P)$, if minimal allowable distances are specified and $\tau = (t)$, if there are no distance constraints. Here $t = \left(t_1^1, t_2^1, \ldots, t_1^m, t_2^m, t_1'^1, t_2'^1, \ldots, t_1'^n, t_2'^n \right)$, where t_1^k, t_2^k are additional variables for the kth pair of ellipses, according to (13.9), $k = 1, \ldots, m$, $m = \frac{(n-1)n}{2}$, and $t_1'^i, t_2'^i$ are additional variables for each ellipse E_i, $i \in I_n$, according to (13.12). If minimal allowable distances are specified, we have to use adjusted quasi-phi-functions (13.11) and (13.13), instead of quasi-phi-functions (13.9) and (13.12). In that case $u_P = \left(u_P^1, \ldots, u_P^m \right)$, $u_P^k = \left(\theta_P^k, \mu_P^k \right)$.

We define the number of the problem variables $\sigma = 2 + 3n + n(n-1) + 2n = n^2 + 4n + 2$ if there are no distance constraints, and $\sigma = 2 + 3n + 2n(n-1) + 2n = 2n^2 + 3n + 2$ if minimal allowable distances are given.

In mathematical model (13.21)–(13.22) for ellipse packing problem we set: $F(u) = l \cdot w$, $\widehat{\Phi}_{ij}'$ is an adjusted quasi-phi-function (13.11) defined for the pair of ellipses E_i and E_j, taking into account minimal allowable distance ρ_{ij}^-, $\widehat{\Phi}_i'$ is an adjusted quasi-phi-function (13.13) defined for the ellipse E_i and the object Ω^* (to hold the *containment* constraint), taking into account minimal allowable distance ρ_i^-. If $\rho_{ij}^- = 0$ and $\rho_i^- = 0$, we replace an adjusted quasi-phi-function $\widehat{\Phi}_{ij}'$ by a quasi-phi-function Φ_{ij}' defined by (13.9) for each pair of ellipses to enforce the *non-overlapping* constraint and $\widehat{\Phi}_i'$ with quasi-function Φ_i' defined by (13.12) for each ellipse and the domain Ω^* to enforce the *containment* constraint.

Due to the forms of quasi-phi-functions in (13.9)–(13.13), the solution space W is now described by a system of inequalities with smooth functions, therefore problem (13.21)–(13.22) becomes a multiextremal nonlinear programming problem.

We follow here the solution strategy introduced in Sect. 13.4.

It is due to the LOFRT procedure our strategy can process large sets of non-identical ellipses (100 and more, see examples below). The reduction scheme used by our LOFRT algorithm is described below.

13.5.1 Starting Point Algorithm for Optimal Ellipse Packing Problem

In order to find a starting point u^0 that belongs to the feasible region W we apply the following algorithm based on homothetic transformation of ellipses. We assume here that homothetic coefficients h_i are variable provided that $h_i = h$, for $i = 1, 2, \ldots, n$, and $0 \le h \le 1$.

The algorithm consists of the following steps:

1. First we choose starting length and width for the container Ω^0. They must be sufficiently large to allow for a placement of all our ellipses with required distance constraints within Ω^0. For example, we can choose $l^0 = w^0 =$

$$2\sum_{i=1}^{n} a_i + (n-1)\rho^-, \rho^- = \max_{i,j \in I_n} \rho_{ij}^-.$$

2. Then we set $h = h^0 = \delta/\max_i a_i$, where $\delta = 0.01\left(\min_i b_i\right)$.

3. Then we generate randomly, within Ω^0, a set of n non-overlapping equal circles of radius δ with randomly chosen centers (x_i^0, y_i^0), $i \in I_n$.

4. Next we generate, randomly, a set of rotation parameters $\theta_i^0 \in [0, 2\pi)$, $i \in I_n$.

5. Then we find starting values for the additional variables τ^0 by a special optimization procedure that solves auxiliary problems of finding $\max_{u_i' \in R^2} \Phi_{i'}\left(u_i^0, u_i'\right)$

(or $\max_{u_i' \in R^2} \widehat{\Phi'}_i\left(u_i^0, u_i'\right)$) and $\max_{u_{ij}' \in R^2} \Phi_{ij'}\left(u_i^0, u_j^0, u_{ij}'\right)$ (or $\max_{u_{ij}' \in R^4} \widehat{\Phi'}_{ij}\left(u_i^0, u_j^0, u_{ij}'\right)$) for each quasi-phi-function (or, respectively, an adjusted phi-function) that is involved in (13.22), under fixed parameters $u_i = (x_i^0, y_i^0, \theta_i^0, \lambda^0)$ for each ellipse.

To solve the above auxiliary problems we use the following model:

$$\max \kappa, \quad s.t. \; u' \in W_{\kappa}',$$

where $W_{\kappa}' = \left\{(u', \kappa) : \Phi'\left(u^0, u'\right) \ge \kappa\right\}$, $\kappa \in R^1$ is a new auxiliary variable, function $\Phi'(u^0, u')$ may take form of $\Phi_i'(u_i^0, u_i')$ (or $\widehat{\Phi'}_i\left(u_i^0, u_i'\right)$) and $\Phi_{ij}'(u_i^0, u_j^0, u_{ij}')$ (or $\widehat{\Phi'}_{ij}\left(u_i^0, u_j^0, u_{ij}'\right)$), u' is the vector of auxiliary variables and u^0 is the vector of fixed parameters for our quasi-phi-functions (respectively, adjusted phi-functions).

Thus all our quasi-phi-functions (or normalized quasi-phi-functions) at the point $u^0 = \left(l^0, w^0, u_1^0, u_2^0, \ldots, u_n^0, \tau^0\right)$ take non-negative values, where $\tau^0 = \left(t^0\right)$ (or, respectively, $\tau^0 = \left(t^0, u_P^0\right)$).

6. Now we take the starting point u^0 under fixed $l = l^0$ and $w = w^0$, and solve the following optimization problem:

$$\max h, \quad s.t. \; u' \in W', \tag{13.23}$$

$$W' = \left\{ u' \in R^{\sigma+1} : \widehat{\Phi}_{ij}' \geq 0, i<j\in I_n, \widehat{\Phi}_i' \geq 0, i\in I_n, \right.$$

$$\left. l=l^0, w=w^0, 0 \leq h \leq 1 \right\}, \tag{13.24}$$

where $u' = (u, h)$ denotes an extended vector of variables and u denotes the original vector of variables for the problem (13.21)–(13.22).

We note that if an optimal global solution is found, then $h = 1$. The solution automatically respects all the non-overlapping and containment constraints.

Thus, the point $u'^0 = \left(l^0, w^0, u_1'^0, u_2'^0, \ldots, u_n'^0, \tau'^0, 1 \right)$ of global maximum of the problem (13.23)–(13.24) guarantees that point $u^0 = (l^0, w^0, u_1'^0, u_2'^0, \ldots, u_n'^0, \tau'^0)$ belongs to feasible region W of problem (13.21)–(13.22).

It should be noted that our algorithm by construction always finds the global solution of the problem (13.21)–(13.22). It is clear that the optimal solution of the above problem will automatically comply with all the non-overlapping and containment constraints.

7. Lastly, our algorithm returns the vector $u^0 = \left(l^0, w^0, u_1'^0, u_2'^0, \ldots, u_n'^0, \tau'^0 \right)$ as a starting point for a subsequent search for a local minimum of the problem (13.21)–(13.22).

13.5.2 Algorithm of Local Optimization with Feasible Region Transformation in the Optimal Ellipse Packing Problem

Let $u^{(0)} \in W$ be one of the starting points found by the previous method. The main idea of the LOFRT algorithm consists in the following.

First we circumscribe a circle C_i of radius a_i around each ellipse E_i, $i = 1, 2, \ldots, n$. Then for each circle C_i we construct an "individual" rectangular container $\Omega_i \supset C_i \supset E_i$ with equal half-sides of length $a_i + \varepsilon$, $i \in I_n$, so that C_i, E_i and Ω_i have the same center (x_i^0, y_i^0) subject to the sides of Ω_i being parallel to those of Ω. Here ε is a predefined fixed constant.

Further we fix the position of each individual container Ω_i and let the local optimization algorithm move the corresponding ellipse E_i only within the container Ω_i. It is clear that if distance between two individual containers Ω_i and Ω_j exceeds ρ_{ij}^- (i.e. $\Phi^{\frown\Omega_i\Omega_j} \geq 0$), then we do not need to check the distance constraint for the corresponding pair of ellipses E_i and E_j.

The above key idea allows us to extract subregions of our feasible region W of the problem (13.21)–(13.22) at each step of our optimization procedure as follows.

We create an inequality system of additional constraints on the translation vector v_i of each ellipse E_i in the form: $\Phi^{C_i\Omega_i^*} \geq 0$, $i \in I_n$, where $\Phi^{C_i\Omega_i^*} =$

$\min\{-x_i + x_i^0 + \varepsilon, -y_i + y_i^0 + \varepsilon, x_i - x_i^0 + \varepsilon, y_i - y_i^0 + \varepsilon\}$ is the phi-function for the circle C_i and $\Omega_i^* = R^2 \setminus \mathrm{int}\,\Omega_i$.

The inequality $\Phi^{C_i\Omega_i^*} \geq 0$ is equivalent to the system of four linear inequalities $-x_i + x_i^0 + \varepsilon \geq 0, -y_i + y_i^0 + \varepsilon \geq 0, x_i - x_i^0 + \varepsilon \geq 0, y_i - y_i^0 + \varepsilon \geq 0$.

Then we form a new region defined by

$$W_1 = \left\{ u \in R^{\sigma - \sigma_1} : \widehat{\Phi'}_{ij} \geq 0, (i,j) \in \Xi_1, \widehat{\Phi'}_i \geq 0, \Phi^{C_i\Omega_i^*} \geq 0, i \in I_n \right\},$$

where $\Xi_1 = \{(i,j) : \Phi^{\Omega_i\Omega_j} < 0, i < j \in I_n\}$.

In other words, we *delete* from the system, which describes W, such quasi-phi-function inequalities for all pairs of ellipses whose individual containers do not overlap and we *add* additional inequalities $\Phi^{C_i\Omega_i^*} \geq 0$, which describe the containment of the circles C_i in their individual containers Ω_i, $i \in I_n$. Thus, we reduce the number of additional variables by σ_1. Then our algorithm searches for a point of local minimum $u_{w_1}^*$ of the subproblem

$$\min F(u_{w_1}) \quad \text{s.t.} \quad u_{w_1} \in W \subset R^{\sigma - \sigma_1}.$$

When the point $u_{w_1}^*$ is found, it is used to construct a starting point $u^{(1)}$ for the second iteration of our optimization procedure (note that the σ_1 previously deleted additional variables τ_1 have to be redefined by a special procedure used in SPA; see step 5, assuming $h^0 = 1$).

At that iteration we again identify all the pairs of ellipses with non-overlapping individual containers, form the corresponding subregion W_2 (analogously to W_1) and let our algorithm search for a local minimum $u_{w_2}^* \in W_2$. The resulting local minimum $u_{w_2}^*$ is used to construct a starting point $u^{(2)}$ for the third iteration, etc.

We stop our iterative procedure when $F\left(u_{w_k}^*\right) = F\left(u_{w_{k+1}}^*\right)$, where $u_{w_k}^*$ is a point of local minimum of the problem

$$\min F(u_{w_k}) \quad \text{s.t.} \quad u_{w_k} \in W \subset R^{\sigma - \sigma_k},$$

where $W_k = \left\{ u \in R^{\sigma - \sigma_k} : \widehat{\Phi'}_{ij} \geq 0, (i,j) \in \Xi_k, \widehat{\Phi'}_i \geq 0, \Phi^{C_i\Omega_{ki}^*} \geq 0, i \in I_n \right\}$, and $\Xi_k = \{(i,j) : \Phi^{\Omega_{ki}\Omega_{kj}} < 0, i < j \in I_n\}$.

We claim that the point $u^* = u^{(k)*} = \left(u_{w_k}^*, \tau_k\right) \in R^\sigma$ is a point of local minimum of the problem (13.21)–(13.22), where $u_{w_k}^* \in R^{\sigma - \sigma_k}$ is the last point of our iterative procedure and $\tau_k \in R^{\sigma_k}$ is a vector of the previously deleted additional variables (the variables can be redefined by the special procedure used in SPA; see step 5). The assertion comes from the fact that any arrangement of each pair of ellipses E_i and E_j subject to $(i,j) \in \Xi \setminus \Xi_k$ guarantees that there always exists a vector τ_k of additional variables such that $\widehat{\Phi'}_{ij} \geq 0, (i,j) \in \Xi \setminus \Xi_k$ at the point $u^{(k)*}$. Here

$\Xi = \{(i,j), i < j \in I_n\}$. Therefore the values of additional variables of the vector τ_k have no effect on the value of our objective function, i.e. $F\left(u_{w_k}^*\right) = F\left(u^{(k)*}\right)$. That is why, indeed, we do not need to redefine the deleted additional variables of the vector τ_k at the last step of our algorithm.

So, while there are $O(n^2)$ pairs of ellipses in the container, our algorithm may in most cases only actively controls $O(n)$ pairs of ellipses (this depends on the sizes of ellipses and the value of ε), because for each ellipse only its nearest neighbors have to be monitored. Thus our LOFRT algorithm allows us to reduce the problem (13.21)–(13.22) with $O(n^2)$ inequalities and a $O(n^2)$-dimensional solution space W to a sequence of subproblems, each with $O(n)$ inequalities and a $O(n)$-dimensional solution subspace W_k. This reduction is of a paramount importance, since we deal with nonlinear optimization problems.

13.6 An Application of Quasi-Phi-Function for the Optimal Packing of 3D-Objects

In the subsection we follow work [34]. Let $O_i \in \left\{\mathbb{P}, \mathbb{S}, \widehat{\mathbb{T}}, \mathbb{T}, \mathbb{C}, \mathbb{D}, \widehat{\mathbb{C}}, \widehat{\mathbb{T}}, \mathbb{E}\right\}$, $i \in I = \{1, 2, \ldots, n\}$ be a collection of 3D-objects, where $I = \bigcup_{j=1}^{9} I_j$, $\mathbb{P}_i$ is a cuboid, $i \in I_1 = \{1, 2, \ldots, k_1 = n_1\}$; $\mathbb{S}_i$ is a sphere, $i \in I_2$; $\widehat{\mathbb{T}}$ is a cone, $i \in I_3$; $\mathbb{T}_i$ is a truncated cone, $i \in I_4$; $\mathbb{C}_i$ is a straight circular cylinder, $i \in I_5$; $\mathbb{D}_i$ is a spherical segment, $i \in I_6$; $\widehat{\mathbb{C}}_i$ is a spherocylinder $i \in I_7$; $\widehat{\mathbb{T}}_i$ is a spherocone, $i \in I_8$, $\mathbb{E}_i$ is a spherical disk, $i \in I_9$, where $I_j = \{k_{j-1} + 1, k_{j-1} + 2, \ldots, n = k_{j-1} + n_j\}$ for $j = 2, \ldots, 9$.

And let $\Omega = \{(x, y, z) \in R^3 : 0 \le x \le w, 0 \le y \le l, \eta_1 \le z \le \eta_2\}$ be a container of height $\eta = \eta_2 - \eta_1$. We denote container Ω of variable height η by $\Omega(\eta)$.

Optimal 3D-object packing problem. Pack the given set of 3D-objects into container Ω of the minimal height.

We use mathematical model (13.21)–(13.22). Now the components of vector $u = (u_\Omega, q, \tau) \in R^\sigma$ for the optimal 3D-object packing problem take the form:

$$u_\Omega = \eta = (\eta_1, \eta_2) \in R^2; \quad q = (u_1, \ldots, u_n) \in R^m, \ m = 6n_1 + 3n_2 + 5\sum_{j=3}^{9} n_j;$$

$\tau = u_p = \left(u_{p_{12}}, u_{p_{13}}, \ldots, u_{p_{1n}}, \ldots, u_{p_{(n-1)n}}\right)$, where $u_{p_{ij}} \in R^3$, $u_p \in R^{\frac{3n(n-1)}{2}}$. The number of the problem variables is defined as $\sigma = 2 + m + \frac{3n(n-1)}{2}$. We set the objective function: $F(\eta) = \eta_2 - \eta_1$ in problem (13.21)–(13.22). To define the feasible region W we use: quasi-phi-functions derived in Sect. 13.3 (see formulas (13.5), (13.7), (13.14)–(13.20)) for non-overlapping constraints and phi-functions derived in [34] for containment constraints. To solve the problem we follow the general solution strategy introduced in Sect. 13.4.

13.6.1 Starting Point Algorithm for the Optimal 3D-Object Packing Problem

In order to find a starting point u^0 that belongs to the feasible region W we apply the following algorithm based on homothetic transformations of 3D-objects.

The algorithm consists of the following steps:

1. We cover each object O_i by sphere $\mathbb{S}_i$ of minimal radius $\mathfrak{R}_i^0$, assuming that local coordinate systems of O_i and $\mathbb{S}_i$ coincide, $i \in I$. Then we assume that $\mathfrak{R}_i$, $i \in I$, are variable and form a vector $\mathfrak{R} = (\mathfrak{R}_1, \mathfrak{R}_2, \ldots, \mathfrak{R}_n)$.
2. Values of components of vector $\eta^0 = (\eta_1^0, \eta_2^0)$ are chosen such that $O_i \subset \Omega(\eta^0)$ for $i \in I$. We suppose that η_1^0, η_2^0 are constants.
3. We take $\mathfrak{R}_i = 0$, $i \in I$, and generate randomly a vector $v = (v_1, \ldots, v_n)$, so that $v_i \in \Omega(\eta^0)$, $i \in I$. As a result we form a point $X^\circ = (v, \mathfrak{R}) = (v, 0)$.
4. Taken a starting point X° we solve the problem

$$\kappa_1\left(\widehat{\mathfrak{R}}\right) = max\kappa_1(\mathfrak{R}), \quad \text{s.t. } X = (v, \mathfrak{R}) \in W_1 \subset R^{4n}, \tag{13.25}$$

$$W_1 = \left\{ X \in R^{4n} : \Phi_{ij}(v_i, v_j, \mathfrak{R}_i, \mathfrak{R}_j) \geq 0, i < j \in I, \Phi_i(v_i, \mathfrak{R}_i) \geq 0, \right.$$
$$\left. \varphi_i(\mathfrak{R}_i) = \mathfrak{R}_i^0 - \mathfrak{R}_i \geq 0, \mathfrak{R}_i \geq 0, i \in I \right\}, \tag{13.26}$$

where $\kappa_1(\mathfrak{R}) = \sum_{i=1}^{n} \mathfrak{R}_i$, $\Phi_{ij}(v_i, v_j, \mathfrak{R}_i, \mathfrak{R}_j)$ is a phi-function of $\mathbb{S}_i$ and $\mathbb{S}_j$, $\Phi_i(v_i, \mathfrak{R}_i)$ is a phi-function of $\mathbb{S}_i$ and Ω^*. We denote a local minimum point of problem (13.25)–(13.26) by $\widehat{X} = \left(\widehat{v}, \widehat{\mathfrak{R}}\right)$.

5. To construct starting point $u^\bullet \in W$ for problem (13.21)–(13.22): we assume $v^\bullet = \widehat{v}$; generate $\theta_{x_i}^\bullet, \theta_{y_i}^\bullet, \theta_{z_i}^\bullet \in [0, 2\pi]$, $i \in I$, randomly; define vector $u_p^\bullet$. In order to derive components $u_{p_{ij}}^\bullet$ of vector $u_p^\bullet$ we construct separating planes for each pair of spheres $S_i(v_i^\bullet)$ and $S_j(v_j^\bullet)$, $i < j \in I$.

6. If $\kappa_1\left(\widehat{\mathfrak{R}}\right) = \sum_{i=1}^{n} \mathfrak{R}_i^0$, then point $u^\bullet = (\eta^0, q^\bullet, u_p^\bullet) \in W$ is taken as a starting point. If $\kappa_1\left(\widehat{\mathfrak{R}}\right) < \sum_{i=1}^{n} \mathfrak{R}_i^0$, then we use the following optimization procedure to define a starting point $u^\bullet \in W$.

Assuming that each object O_i undergo the homothetic transformations with variable homothetic coefficient h_i, $i \in I$, we solve the problem

$$\kappa_2\left(h^*\right) = max\kappa_2(h), \quad \text{s.t. } u^{'} = (u, h) \in W_2 \subset R^{\sigma+n-2}, \tag{13.27}$$

$$W_2 = \left\{ u' \in R^{\sigma+n-2} : \; \Phi_{ij}{}' \left(u_i, u_j, u_{p_{ij}}, h_i, h_j \right) \geq 0, i < j \in I, \Phi_i \left(u_i, h_i \right) \geq 0, \right.$$

$$\left. \varphi_i \left(h_i \right) = 1 - h_i \geq 0, h_i \geq 0, i \in I \right\}, \tag{13.28}$$

where $\kappa_2(h) = \sum_{i=1}^{n} \rho_i h_i$, ρ_i is a sum of metric characteristics (sizes) of O_i, $i \in I$, $h = (h_1, h_2, \ldots, h_n) \in R^n$. We denote a local maximum point of problem (13.27)–(13.28) by $u'^* = (u^*, h^*)$ and take point $u^{\bullet} = \left(\eta^0, q^*, u_p^* \right) \in W$ as a starting point for problem (13.21)–(13.22).

13.6.2 Algorithm of Local Optimization with Feasible Region Transformation in the Optimal 3D-Object Packing Problem

Local optimization. Taking a starting point $u^{\bullet} \in W$, we can extract from the system of phi-inequalities in (13.22) a system of inequalities, describing a nonempty subregion of feasible region W of problem (13.21)–(13.22) and search for a local minimum of the problem. However in this case we deal with a huge number of inequalities in the system. We propose here the algorithm, which reduces the problem (13.21)–(13.22) to a sequence of nonlinear programming subproblems of smaller dimensions. The solution space of each subproblem is specified by the incomparably smaller number of inequalities. This allows us to decrease essentially the computational time. Our algorithm is based on two related ideas: constructing of subregions of feasible region W and decreasing of the number of inequalities, specifying the subregions. The first idea is described in, e.g., [31] and the second idea is introduced in Sect. 13.5 for the optimal ellipse packing problem.

Transition from a local minimum point to another one. Let u^{0*} be a local minimum point of problem (13.21)–(13.22). In order to obtain next local minimum point $u^{1*} \neq u^{0*}$ of problem (13.21)–(13.22) we may generate a new starting point $u^{\bullet} \in W$ for problem (13.21)–(13.22) (see Sect. 13.6.1) and solve the problem using the local optimization algorithm mentioned above.

The other way is to apply a special algorithm to transit from the local minimum point u^{0*} to a local minimum point u^{1*} so that $F \left(\eta^{1*} \right) < F \left(\eta^{0*} \right)$. Let us consider the algorithm.

We solve the following problem:

$$F \left(\eta^* \right) = \min F \left(\eta \right), \quad \text{s.t.} \;\; u'' \in W_3 \subset R^{\sigma+n}, \tag{13.29}$$

$$W_3 = \left\{ u'' \in R^{\sigma+n} : \Phi'_{ij}\left(u_i, u_j, u_{p_{ij}}, h_i, h_j\right) \geq 0, i < j \in I, \right.$$

$$\left. \Phi_i\left(u_i, \eta, h_i\right) \geq 0, h_i \geq 0, i \in I, \sum_{i=1}^{n} V_i h_i^3 - \sum_{i=1}^{n} V_i \geq 0 \right\}, \tag{13.30}$$

where V_i is a volume of $O_i, i \in I$. Here components of vector η are variable.

Then we assume that $h_i^0 = 1, i \in I$. Let u^{0*} be a local minimum point of problem (13.21)–(13.22). We form a point $u''^{0*} = \left(u^{0*}, 1\right)$ and compute the steepest descent vector Z^0 at the point u''^{0*} for problem (13.29)–(13.30), using the iterative procedure

$$u''^k = u''^{0*} + 0.5^{k-1} Z^0, \quad k \in M = \{1, 2, \dots\}.$$

If $h_i^k > 1, i \in J_1 \subset I$, then the appropriate object is expanded and, therefore, a free space around the true object occurs; if $h_i^k < 1, i \in J_2$, the appropriate object is shrunk.

It is evident that $F\left(\eta^k\right) < F\left(\eta^{0*}\right)$ for any $k \in M$ and, in the general case, $u''^k \notin W_3$. This allows us to define m such that: if $k \geq m$, then $u''^k \in W_3$.

Assuming $k = m$, we take point u'''^m and define point $u'''^m = \left(u^m, h^{0m}\right)$, where $h^{0m} = \left(h_1^{0m}, h_2^{0m}, \dots, h_n^{0m}\right)$ and $h_i^{0m} = 1$, if $i \in I \backslash J_2, h_i^{0m} = h_i^m$, if $i \in J_2$. Whence,

$$\sum_{i=1}^{n} V_i \left(h_i^{0m}\right)^3 - \sum_{i=1}^{n} V_i < 0 \text{ if } J_2 \neq \{\varnothing\}. \text{ Let } \eta = \eta^m, \text{ i.e. } F\left(\eta^m\right) < F\left(\eta^{0*}\right).$$

Then we try to "change over" objects of collections $\left\{O_i\left(u_i^m, h_i^{0m}\right), i \in J_1\right\}$, and $\left\{O_i\left(u_i^m, h_i^{0m}\right), i \in J_2\right\}$, so that the value of $\kappa_2(h)$ in (13.27)–(13.28) increases with respect to point u'^m.

For the sake of simplicity, we assume that each object O_i is covered by a circular cylinder $\mathbb{C}_i \supseteq O_i, i \in I$. Taking point u'^m, we generate a point $\tilde{u}'$ as follows.

First we form index subsets $J_{11} \subset J_1$ and $J_{22} \subset J_2$ for which

$$r_j^0 h_j^m < r_i^0 h_i^m, e_j^0 h_j^m < e_i^0 h_i^m, r_i^0 \leq r_j^0 h_j^m, e_i^0 \leq e_j^0 h_j^m, i \in J_1, j \in J_2. \tag{13.31}$$

Then we set $\tilde{h}_i = 1, \tilde{u}_i = u_j^m, \tilde{u}_j = u_i^m, \tilde{h}_j = \min\left\{h_j^m + \varepsilon_j, 1\right\}, \varepsilon_j = \min\left\{\varepsilon_{1j}, \varepsilon_{2j}\right\}$, $\varepsilon_{1j} = \frac{r_i^0 h_i^m}{r_j^0} - h_j^m, \varepsilon_{2j} = \frac{e_i^0 h_i^m}{e_j^0} - h_j^m$.

In order to find the values of components of vector $\tilde{u}_p = \left(\tilde{u}_{p_{12}}, \tilde{u}_{p_{13}}, \dots, \tilde{u}_{p_{1n}}, \dots, \tilde{u}_{p_{it}}, \dots, \tilde{u}_{p_{jk}}, \dots, \tilde{u}_{p_{n(n-1)}}\right)$ we solve the following problems:

$$\max \Phi'_{il}\left(u_i, u_l, \tilde{u}_{p_{it}}, h_i, h_l\right), \quad \text{s.t. } \tilde{u}_{p_{it}} \in R^3 \text{ for } i \in J_{11}, l \in I,$$

$$\max \Phi'_{jk}\left(u_j, u_k, \tilde{u}_{p_{jk}}, h_j, h_k\right), \quad \text{s.t. } \tilde{u}_{p_{jk}} \in R^3 \text{ for } i \in J_{22}, k \in I.$$

If at least one of inequalities (13.31) is not fulfilled for $i \in J_1$ or $j \in J_2$, then we set $\tilde{h}_i = h_i^{0m}, \tilde{u}_i = u_i^{0m}, \tilde{h}_j = h_j^{0m}, \tilde{u}_j = u_j^{0m}, \tilde{u}_{p_{ij}} = u_{p_{ij}}^{0m}$. Note that if $\tilde{u}' \neq u'^m$ then points $\tilde{u}'$ and u'^m are in attraction zones of different local maximum points. We prove in [31] that: if $\tilde{u}' \neq u'^m$ then $\kappa_2(\tilde{h}) > \kappa_2(h^{0m})$.

Starting from point $\tilde{u}' \in W_2$ we can obtain a new local maximum point $\tilde{u}'^*$ of problem (13.27)–(13.28) such that $\kappa_2(\tilde{h}^*) > \kappa_2(\tilde{h})$. If $\kappa_2(\tilde{h}^*) = n$, then $\tilde{u}^* = (\eta^m, \tilde{q}^*, \tilde{u}_p^*) \in W$ and $F(\eta^m) < F(\eta^{0*})$. Since point $\tilde{u}^*$ may not be a local minimum point of problem (13.21)–(13.22), we take the point as a starting point to solve problem (13.21)–(13.22). Then we obtain a local minimum point u^{1*}. Evidently, $F(\eta^{1*}) \leq F(\eta^m) < F(\eta^{0*})$. The approach is described in detail in [31] for optimal packing problem of non-oriented parallelepipeds and spheres.

13.7 Computational Results

Here we present a number of examples to demonstrate the high efficiency of our methodology. We have run our experiments on an AMD Athlon 64 X2 5200+ computer. For local optimization we used the IPOPT code (https://projects.coin-or.org/Ipopt) developed by [36].

13.7.1 Examples for the Optimal Ellipse Packing Problem

First we give a new benchmark instances. We set the computational time limit for each example to search for at least 10 local minima. For our computational experiments we take $\varepsilon = \sum_{i=1}^{n} b_i/n$.

Example 1 $n = 28$, $\{(a_i, b_i) = (2.2, 1.80), i = 1, \ldots, 7\}$, $\{(a_i, b_i) = (2.60, 1.70), i = 8, \ldots, 14\}$, $\{(a_i, b_i) = (3.5, 0.7), i = 15, \ldots, 21\}$, $\{(a_i, b_i) = (3.6, 2.7), i = 22, \ldots, 28\}$. Figure 13.1a shows the packing of ellipses into a rectangular container, which corresponds to the local minimum point u^*. Container has sizes $(l^*, w^*) = (22.273763, 24.126932)$ and area $F(u^*) = 537.397581$.

Figure 13.1b shows the packing of ellipses into a rectangular container taking into account minimal allowable distance ($\rho^- = 0.5$ between each pair of ellipses), which corresponds to the local minimum point u^*. Container has sizes $(l^*, w^*) = (25.984532, 25.024524)$ and area $F(u^*) = 650.250548$. The computational time limit is 1 h.

Example 2 $n = 36$, $\{(a_i, b_i) = (2.2, 1.80), i = 1, \ldots, 9\}$, $\{(a_i, b_i) = (2.60, 1.70), i = 10, \ldots, 18\}$, $\{(a_i, b_i) = (3.5, 0.7), i = 19, \ldots, 27\}$, $\{(a_i, b_i) = (3.6, 2.7), i = 28, \ldots, 36\}$. Figure 13.2a shows the placing of ellipses into a rectangular container,

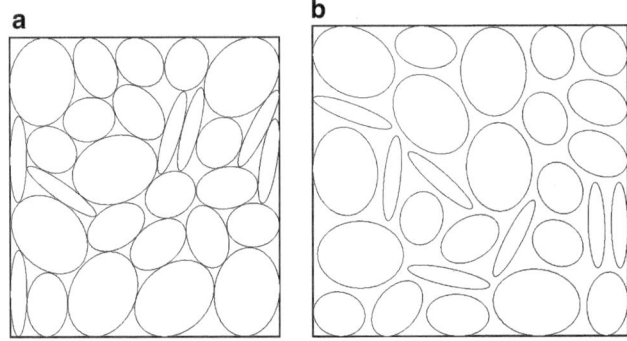

Fig. 13.1 Local optimal packing of ellipses in Example 1: (**a**) no distance constraints, (**b**) with distance constraints

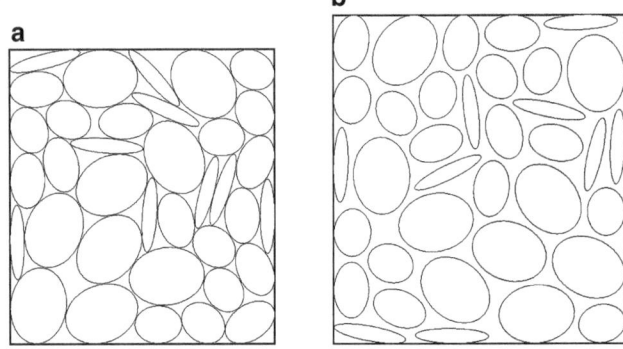

Fig. 13.2 Local optimal placement of ellipses in Example 2: (**a**) no distance constraints, (**b**) with distance constraints

which corresponds to the local minimum point u^*. Container has sizes $(l^*, w^*) = (25.176786, 27.380105)$ and area $F(u^*) = 689.343044$.

Figure 13.2b shows the placing of ellipses into a rectangular container taking into account minimal allowable distance ($\rho^- = 0.5$ between each pair of ellipses), which corresponds to the local minimum point u^*. Container has sizes $(l^*, w^*) = (27.498755, 30.282542)$ and area $F(u^*) = 832.732196$.

Further we give a couple of examples with our records to place a large number of ellipses. Time limit for these large example was set to 48 h.

Example 3 $n = 140$, $\left\{(a_i, b_i) = (222, 180), i = 1, \ldots, 50\right\}$, $\left\{(a_i, b_i) = (260, 170), i = 51, \ldots, 90\right\}$, $\left\{(a_i, b_i) = (350, 70), i = 91, \ldots, 120\right\}$, $\left\{(a_i, b_i) = (360, 270), i - 121, \ldots, 140\right\}$.

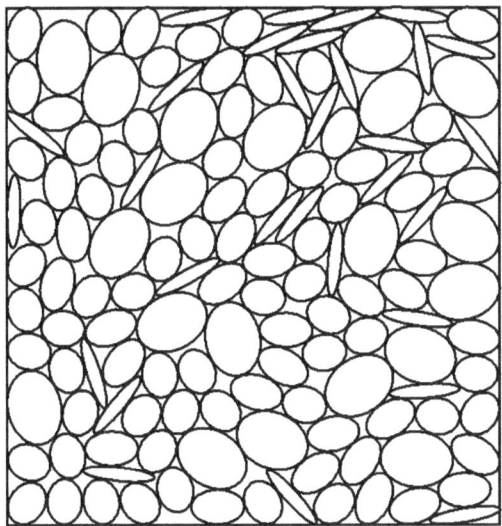

Fig. 13.3 Local optimal packing of ellipses in Example 3

The local optimal ellipse packing is shown in Fig. 13.3, the container has sizes $(l^*, w^*) = (4854.0329, 4970.3722)$ and area $F(u^*) = 24126350.3955$.

Example 4 $n = 150$, $\{(a_i, b_i), i = 1, \ldots, 6 = (2, 1.5, 1.5, 1, 1, 0.8, 0.9, 0.75, 0.8, \}$, $0.6, 0.7, 0.3)$ $\{(a_i, b_i) = (1, 0.8), i = 7, \ldots, 50\}$, $\{(a_i, b_i), i = 51, \ldots, 56 = (2, 1.5, 1.5, 1, 1, 0.8, 0.9, 0.75, 0.8, 0.6, 0.7, 0.3)\}$, $\{(a_i, b_i) = (1, 0.8), i = 57, \ldots, 100\}$, $\{(a_i, b_i), i = 101, \ldots, 106 = (2, 1.5, 1.5, 1, 1, 0.8, 0.9, 0.75, \}$, $0.8, 0.6, 0.7, 0.3)$ $\{(a_i, b_i) = (1, 0.8), i = 107, \ldots, 150\}$.

The local optimal packing is shown in Fig. 13.4, the container has sizes $(l^*, w^*) = (19.865110, 22.839405)$ and area $F(u^*) = 453.70729$. Time limit is 48 h.

We applied our method to some instances used in paper [23] and compare our local optimal solutions to theirs. Table 13.1 lists the examples. For each example the minimal area of the container found by our method happens to be smaller than the best solution reported in [23]. The improvement is not so big (1–2 %) for smaller sets of ellipses, but it becomes significant (8–9 %) for larger sets of ellipses. It should be noted that for examples TC02, TC03, and TC04 presented in [23] our method found the same optimal results.

We set the computational time for the group of instances: up to 20 objects—time limit 2 h, up to 50—time limit 5 h, 100 objects—time limit 12 h.

Our ellipse packing instances are available at https://app.box.com/s/mo7xjvjve7v 52p9movfi.

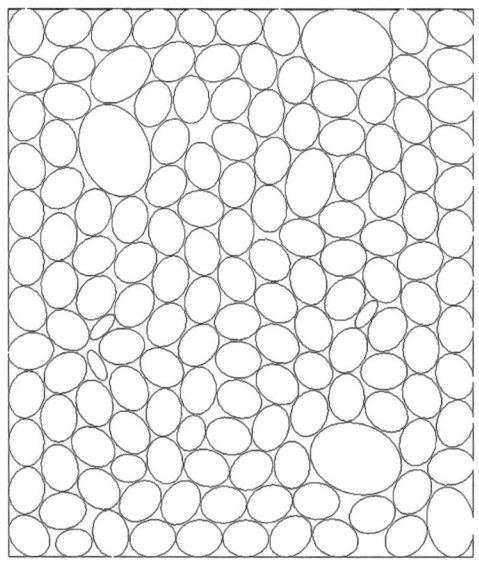

Fig. 13.4 Local optimal packing of ellipses in Example 4

Table 13.1 Comparison of our results to those in [23]

Number of ellipses	Name of instance	Our result	The best result from [23]	Improvement (%)
5	TC05a	25.0206	25.29557	1.0990
5	TC05b	30.84870	31.28873	1.4264
6	TC06	25.47173	25.51043	0.1520
11	TC11	57.1783	57.24034	0.1085
14	TC14	24.25099	24.84634	2.4550
20	TC20	66.13647	67.83459	2.5676
30	TC30	95.36535	103.45212	8.4798
50	TC50	154.47048	166.91505	8.0563
100	TC100	297.73798	322.64663	8.3660

13.7.2 Examples for the Optimal 3D-Object Packing Problem

Example 5 $n = 10, w = 70$ and $l = 70$. Types and sizes of 3D-objects are presented in Table 13.2.

AQ1   Figure 13.5 shows a local optimal packing of 3D-objects.

Placement parameters of objects are given in Table 13.3. Container has height $F(\eta^*) = 26, 192$.

Below we give new nine benchmark instances: packings of 3D-objects (from 10 to 200). The input and output data for the instances are available at http://www.datafilehost.com/d/55384293.

Table 13.2 Types and sizes of 3D-objects

i	Type	e_i	r_{1i}	r_{2i}	$\overline{\omega}_{1i}$	$\overline{\omega}_{2i}$	l_i	w_i	g_i
1	Cuboid	–	–	–	–	–	11.45	5.547	4.133
2	Sphere	–	8.387	–	–	–	–	–	–
3	Cone	8.691	8.823	–	–	–	–	–	–
4	Truncated cone	8.608	9.008	4.124	–	–	–	–	–
5	Cylinder	5.175	8.102	8.102	–	–	–	–	–
6	Segment	–	9.452	–	3.193	–	–	–	–
7	Spherocylinder	8.344	5.376	5.376	5.322	3.295	–	–	–
8	Spherocylinder	7.644	7.822	7.822	7.014	2.281	–	–	–
9	Spherocone	6.6	7.037	6.899	4.513	4.19	–	–	–
10	Disk	–	8.597	8.597	2.696	4.202	–	–	–

Fig. 13.5 Packing of 3D-objects in Example 5

Table 13.3 Placement parameters of 3D-objects in Example 5

i	Type	x_i	y_i	z_i	θ_{x_i}	θ_{y_i}	θ_{z_i}
1	Cuboid	8.744	7.487	−8.96	−3.14	0	5.347
2	Sphere	11.61	−11.6	−4.71	–	–	–
3	Cone	−6.2	−11.1	−4.4	1.556	0.005	–
4	Truncated cone	9.524	5.837	6.085	−0.46	1.807	–
5	Cylinder	1.454	−6.3	3.571	−3.32	0.853	–
6	Segment	11.18	−11.5	7.016	0.544	0.428	–
7	Spherocylinder	−14.6	−1.37	−7.72	1.571	3.142	–
8	Spherocylinder	−5.84	12.18	0.657	0	5.451	–
9	Spherocone	−13	−6.64	5.648	4.625	0.01	–
10	Disk	11.41	15.72	4.593	1.719	−0.05	–

Figure 13.6 illustrates local optimal packings of 3D-objects into a cuboid container of minimal height.

The number and types of the objects are given in Table 13.4.

Figure 13.7 demonstrates a diagram of the dependence of the computational time on the number of objects to be packed.

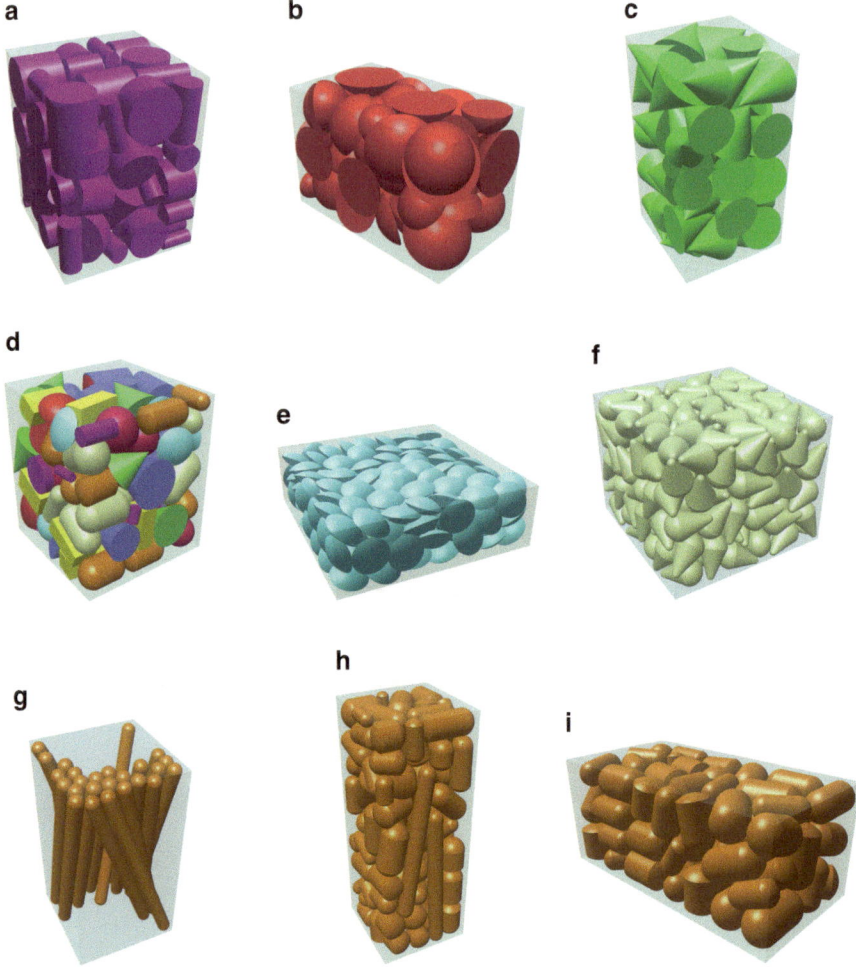

Fig. 13.6 Local optimal 3D-packings: new nine instances (**a–i**)

13.8 Conclusions

In this chapter we introduce new functions, *quasi-phi-functions*, which we use for analytical description of non-overlapping, containment, and distance constraints. We employ the function for extended class of 2D- and 3D-objects, involving new shapes of objects, such as ellipses, spherocones, and spherocylinders for which phi-functions could not be constructed. In addition, these functions (in common with phi-functions) take into account continuous translations and rotations of objects as well as variable sizes of objects. Our quasi-phi-functions are defined by simple enough formulas, which allow us to use nonlinear programming. We propose

Table 13.4 The number and types of 3D-objects

	a	b	c	d	e	f	g	h	i
Cuboid	–	–	–	11	–	–	–	–	–
Sphere	–	–	–	11	–	–	–	–	–
Cone	–	–	60	11	–	–	–	–	–
Truncated cone	–	–	–	11	–	–	–	–	–
Cylinder	60	–	–	11	–	–	–	–	–
Segment	–	60	–	11	–	–	–	–	–
Spherocylinder	–	–	–	12	–	–	25	100	80
Spherocone	–	–	–	11	–	200	–	–	–
Disk	–	–	–	11	200	–	–	–	–

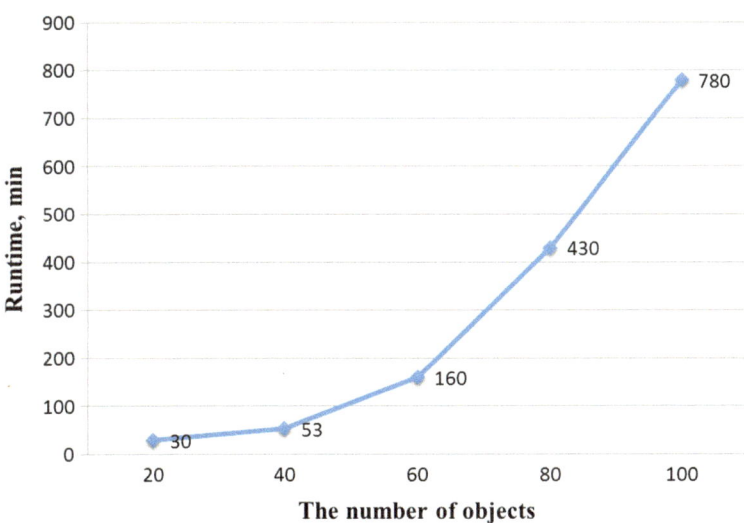

Fig. 13.7 Dependence of the computational time on the number of objects

also fast algorithms to construct feasible starting points based on object homothetic transformations, as well as efficient optimization procedures to search for local extrema in optimal packing problems. We apply our quasi-phi-functions and the algorithms to 2D- and 3D-packing problems and demonstrate the high efficiency of our methodology.

References

1. Wascher, G., Hauner, H., Schumann, H.: An improved typology of cutting and packing problems. Eur. J. Oper. Res. **183**(3), 1109–1130 (2007)
2. Chazelle, B., Edelsbrunner, H., Guibas, L.J.: The complexity of cutting complexes. Discrete Comput. Geom. **4**(2), 139–181 (1989)
3. Aladahalli, C., Cagan, J., Shimada, K.: Objective function effect based pattern search – theoretical framework inspired by 3D component layout. J. Mech. Des. **129**, 243–254 (2007)
4. Burke, E., Hellier, R., Kendall, G., Whitwell, G.: Irregular packing using the line and arc no-fit polygon. Oper. Res. **58**(4), 948–970 (2010)
5. Cagan, J., Shimada, K., Yin, S.: A survey of computational approaches to three-dimensional layout problems. Comput.-Aided Des. **34**, 597–611 (2002)
6. Costa, M.T., Gomes, A.M., Oliveira, J.F.: Heuristic approaches to large-scale periodic packing of irregular shapes on a rectangular sheet. Eur. J. Oper. Res. **192**, 29–40 (2009)
7. Egeblad, J.: Heuristics for multidimensional packing problems. PhD Thesis (2008)
8. Egeblad, J., Nielsen, B.K., Odgaard, A.: Fast neighborhood search for two- and three-dimensional nesting problems. Eur. J. Oper. Res. **183**(3), 1249–1266 (2007)
9. Fasano, G.: MIP-based heuristic for non-standard 3D-packing problems. 4OR: Quart. J. Belgian, French and Italian Oper. Res. Soc. **6**(3), 291–310 (2008)
10. Gan, M., Gopinathan, N., Jia, X., Williams, R.A.: Predicting packing characteristics of particles of arbitrary shapes. KONA **22**, 82–93 (2004)
11. Hifi, M., M'Hallah, R.: A literature review on circle and sphere packing problems: models and methodologies. Adv. Oper. Res. (2009). doi:10.1155/2009/150624
12. Jia, X., Gan, M., Williams, R.A., Rhodes, D.: Validation of a digital packing algorithm in predicting powder packing densities. Powder Technol. **174**, 10–13 (2007)
13. Korte, A.C.J., Brouwers, H.J.H.: Random packing of digitized particles. Powder Technol. **233**, 319–324 (2013)
14. Li, S.X., Zhao, J.: Sphere assembly model and relaxation algorithm for packing of non-spherical particles. Chin. J. Comp. Phys. **26**(3), 167–173 (2009)
15. Li, S.X., Zhao, J., Lu, P., Xie, Y.: Maximum packing densities of basic 3D objects. Chin. Sci. Bull. **55**(2), 114–119 (2010)
16. Sriramya, P., Varthini, P.B.: A state-of-the-art review of bin packing techniques. Eur. J. Sci. Res. **86**(3), 360–364 (2012)
17. Birgin, E.G., Martinez, J.M., Nishihara, F.H., Ronconi, D.P.: Orthogonal packing of rectangular items within arbitrary convex regions by nonlinear optimization. Comput. Oper. Res. **33**, 3535–3548 (2006)
18. Birgin, E., Martínez, J., Ronconi, D.: Optimizing the packing of cylinders into a rectangular container: a nonlinear approach. Eur. J. Oper. Res. **160**(1), 19–33 (2005)
19. Egeblad, J., Nielsen, B.K., Brazil, M.: Translational packing of arbitrary polytopes. Comput. Geom. **42**(4), 269–288 (2009)
20. Fasano, G.A.: Global optimization point of view for non-standard packing problems. J. Glob. Optim. **55**(2), 279–299 (2013)
21. Gomes, A.M., Oliveira, J.F.: Solving irregular strip packing problems by hybridising simulated annealing and linear programming. Eur. J. Oper. Res. **171**, 811–829 (2006)
22. Kallrath, J.: Cutting circles and polygons from area-minimizing rectangles. J. Glob. Optim. **43**, 299–328 (2009)
23. Kallrath, J., Rebennack, S.: Cutting ellipses from area-minimizing rectangles. J. Glob. Optim. **59**(2-3), 405–437 (2014)
24. Petrov, M.S., Gaidukov, V.V., Kadushnikov, R.M.: Numerical method for modelling the microstructure of granular materials. Powder Metall. Met. Ceram. **43**(7-8), 330–335 (2004)
25. Torquato, S., Jiao, Y.: Dense polyhedral packings: Platonic and Archimedean solids. Phys. Rev. **80**, 041104 (2009)

26. Bennell, J.A., Scheithauer, G., Stoyan, Y., Romanova, T.: Tools of mathematical modelling of arbitrary object packing problems. J. Ann. Oper. Res. **179**(1), 343–368 (2010)
27. Chernov, N., Stoyan, Y., Romanova, T.: Mathematical model and efficient algorithms for object packing problem. Comput. Geom.: Theory Appl. **43**(5), 535–553 (2010)
28. Stoyan, Y., Romanova, T.: Mathematical models of placement optimisation: two- and three-dimensional. In: Fasano, G., Pintér, J. (eds.) Modeling and Optimization in Space Engineering, pp. 363–388. Springer, Problems and Applications, New York (2013)
29. Chernov, N., Stoyan, Y., Romanova, T., Pankratov, A.: Phi-functions for 2D-objects formed by line segments and circular arcs. Adv. Oper. Res. (2012). doi:10.1155/2012/346358
30. Bennell, J.A., Scheithauer, G., Stoyan, Y., Romanova, T., Pankratov, A.: Optimal clustering of a pair of irregular objects. J. Glob. Optim. (2014). doi:10.1007/S10898-014-0192-0
31. Stoyan, Y., Chugay, A.: Packing different cuboids with rotations and spheres into a cuboid. Adv. Decis. Sci. (2014). doi:10.1155/2014/571743
32. Stoyan, Y., Chugay, A.: Mathematical modeling of the interaction of non-oriented convex polytopes. Cyber. Syst. Anal. **48**(6), 837–845 (2012)
33. Chernov, N., Stoyan, Y., Pankratov, A., Romanova, T.: Quasi-phi-functions and optimal packing of ellipses. J. Glob. Optim. (2014, submitted)
34. Semkin, V., Chugay, A.: Placement of non-oriented 3D objects taking into account a given minimal allowable distances. Artif. Intell. (Ukraine) **2**, 39–44 (2014, in Russian)
35. Minkovski, H.: Dichteste gitterformige. Lagerung, in Nachr. Ges. Wiss., Gottingen (1904)
36. Wachter, A., Biegler, L.T.: On the implementation of an interior-point filter line-search algorithm for large-scale nonlinear programming. Math. Program. **106**(1), 25–57 (2006)

AUTHOR QUERIES

AQ1. Note that "Figs. 13.6, 13.7, and 13.8" have been changed to "Figs. 13.5, 13.6, and 13.7" in captions and citations. Please check.
AQ2. The decimal comma has been changed to a decimal point in Tables 13.2 and 13.3. Please check, and correct if necessary.
AQ3. Please update the Refs. [33].

Chapter 14
Graph Coloring Models and Metaheuristics for Packing Applications

Nicolas Zufferey

Abstract On the one hand, in the famous graph coloring problem, each vertex of the considered graph has to get a single color. If two vertices are connected with an edge, then their colors have to be different. The goal consists in coloring the graph with the smallest number of colors. On the other hand, consider the packing problem where items have to be loaded in a container. For each item, we have to decide in which container it will be assigned. As some pairs of items are incompatible, they cannot be loaded in the same container. The goal is to load all the items in a minimum number of containers. Even if the correspondence between these two problems is obvious (a vertex is an item, a color is a container, and an edge represents an incompatibility), there is no obvious bridge between the packing and the graph coloring literatures. In this chapter, some packing problems will be modeled and solved with graph coloring models and methods.

Keywords Graph coloring • Packing with incompatibilities • Metaheuristics

14.1 Introduction

On the one hand, consider the problem *PACK* where n items have to be packed. For each item, we have to decide in which container (of a boat) it will be assigned. The $m \geq n$ containers are assumed to be identical. However, for some reasons (security, volume, weight, etc.), some pairs of items are *incompatible*, as they cannot be loaded in the same container. The goal is to load all the items in a minimum number of containers.

On the other hand, consider the classical *graph coloring* problem denoted *COL*. $G = (V, E)$ is a graph where $V = \{1, \ldots, n\}$ is the set of n vertices and E is the set of edges. *COL* consists in assigning a color (i.e., an integer between 1 and n) to each vertex in V such that two adjacent vertices have different colors, while

N. Zufferey (✉)

Geneva School of Economics and Management (GSEM), University of Geneva,
Geneva, Switzerland

e-mail: n.zufferey@unige.ch

© Springer International Publishing Switzerland 2015

G. Fasano, J.D. Pintér (eds.), *Optimized Packings with Applications*, Springer Optimization and Its Applications 105, DOI 10.1007/978-3-319-18899-7_14

295

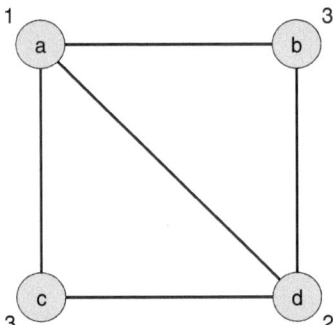

Fig. 14.1 Incompatibility graph representing n = 4 items

minimizing the number of different colors used. *COL* can be used to model *PACK* as follows: a vertex x represents an item x, an edge $[x, y]$ indicates that items x and y are incompatible, and a color t corresponds to a container t. The incompatibility graph in Fig. 14.1 represents the loading of $n = 4$ items a, b, c and d, where items a and d are incompatible with all the other items. The optimal solution shown (where vertex a gets color 1) indicates the loading of item a in container 1, item d in container 2, and items b and c in container 3.

There are numerous metaheuristics for *COL* [23], and some of them will be described in Sects. 14.3 (whose main references include [8, 9, 18]) and 14.4. Three variations of *PACK* will be tackled in this chapter from the graph coloring perspective.

- When the number of containers is limited (i.e., $m < n$), the objective is to minimize the number of unloaded items. This problem is studied in Sect. 14.4, relying on references [2, 30].
- If precedence constraints between specific pairs of items have to be satisfied, the resulting problem can be represented using the *mixed graph coloring* model. This problem is tackled in Sect. 14.5, relying on reference [25].
- Another problem authorizes that incompatible items can be loaded in the same container, but incompatibility costs are encountered. This problem is considered in Sect. 14.6, relying on reference [37].

The objective of this chapter is to study the above packing problems using metaheuristics and graph coloring models and methods. Note that a similar study was conducted in [35] for production scheduling problems. For practical reasons, the packing terminology (e.g., item, container, etc.) and the graph coloring terminology (e.g., vertex, color, etc.) will be indifferently used in this work, which will start by an introduction to metaheuristics in Sect. 14.2.

Even if the focus of this chapter is on what graph coloring models and metaheuristics can bring to packing and container loading problems, the reader interested in metaheuristics for container loading problems could refer to [4, 11, 13, 22, 27, 29, 33].

14.2 Introduction to Metaheuristics

As presented in [34], an *exact method* guarantees the optimality of the provided solution. Among the exact methods are branch-and-bound, dynamic programming, Lagrangian relaxation based methods, and linear and integer programming based methods (e.g., branch-and-cut, branch-and-price, branch-and-cut-and-price). However, for a large number of applications and most real-life optimization problems, such methods need a prohibitive amount of time to find an optimal solution, because such problems are NP-hard [10]. For these difficult problems, one should prefer to quickly find a satisfying solution, which is the goal of heuristic solution methods. A *heuristic* can be defined as an optimization method which produces a satisfying but non-necessarily optimal solution in a reasonable amount of time. The word heuristic is from the Greek and means "to find." The term metaheuristic was first introduced in [14], where "meta" is also from the Greek and means "beyond, in an upper level." Many definitions of metaheuristics can be found in the literature [3]. In [28], it is presented as follows: "a *metaheuristic* is formally defined as an iterative generation process which guides a subordinate heuristic by combining intelligently different concepts for exploring and exploiting the search space, learning strategies are used to structure information in order to find efficiently near-optimal solutions."

There exist several ways of classifying the metaheuristics (e.g., [3, 5, 31, 32]), but the focus will be made on the classification of the heuristics in two classes, namely local search methods and population based methods. A *local search* method starts with an initial solution and tries to improve it iteratively. At each iteration, a modification, called *move*, of the current solution is performed in order to generate a neighbor solution. The definition of a move, i.e. the definition of the neighborhood structure, depends on the considered problem. In contrast, *population based* methods, also called *evolutionary* algorithms, can be defined as iterative procedures that use a central memory where information is collected during the search process. Each iteration, called *generation*, involves of two complementary ingredients: cooperation and self-adaptation. In the *cooperation* effort, the central memory is used to build new offspring solutions, whereas *self-adaptation* consists of individually modifying the offspring solutions. The output solutions of the self-adaptation phase are used for updating the content of the central memory. The most popular local search methods are simulated annealing, tabu search, variable neighborhood search, guided local search, and threshold algorithms, whereas the most used population based methods are genetic algorithms, ant colonies, adaptive memory algorithms, and memetic search which can be seen as a generalization of genetic algorithms, and scatter search. The reader interested in more information on metaheuristics is referred to [12].

Theoretically, there exist some convergence theorems associated with the use of metaheuristics (e.g., [1, 7, 15, 17]). Basically, the theorems state that the search has a high probability to find an optimal solution, but in a very large amount of time, which is likely to be larger than the time needed for a complete enumeration. Therefore, such theoretical results do not have any impact in practice, and, moreover, do not help to efficiently design a metaheuristic. The performance of a metaheuristic can be evaluated according to several criteria. The most relevant criteria are: (1) quality: value of the obtained results, according to a given objective function f; (2) quickness: time needed to get good results; (3) robustness: sensitivity to variations in problem characteristics and data quality; (4) facility of adaptation of the method to a problem; (5) possibility to incorporate properties of the problem (it is widely admitted that an efficient metaheuristic should incorporate knowledge from the considered problem [16]).

14.3   Minimizing the Number of Containers

COL (the classical graph coloring problem) is one of the most studied combinatorial optimization problem, and it has been the focus of many studies. Let $y_c = 1$ if color c is used, and $y_c = 0$ otherwise (for $c \in \{1, \ldots, n\}$). In addition, let $x_{ic} = 1$ if color c is given to vertex i, and 0 otherwise. Thus, the y_c's and the x_{ic}'s are in $\{0, 1\}$. The resulting integer linear program is described below. Constraint (14.1) gives a color to each vertex. Constraint (14.2) is the linking constraint (vertex i gets color c only if c is used). Constraint (14.3) forbids two adjacent vertices to get the same color. Constraints (14.4) and (14.5) are domain constraints.

$$\min \sum_{c=1}^{n} y_c \qquad \text{s.t.} \sum_{c=1}^{n} x_{ic} = 1 \qquad \forall\, i \in \{1, \ldots, n\} \tag{14.1}$$

$$x_{ic} - y_c \leq 0 \qquad \forall\, i, c \in \{1, \ldots, n\} \tag{14.2}$$

$$x_{ic} + x_{jc} \leq 1 \qquad \forall\ \text{edge } [i,j] \in E, \forall\, c \in \{1, \ldots, n\} \tag{14.3}$$

$$0 \leq x_{ic}, y_c \leq 1 \qquad \forall\, i, c \in \{1, \ldots, n\} \tag{14.4}$$

$$x_{ic}, y_c \in \mathbb{Z} \qquad \forall\, i, c \in \{1, \ldots, n\} \tag{14.5}$$

Given that COL is an NP-hard problem [10], exact methods are not appropriate to tackle large instances (above one hundred vertices). It is therefore not surprising that the most efficient algorithms are metaheuristics. The reader is referred to [23] for an accurate literature review.

The most efficient metaheuristics for PACK generally work with a fixed number k of containers. This therefore raises the k-PACK problem, which consists in assigning a container between 1 and k to each item, so as not to generate conflicts (a *conflict* occurs if two incompatible items belong to the same container). If a feasible solution is found (also known as a k-packing without conflict), we restart the process with $k-1$ containers, and so on until the used metaheuristic can no longer find a feasible solution (i.e., without conflict). PACK is therefore tackled by solving a series of k-PACK problems, beginning, for example, with $k = n$, where there is definitely a feasible solution (which consists in assigning a different container to each item).

We first present two metaheuristics for *k-PACK*, namely a tabu search and a hybrid genetic algorithm. Subsequent methods will then be discussed.

14.3.1 Tabu Algorithm

A popular metaheuristic for *k-PACK* is *tabu search*. It is a local search method where a *neighbor* solution s' is generated at each iteration from a *current* solution s by slightly modifying the latter, with respect to a predefined rule. We can therefore say that we generate s' from s by performing a *move*. In order to avoid cycling (i.e., returning to a previously visited solution in the recent past), when a move is performed, its reverse is tabu (forbidden) during *tab* (parameter) iterations. At each iteration, tabu search performs the best possible non tabu move (a tabu move is however allowed if it reaches a solution which is strictly better than all the previously visited ones). The method is stopped, for example, when a predefined time limit is reached, or when an optimal solution is encountered (assuming the value of an optimal solution is known). For further details on tabu search and more generally on metaheuristics, please refer to [12].

The most well-known tabu algorithm for *k-PACK* is *TabuCol*, which was firstly proposed in [18] and then improved in [8]. The best version of *TabuCol* works as follows. A solution is modeled by $s = (C_1, C_2, \ldots, C_k)$, where each C_t contains the set of items loaded in container t. Given that a solution s is a simple partition of all the items into k sets (also known as classes), it may contain conflicts. The objective function f to minimize is therefore the number of conflicts, and the algorithm can thus stop if a solution s such as $f(s) = 0$ is found. A move consists in changing the container of a conflicting item. Let us suppose that to move from the current solution s to the neighbor solution s', the move $(j, C_t, C_{t'})$ is performed, where container t' is assigned to item j instead of container t. It is then tabu (forbidden) to reallocate container t to item j during *tab* iterations, where *tab* depends on the number of conflicts $n_c(s)$ in s. More precisely, $tab = U(0, 9)+0, 6 \cdot n_c(s)$, where $U(a, b)$ returns a randomly selected integer between a and b (bounds included). It is clearly easy to evaluate the quality of a move $(j, C_t, C_{t'})$: it is the number of items in $C_{t'}$ which are incompatible with j minus the number of items in C_t which were incompatible with j. This powerful incremental computation is part of the effectiveness of *TabuCol*. Another fundamental aspect is the fact that the focus is only put on conflicting items (which thus directly contributes to the objective function) during each iteration: the size of the explored neighborhood is therefore drastically reduced.

14.3.2 Hybrid Genetic Algorithm

At the time of publication in 1987, *TabuCol* was the best algorithm for *k-PACK* (and thus for *PACK*). Today, *TabuCol* is still used as an intensification procedure in some evolutionary methods, which are among the best algorithms for this problem

---

Algorithm 1 Hybrid genetic algorithm

While a stopping criterion is not satisfied, do

1. *recombine*: construct an offspring solution $s_{(off)}$ from two parent solutions of P;
2. *intensify*: improve $s_{(off)}$ using a local search, and let s be the resulting solution;
3. *update P*: s replaces a solution of P.

[8, 9, 20, 21]. Such methods are often based on a population P of solutions, which cooperate (information exchange phase) and adapt individually. Some of the best known evolutionary methods include genetic algorithms, ant colonies, scatter search and the adaptive memory algorithm. For further information on such metaheuristics, please refer to [12].

In 1999, a metaheuristic outperformed all the others for *k-PACK*, which is the hybrid genetic algorithm proposed in [8]. Its generic version is presented in Algorithm 1. Steps (1) to (3) constitute a *generation*. The cooperation phase occurs in step (1) with the recombination operator. The individual adaptation phase appears in step (2). Finally, the solution removed from P at the end of a generation is, for example, the worst solution of P (as it is the case in [8]).

The adaptation of the above method to *k-PACK* will be further examined below. Firstly, the search space is the same as the one used by *TabuCol*. The intensification operator is *TabuCol*, and the population P contains ten solutions.

The recombination operator, denoted *X-GH*, constructs an offspring solution $s_{(off)}$ class by class from two parent solutions s_1 and s_2 randomly selected from P. At step t of this operator (with $1 \leq t \leq k$), the offspring solution under construction contains classes $C_1, C_2, \ldots, C_{t-1}$, and the set C_t of items loaded in container t is determined. This set corresponds to the class C in s_i (where $i = 1$ if t is odd and $i = 2$ if t is even, allowing s_1 and s_2 to be used alternately) maximizing the number of additional items which could be added to the solution $s_{(off)}$. In other words, $C \in s_i$ maximizes $g(C') = |C_1 \cup C_2 \cup \ldots \cup C_{t-1} \cup C'|$. When k such steps have been performed, a random container is assigned to each unloaded item. *X-GH* is now illustrated with the set of items $\{a, b, c, \ldots, i, j\}$, and $k = 3$. We therefore want to build $s_{(off)} = (C_1, C_2, C_3)$. We assume that $s_1 = (C_1^{(1)}, C_2^{(1)}, C_3^{(1)}) = (\{a, b, c\}, \{d, e, f, g\}, \{h, i, j\})$ and $s_2 = (C_1^{(2)}, C_2^{(2)}, C_3^{(2)}) = (\{c, d, e, g\}, \{a, f, i\}, \{b, h, j\})$. Since $C_1^{(1)}$ is the largest class of s_1, we have $C_1 = C_2^{(1)} = \{d, e, f, g\}$. We can then remove items d, e, f and g from s_1 and s_2, and we are left with $s_1 = (\{a, b, c\}, \{\}, \{h, i, j\})$ and $s_2 = (\{c\}, \{a, i\}, \{b, h, j\})$. The class $C_3^{(2)}$ is thus the largest in s_2, and we therefore have $C_2 = C_3^{(2)} = \{b, h, j\}$. We then obtain $s_1 = (\{a, c\}, \{\}, \{i\})$ and $s_2 = (\{c\}, \{a, i\}, \{\})$, and finally we have $C_3 = C_1^{(1)} = \{a, c\}$. At the end of this class by class construction phase, the item i has not received any container. We can then randomly insert i in C_3, which leads to $s_{(off)} = (\{d, e, f, g\}, \{b, h, j\}, \{a, c, i\})$.

The superiority of algorithm *GH* strongly relies on its recombination operator, which is obviously not a simple crossover of two parent solutions s_1 and s_2 (taking half of each parent). The transmitted information is therefore not a couple (item, container), as it was the case in genetic algorithms prior to *GH*, but rather the belonging of some items to a common container.

14.3.3   Methods Developed After GH

In [9] is proposed an adaptive memory algorithm whose pseudo-code is described in Algorithm 1. A major difference lies in the recombination operator, where all (and therefore not only two) the solutions of the population P can contribute to construct the offspring solution $s_{(off)}$. The adaptive memory algorithm for *k-PACK* is denoted by *AmaCol*, and its recombination operator is an extension of *X-GH*. All the parent solutions can be used to provide classes to $s_{(off)}$, although it is forbidden for the same solution from P to consecutively provide two classes, so that $s_{(off)}$ does not resemble too much to a solution of P. A numerical comparison between *TabuCol* and *AmaCol* can be found in [9]. It is showed that *AmaCol* and *GH* have a comparable performance, and significantly outperform *TabuCol*. However, *AmaCol* has an advantage over *GH* due to its relative simplicity.

More recently, a method from the same family as *GH* and *AmaCol* has been proposed in [21]. It can also be considered as an adaptive memory algorithm, and it is currently the most efficient approach for *k-PACK*, because its average performance is averagely the best among eleven other well-known algorithms (including *TabuCol*, *GH*, and *AmaCol*). Its recombination operator is also a generalization of *X-GH*. A force of this method is the population update mechanism, which is based on a distance function between solutions. The distance $d(s, s')$ between two solutions s and s' measures the structural difference between s and s'. In other words, the larger is $d(s, s')$, the less s and s' are similar. The idea is then to remove from P the solutions which do not provide very much diversity to P, and to replace them with offspring solutions (which are improved using a procedure similar to *TabuCol*).

14.4   Maximizing the Number of Loaded Items

Let us consider the *k-PACK* problem where the objective is to maximize the number of loaded items if the number of containers is limited to $k < n$. In practical situations, the unloaded items could be loaded later in a different boat, which might result in late deliveries to the final clients.

It Sect. 14.3, we mentioned that the most efficient strategy for tackling *PACK* is to solve a decreasing series of *k-PACK* problems, beginning for example with $k = n$. If k is fixed, a solution can be modeled with $s = (C_1, C_2, \ldots, C_k)$ (it is thus a partition of the items into k classes), and the goal consists in minimizing the number of conflicts (a conflict occurs if two adjacent vertices have the same color). If this number reaches zero, the problem is solved. Another powerful approach to tackle *k-PACK* consists in working with solutions modeled with $s = (C_1, C_2, \ldots, C_k; O)$, which is a partition of the n items into $k + 1$ classes. Each class C_t is conflict-free and contains items loaded in container t. In addition, the set O contains the set of unloaded items and can contain conflicts. The objective function to minimize is simply $|O|$. If this number reaches zero, the problem is solved. As a result, there are mainly two efficient solution spaces for *k-PACK*:

- $\mathscr{E}^{(C)}$ of *k*-complete loadings, where the number of conflicts has to be minimized;
- $\mathscr{E}^{(P)}$ of *k*-partial loadings without conflict, where the number of unloaded items has to be minimized.

An algorithm A_1 associated with $\mathscr{E}^{(C)}$ can therefore be compared to an algorithm A_2 associated with $\mathscr{E}^{(P)}$: we only need to compare the smallest k where A_1 finds a k-packing without conflict, with the smallest k where A_2 finds a k-packing without unloaded items. The literature on $\mathscr{E}^{(P)}$ related algorithms is more recent and has very convincing results [2, 24, 30].

Even if the search spaces $\mathscr{E}^{(C)}$ and $\mathscr{E}^{(P)}$ are associated with the same *k-PACK* (and therefore *PACK*) problem, an important distinction is made in this chapter. Indeed, from a packing perspective, loading n items while minimizing the number of containers is very different from minimizing the number of unloaded items for a given upper bound on the number of containers. The latter problem is particularly relevant when incompatibilities between items are too numerous, so that it is not possible to load all the n items in the set of available containers. In such an environment, the adequate selection of items to load is an important issue, particularly in a situation where each container delivery only occurs after an order from a client.

Two metaheuristics will be discussed below for this problem, namely a tabu search and an ant colony algorithm.

14.4.1 Tabu Search

A powerful tabu search for this problem is *PartialCol*, which works in $\mathscr{E}^{(P)}$ [2]. A move (j, C_t) is completed in two steps: (1) assign a container t (with $t \leq k$) to an item $j \in O$ (i.e., put item j in class C_t); (2) put in O all the items of C_t which are incompatible with j. It is thus very quick to evaluate the value of a move (j, C_t): it is the number of items in C_t which are incompatible with j (as with *TabuCol*, an efficient incremental computation is therefore used). When a move (j, C_t) is performed, it is tabu (forbidden) for *tab* iterations to reinsert in C_t an item which was just removed in the above step (2) (in order to avoid removing from C_t the item j which has just entered it). The value of *tab* is adjusted in an original and efficient way. It depends on the variability Δf of the objective function $f = |O|$ in the last cycle of iterations (a cycle contains several hundred iterations). More specifically, Δf is defined as the gap between the largest and smallest value of f during the last cycle of iterations. The larger Δf is (which indicates that the visited solutions are likely to be different), the smaller *tab* is (which enables a more in-depth exploration of the search space zone under examination). In contrast, the smaller Δf is (which indicates that the algorithm is blocked in a specific zone of the solution space), the larger *tab* is (a large number of forbidden moves favors the exploration of new search space zones). The intensification and diversification phases of *PartialCol* are therefore regulated by the dynamic management of Δf during the search.

A detailed comparison between *TabuCol* and *PartialCol* is presented in [2] for *k-PACK*, when the objective is to find the smallest k such that all the items can be loaded without conflict. It is showed that *PartialCol* is on average slightly better than *TabuCol*. At the time of publication (in 2008), *PartialCol* beat a record on an instance with $n = 300$ items (labeled *flat300280*), using 28 containers (which is optimal), whereas no other algorithm was able to use less 31 containers. As it was the case for *TabuCol*, *PartialCol* has then often been used as an intensification procedure for *k-PACK* in the best evolutionary methods (e.g., an adaptive memory algorithm [24], a variable space search [20], an ant colony algorithm [30]). An unconventional but successful ant algorithm is presented below.

14.4.2   The Ant Colony Algorithm

An ant colony algorithm is generally based on a population of N ants. At each generation, each of the N ants provides a solution. At the end of each generation, a central memory (the *trail system*) is updated. Starting from an "empty" solution, the role of each ant is to build, step by step, a complete solution for the considered problem. At each step, an ant adds an element to the partial solution under construction. Each move (or decision) u is based on two ingredients: (1) the *greedy force* $GF(u)$ (short-term profit of the considered ant) which represents the individual adaptation of each ant, (2) the *trail* $Tr(u)$ (information obtained from other ants) which represents the collaborative phase of the algorithm. U represents all the possible moves at the step being considered. The probability $p_i(u)$ that ant i performs move u is given by Eq. (14.6), where α and β are parameters and $U_i(adm)$ is the set of allowed moves that ant i can perform.

$$p_i(u) = \frac{GF(u)^\alpha \cdot Tr(u)^\beta}{\sum\limits_{u' \in U_i(adm)} GF(u')^\alpha \cdot Tr(u')^\beta} \qquad (14.6)$$

When each ant of the population has built its solution (i.e., at the end of a generation), the trails are generally updated as follows: $Tr(u) = \rho \cdot Tr(u) + \Delta Tr(u)$, $\forall u \in U$, where $\rho \in]0, 1[$ is a parameter (often close to 0.9) representing the evaporation of trails, and $\Delta Tr(u)$ is a term which reinforces the trails left on move u by ants from the very last generation. This quantity is generally proportional to the number of times that the ants have performed move u, as well as the quality of the solutions which have been obtained thanks to move u. More precisely,

---

Algorithm 2 *ALS* algorithm

While a stopping criterion is not satisfied, do

1. for $i = 1$ to N: apply the local search associated with the ant i, and let s_i be the resulting solution;
2. update the trails using a subset of $\{s_1, \ldots, s_N\}$.

$\Delta Tr(u) = \sum_{i=1}^{N} \Delta Tr_i(u)$, where $\Delta Tr_i(u)$ is proportional to the quality of the solution provided by the ant i which performed move u. The reader is referred to [6] for more information on ant algorithms.

However, in order to obtain competitive results, it is often necessary to apply a local search method to the solutions provided by the ants [6]. An alternative proposed in [30] gives a more significant role to each ant. An ant is no longer considered as a constructive algorithm, but rather as a local search procedure. The resulting method is known as *ALS* (for Ant Local Search) and is summarized in Algorithm 2, where each generation requires steps (1) and (2).

In most of the ant algorithms, selecting a move based on Eq. (14.6) is very costly in terms of computational effort. For this reason, a quicker and more efficient technique, also based on the greedy forces and the trails, will be briefly presented below. At each iteration of the local search associated with the ant being considered, A is the set of $|A|$ moves which have the largest greedy forces. The selected move is the one of A which has the largest trail value (ties are broken randomly). One can easily remark that the size of A is an important and sensitive parameter. The advantages of this technique are accurately explained in [30]. Globally, it is obvious that *ALS* is quicker, more efficient, and easier to calibrate and manage than a classical ant colony algorithm.

There are around a dozen ant colony algorithms for *PACK* [19]. The best is *ALS* by far, whose characteristics are summarized below. Firstly, each ant is a tabu search very close to *PartialCol* (see Sect. 14.3). The greedy force $GF(j, C_t)$ of a move (j, C_t) is defined as the opposite of the number of items which are incompatible with j in C_t (if this number is zero, GF is fixed to an arbitrarily large number). The trail $Tr(j, C_t)$ associated with (j, C_t) is defined as follows. Let j and j' be two items, and let $s_i = (C_1, \ldots, C_k; O)$ be a solution provided by the ant i from the population in a specific generation. If the ant i gives the same container t to items j and j' in the solution s_i (i.e., $j, j' \in C_t$), this information must be transmitted to the ants of the next generations, and the intensity of the information must be even more important if j and j' are in a container with a lot of items. During the search, an unloaded item $j \in O$ will tend to be inserted into a container C_t containing the items with which the item j is used to share the same container. More formally, this can be expressed as:

$$\Delta Tr_i(j, j') = \begin{cases} |C_t|^2 & \text{if } j \text{ and } j' \text{ have the same container } t \text{ in } s_i; \\ 0 & \text{if } j \text{ and } j' \text{ do not share the same container in } s_i. \end{cases}$$

Subsequently, as in many ant colony algorithms, at the end of each generation, the trail is reinforced with $\Delta Tr(j,j') = \sum_{i=1}^{N} \Delta Tr_i(j,j')$, and the trails are globally updated as follows: $Tr(j,j') = 0.9 \cdot Tr(j,j') + \Delta Tr(j,j')$. During the tabu search associated with each ant, the trail of a move (j, C_t) is therefore $Tr(j, C_t) = \sum_{j' \in C_t} Tr(j,j')$.

14.5 Precedence Constraints

Let us now consider the *PACK* problem, but including precedence constraints, which has been tackled in [25]. For each item j in the set V of n items to be loaded, we know the set $R(j) \subset V$ of its immediate predecessors. In other words, if $j' \in R(j)$, it indicates that item j' (reps. j) must be loaded in a container i' (resp. i) such that $i' < i$. This kind of precedence constraint is expressed as (j',j). Let us suppose, for example, that $V = \{a,b,c\}$. If we have constraints (a,b) and (b,c), then we have $R(a) = \emptyset, R(b) = \{a\}$ and $R(c) = \{b\}$, and not $R(c) = \{a,b\}$. In other words, a is a non immediate predecessor of c. The objective is to assign a container t to each item j while minimizing the number of containers and satisfying the incompatibility and precedence constraints. Let *PACK-PREC* denote this problem, whose version with a number of containers fixed to k is expressed as k-*PACK-PREC*. As it was the case for *PACK*, *PACK-PREC* can be approached by solving a series of k-*PACK-PREC* problems, beginning, for example, with $k = n$.

From a practical perspective, the label i of each container is associated with its unloading time at the corresponding destination point. If $i' < i$ for containers i and i', it means, for example, that i and i' have to be sequentially unloaded in two different delivery points. In other words, the unloading time of container i' is larger than the one of i.

PACK-PREC can be represented by the *mixed graph coloring model*, denoted *MCOL*. A mixed graph $G = (V,E,A)$ is a graph with a set of vertices V, a set of edges E, and a set of arcs A. By definition, an edge is undirected, and an arc is a directed edge. An edge linking vertices x and y is denoted $[x,y]$, whereas an arc from x to y is expressed as (x,y). *MCOL* consists in giving a color to each vertex in order to minimize the number of different colors used, while satisfying incompatibility (if the edge $[x,y]$ exists in E, the vertices x and y must receive two different colors) and precedence constraints (if the arc (x,y) exists in A, the color of the vertex x must be strictly smaller than the color of the vertex y).

For further information on this problem, please refer to [25]. We can easily see the equivalence between *MCOL* and *PACK-PREC*: a vertex represents an item j, an edge $[j,j']$ indicates that items j and j' are incompatible, an arc (j'',j) corresponds to a precedence constraint, and a color t represents a container t. The incompatibility and precedence constraints in Fig. 14.2 represent the items in the set $\{a,b,c,d,p,s\}$, with the precedence constraints $(s,a),(s,c),(a,d),(b,p)$ and (d,p), as well as the incompatibility constraints $[a,b],[a,c],[c,d]$ and $[b,d]$. The non-optimal solution shown is: $C_1 = \{s\}, C_2 = \{a\}, C_3 = \{b,c\}, C_4 = \{d\}, C_5 = \{p\}$.

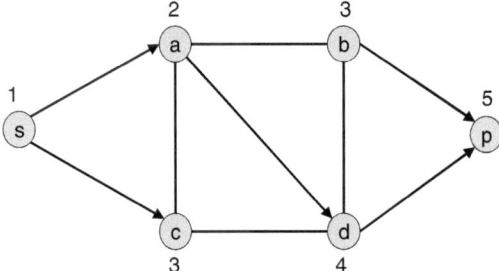

Fig. 14.2 Graph of incompatibilities and precedences representing n = 6 items

As proposed in [25], the integer linear model associated with *MCOL* is the following. Let $G = (V, E, A)$ be a mixed graph with $V = \{v_1, \ldots, v_n\}$, E being the edge set and A being the arc set. Let $C = \{1, \ldots, k\}$ be the set of available colors. For all $i \in \{1, \ldots, n\}$ and $j \in \{1, \ldots, k\}$, $x_{ij} = 1$ if vertex v_i gets color j ($x_{ij} = 0$ otherwise). For all $j \in \{1, \ldots, k\}$, $z_j = 1$ if at least one vertex gets color j ($z_j = 0$ otherwise). The objective function to minimize is $\sum_{i=1}^{k} z_i$, and the constraints to satisfy are:

$$x_{i_1 j} + x_{i_2 j} \leq 1 \quad \forall \, [v_{i_1}, v_{i_2}] \in E, \forall \, j \in \{1, \ldots, k\} \tag{14.7}$$

$$\sum_{j=1}^{k} x_{ij} = 1 \quad \forall \, v_i \in V \tag{14.8}$$

$$x_{ij} \leq z_j \quad \forall \, v_i \in V, \forall \, j \in \{1, \ldots, k\} \tag{14.9}$$

$$x_{i_1 j_1} + x_{i_2 j_2} \leq 1 \quad \forall \, (v_{i_1}, v_{i_2}) \in A, \forall \, j_1 \geq j_2, j_1, j_2 \in \{1, \ldots, k\} \tag{14.10}$$

$$x_{ij}, z_j \in \{0, 1\} \quad \forall \, v_i \in V, \forall \, j \in \{1, \ldots, k\} \tag{14.11}$$

Constraints (14.7) impose that two vertices linked with an edge must get different colors. Constraints (14.8) impose that each vertex must get exactly one color. Constraints (14.9) are linking constraints. Constraints (14.10) forbid to give a larger color to the start vertex of an arc than to the end vertex of an arc. Constraints (14.11) impose integer values for variables x_{ij} and z_j.

As it was the case for *PACK*, *PACK-PREC* is an NP-hard problem and metaheuristics are well suited to tackle it. The two existing metaheuristics for *PACK-PREC* are proposed in [25], namely a tabu search and a variable neighborhood search. These methods have comparable performances and can tackle instances with several hundred of vertices. Note that the above integer linear model based on CPLEX 10.2 is limited to fifty vertices only (and requires several hours of computation). On such small instances, the two above-mentioned metaheuristics can generally find the optimal solution much quicker (only a few minutes are usually necessary).

Before proposing solution methods for k-$PACK$-$PREC$, it is useful to use a technique which reduces the number of possible colors for each vertex j. A path is a series of adjacent arcs $(j_1, j_2), (j_2, j_3), \ldots, (j_{p-2}, j_{p-1}), (j_{p-1}, j_p)$ such that $j_{i_1} \neq j_{i_2}$ if $i_1 \neq i_2$. For example, $(a, b), (b, c)$ is a path but not $(a, b), (c, b)$. Let us suppose that we want to color the mixed graph composed of the path $(a, b), (b, c)$. Using $k = 4$ colors and beginning with an empty solution (no vertex is colored), if we first assign color 4 to vertex a, it is then impossible to find a color in $\{1, 2, 3, 4\}$ for vertices b and c in order to reach a 4-coloring without conflict. As a result, color 4 must never be considered for vertex a. More generally, in [25], it is proposed to reduce the search space as follows. The length of the path from the vertex x to y is the number of arcs belonging to it. $InRank(j)$ (respectively $OutRank(j)$) is the number of vertices belonging to a longest path leading to (respectively starting from) vertex j. It is obvious that k must be larger than the length of the longest path in the graph being considered. Let $FC(x)$ be the set of feasible colors for vertex x. According to the above example (i.e., considering path $(a, b), (b, c)$, with $k = 4$), we have $FC(a) = \{1, 2\}, FC(b) = \{2, 3\}$ and $FC(c) = \{3, 4\}$.

14.5.1   Tabu Search

Tabu search for k-$PACK$-$PREC$ is an extension of $PartialCol$ proposed for k-$PACK$. The following elements must be defined: the way to represent a solution, the neighborhood structure (i.e., the nature of a move), the objective function to be minimized, and the tabu status updating mechanism.

Let t_j denote the container assigned to item j. In this context, there is a conflict between items j and j' if one of the two following conditions is verified: (1) $[j, j'] \in E$ and $t_j = t_{j'}$ (violation of an incompatibility constraint); (2) $(j, j') \in A$ and $t_j \geq t_{j'}$ (violation of a precedence constraint). A solution s can also be modeled by $s = \{C_1, \ldots, C_k; O\}$, where C_t is the set of items loaded in container t (without any conflict). The function $f = |O|$ has to be minimized (all the items which have not received a container are in O). Note, however, that it is not always possible to complete a solution s. Assuming $k = 4$ for example, let us suppose that the considered graph contains a path $(a, b), (b, c)$, as well as a set $\{a, d, e, f\}$ of mutually adjacent vertices, as illustrated in Fig. 14.3. It results that $FC(a) = \{1, 2\}, FC(b) = \{2, 3\}, FC(c) = \{3, 4\}, FC(d) = FC(e) = FC(f) = \{1, 2, 3, 4\}$. In this case, the partial solution $s = \{C_1 = \{e\}, C_2 = \{a\}, C_3 = \{c, d\}, C_4 = \{f\}\}$ does not contain conflicting vertices, but it cannot be completed because it is impossible to find a feasible solution where colors 2 and 3 are, respectively, assigned to vertices a and c.

The neighborhood structure is the same as the one used in $PartialCol$: we assign a color t to an uncolored vertex j, and we then remove the color of the vertices in C_t in conflict with j. When such a move is performed to move from a current solution to a neighbor solution, it is then tabu to remove j from C_t during tab iterations, which is a number depending on the number of conflicts in the current solution. It is obvious that $f = |O|$ can assign the same value to several neighbor solutions, which indicates that there are many equivalent options at each iteration of tabu search.

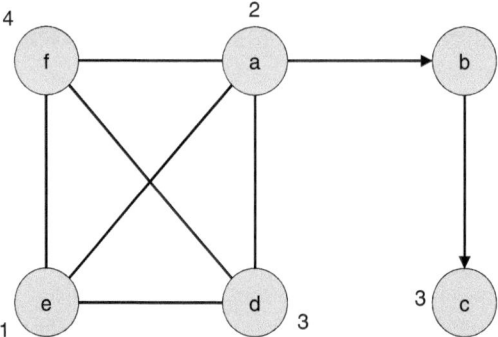

Fig. 14.3 Partial solution without conflict which cannot be completed as a feasible 4-coloring

To break ties, an auxiliary objective function g is used instead of f. Note that a conflict may appear either on an edge (violation of an incompatibility) or on an arc (violation of a precedence). It has been noted that it is better to give different weights to these types of conflict. Consider the solution $s = \{C_1, \ldots, C_k; O\}$, a vertex $j \in O$ and a color $t \in \{1, \ldots, k\}$. We define:

$$A(j, t) = \{j' \in V \mid \exists \text{ edge } [j, j'] \in E \mid t_{j'} = t\}$$

$$B(j, t) = \{j' \in V \mid \{\exists \text{ arc } (j, j') \in A \mid t_{j'} \leq t\} \text{ or } \{\exists \text{ arc } (j', j) \in A \mid t_{j'} \geq t\}\}$$

In other words, $A(j, t)$ (resp. $B(j, t)$) is the set of incompatible items (resp. the set of items involved in some precedence constraints) with j which will become in conflict if the decision $t_j = t$ is taken. At each iteration of tabu search, the function g used to choose among the equivalent options according to f is $g(j, t) = \alpha \cdot |A(j, t)| + \beta \cdot |B(j, t)|$. It has been observed that $\alpha = 4$ and $\beta = 1$ are reasonable values for these parameters. The function g can quickly evaluate a move. Let us for example suppose that, in order to generate s' (neighbor solution) from s (current solution), the vertex j is first moved from O to C_t, then the vertex j_1 is moved from C_t to O due to the violation of an incompatibility constraint, and finally j_2 and j_3 moved from $C_{t'}$ to O due to the violation of precedence constraints. It is easy to evaluate the resulting neighbor solution s' with $g(j, t) = \alpha \cdot 1 + \beta \cdot 2$.

The pseudo-code of such a tabu search is presented in Algorithm 3, which returns the best solution $s^\star$ encountered during the search (its value is $f^\star$).

14.5.2 Variable Neighborhood Search

Usually, a local search methods only use a single type of neighborhood $\mathcal{N}$: a local optimum is therefore defined according to $\mathcal{N}$. To escape from a local optimum, tabu search, for example, relies on the notion of forbidden moves. In contrast, a variable neighborhood search attempts to avoid being trapped in a local optimum by the use of several types of neighborhood: a local optimum for a neighborhood

---

Algorithm 3 Tabu search for *k-PACK-PREC*

Input: set of *n* items, incompatibility and precedence constraints.

Initialization

1. generate an initial solution *s* (at random or by putting all the items in *O*);
2. set $s^* = s$ and $f^* = f(s)$;
3. set *Iter* = 0 (iteration counter).

While a stopping criterion is not satisfied, do

1. update the iteration counter: set *Iter* = *Iter* + 1;
2. generate the set *D* of all the non tabu candidate neighbor solutions by assigning a container to an item $j \in O$, $\forall j$ (exception: *D* can contain tabu solutions if their values are smaller than f^*);
3. let s' be the solution of *D* minimizing *g* (break ties randomly); suppose that s' is obtained from *s* by assigning a container to the item *j*;
4. update the record: if $f(s') < f^*$, set $f^* = f(s')$ and $s^* = s'$;
5. update the tabu status: it is forbidden to reinsert *j* into *O* until iteration *Iter* + *tab*;
6. update the current solution: set $s = s'$.

Output: solution s^* with value f^*.

Algorithm 4 Variable neighborhood search

Input: neighborhood structures $\mathcal{N}^{(i)}$ ($i = i_{min}, \ldots, i_{max}$).

Initialization: generate an initial solution *s* and set $i = i_{min}$.

While a stopping criterion is not satisfied, do

1. generate a solution s' in the ith neighborhood of *s*: $s' \in \mathcal{N}^{(i)}(s)$;
2. apply a local search procedure during *I* iterations, with s' as an initial solution, and let s'' be the resulting solution;
3. if s'' is better than the current solution *s*, set $s = s''$ and continue the search with the first neighborhood $\mathcal{N}^{(i_{min})}$ (i.e. set $i = i_{min}$); otherwise, move to the next neighborhood (i.e. set $i = \max\{i_{min}; (i \bmod i_{max}) + 1\}$).

Output: best solution found during the search.

---

is not necessarily a local optimum for another neighborhood. Let $\mathcal{N}^{(i)}$ (with $i \in \{1, \ldots, i_{max}\}$) be a series of neighborhood structures, where $\mathcal{N}^{(i)}(s)$ is the set of solutions in the ith neighborhood of solution *s*. The variable neighborhood search method, initially proposed in [26], is summarized in Algorithm 4.

For *k-PACK-PREC*, the parameters i_{min}, i_{max} and *I* are, respectively, fixed to 2, 5 and 100,000 in [25]. The local search used at step (2) is the tabu search described above. The different neighborhood structures are now presented, requiring the following additional definitions. Let t_j be the color assigned to the vertex *j*, and let *x* and *y* be two vertices in conflict (which are therefore connected by a conflicting edge or arc). For $i \geq 2$, we say that there is an *i*-conflict between vertices *x* and *y* if at least one of the two following conditions is verified: (1) there is a path with length *i* from *x* to *y* such that $t_x + i > t_y$; (2) there is a path with length *i* from *y* to *x* such that

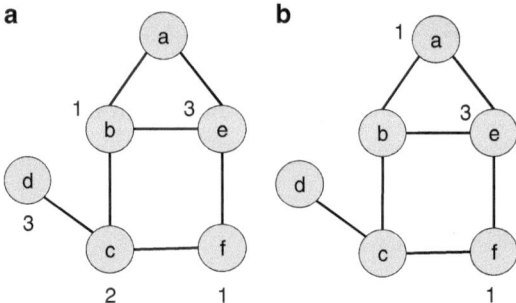

Fig. 14.4 (a) A partial and legal solution s. (b) A neighbor solution $s' \in \mathcal{N}^{(3)}(s)$ obtained by assigning the color 1 to the vertex a

$t_y + i > t_x$. In these two cases, it is impossible to feasibly color the vertices of the involved path. In the neighborhood $\mathcal{N}^{(1)}$, a move consists in assigning a color to an uncolored vertex x, and then to remove the color of the conflicting vertices with x. Such conflicting vertices are necessarily adjacent to x. For $i \geq 2$, we define the neighborhood $\mathcal{N}^{(i)}(s)$ of a current solution s as all the solutions obtained from s by assigning a color to $x \in O$, and then removing the color of all the conflicting vertices and in r-conflict with x (with $2 \leq r \leq i$). The neighborhood $\mathcal{N}^{(3)}(s)$ is illustrated in Fig. 14.4.

14.6 Incompatibility Costs

Let us define a new problem *k-PACK-INC* from *k-PACK* by relaxing the conflict constraints as follows: if two incompatible items j and j' belong to the same container t, an incompatibility cost $c(j,j') = c(j',j)$ is incurred; (2) if an item j is belongs to container t, an assignment cost $a(j,t)$ must be paid. The objective consists in assigning a container to each item in order to minimize the incompatibility and assignment costs. A solution s can thus be denoted $s = (C_1, \ldots, C_k)$, and an incompatibility graph can model this problem: an edge $[j,j']$ between two vertices j and j' indicates that if the same color is given to vertices j and j', then the cost $c(j,j') > 0$ is encountered.

From a practical perspective, the incompatibility cost $c(j,j')$ could be proportional to the risk encountered if two conflicting items belong to the same container. In such a case, a goal would be to minimize the risk. The assignment cost $a(j,t)$ represents, for example, the cost of the resources involved in the loading/unloading of item j in container t. A solution s using k containers can be generated using a function *per* : $V \longrightarrow \{1, \ldots, k\}$ which attributes a container $per(j)$ to each item $j \in V$. The value of a solution $s - (C_1, \ldots, C_k)$ is

$$f(s) = \sum_{t=1}^{k} \sum_{j \in C_t} a(j,t) + \sum_{j=1}^{n-1} \sum_{j' \in \{j+1,...,n\} \cap C_{per(j)}} c(j,j') \qquad (14.12)$$

As proposed in [37], let $x_{jt} = 1$ if container t is assigned to item j ($x_{jt} = 0$ otherwise). Then, we can formulate *k-PACK-INC* as follows.

Objective function: $\min \displaystyle\sum_{t=1}^{k} \sum_{j=1}^{n} a(j,t) \cdot x_{jt} + \sum_{t=1}^{k} \sum_{j=1}^{n-1} \sum_{j'=j+1}^{n} c(j,j') \cdot x_{jt} \cdot x_{j't}$ (14.13)

Constraints: $\displaystyle\sum_{t=1}^{k} x_{jt} = 1, \qquad \forall j \in \{1, \ldots, n\}$ (14.14)

$$x_{jt} \in \{0,1\}, \quad \forall j \in \{1, \ldots, n\}, \forall t \in \{1, \ldots, k\} \qquad (14.15)$$

Equation (14.14) imposes to assign exactly one container to each item. The above formulation can be linearized by using $y_{jj'} = \sum_{t=1}^{k} x_{jt} x_{j't}$ (see the details in [37]). Note that CPLEX 10.0 (used during several hours and the linearized formulation) is not able to optimally solve instances with more than fifty items. For such instances, it has been noted that tabu search performs similarly, but only requires a few minutes.

There are only two metaheuristics for this NP-hard problem: a tabu search and an adaptive memory algorithm [37]. These two methods have comparable performances and can be used for instances with several hundred items. Such algorithms are presented below.

14.6.1   Tabu Search

A move $(j, C_t, C_{t'})$ simply consists in giving container t' instead of container t to item j. However, to avoid evaluating all the possible moves at each iteration, only the most promising moves are examined, which are the ones which contribute the most to the objective function f. To do so, for each item j, its contribution $cost(j)$ to f is computed as follows, where $I(j)$ is the set of items which are incompatible with j:

$$cost(j) = a(j, per(j)) + \frac{1}{2} \sum_{j' \in I(j) \cap C_{per(j)}} c(j,j') \qquad (14.16)$$

Note that the fraction $\frac{1}{2}$ is used to consider the fact that items j and j' contribute equally to $c(j,j')$. As a result, the other half of $c(j,j')$ is taken into account in $cost(j')$.

At each iteration, all the moves involving the modification of a container of an item j are considered, but only for the j's belonging to the $q\%$ (parameter fixed to 40%) most expensive items according to the *cost* function.

In order to save time at each iteration, an incremental computation is used to evaluate a move $(j, C_t, C_{t'})$ associated with the generation of a neighbor solution s' from the current solution s. Rather than computing $f(s')$ from Eq. (14.12), only the variation $\Delta f(s, s') = f(s') - f(s)$ of f is computed as proposed in Eq. (14.17):

$$\Delta f(s, s') = a(j, t') + \sum_{j' \in I(j) \cap C_{t'}} c(j, j') - a(j, t) - \sum_{j' \in I(j) \cap C_t} c(j, j') \qquad (14.17)$$

When a move $(j, C_t, C_{t'})$ is performed, it is tabu to assign container t to item j during *tab* iterations. As a result, at each iteration, the best non-tabu move is performed (a tabu move is, however, allowed if it reaches a solution which is strictly better than all the previously visited ones). The value of *tab* is determined as indicated in Eq. (14.18). The maximum is used to enforce *tab* to be positive. The last term of Eq. (14.18) represents the improvement of the objective function f during the generation of the neighbor solution s' from the current solution s. If s' is better than s, the improvement is positive and the reverse move will be forbidden for a larger number of iterations (when compared to a move with a negative improvement), which is straightforward.

$$tab = \max\left\{1; U(10, 20) + 15 \cdot \frac{f(s) - f(s')}{f(s)}\right\} \qquad (14.18)$$

A more refined management of the tabu status is also used, based on the following idea: if the diversity of the visited solutions is below a specific threshold δ, the value of *tab* must be increased for a couple of iterations in order to favor the exploration of new zones of the solution space (diversification phase). In contrast, if the diversity is larger than δ, the value of *tab* must be reduced for a couple of iterations in order to favor a deeper exploration of the search space in which the current solution lies (intensification phase). Two important points must therefore be considered: (1) how to determine the diversity of the visited solutions, and (2) how to determine the threshold δ. To tackle these two issues, additional definitions are required.

The similarity $sim(s, s')$ between two solutions $s = (C_1, C_2, \ldots, C_k)$ and $s' = (C'_1, C'_2, \ldots, C'_k)$ is defined in Eq. (14.19) (with the following convention: if $\frac{|C_i \cap C'_i|}{|C_i \cup C'_i|} = \frac{0}{0}$, we have $\frac{|C_i \cap C'_i|}{|C_i \cup C'_i|} = 1$, because in this case $C_i = C'_i = \emptyset$):

$$sim(s, s') = \sum_{i=1}^{k} \frac{|C_i \cap C'_i|}{|C_i \cup C'_i|} \qquad (14.19)$$

The distance $d(s, s')$ between two solutions s and s' can therefore be defined by $d(s, s') = k - sim(s, s')$. In addition, the distance $d(s, Z)$ between a solution s and a set Z of solutions can be defined by:

$$d(s, Z) = \frac{\sum_{s' \in Z} d(s, s')}{|Z|} \qquad (14.20)$$

Algorithm 5 Tabu search for *k-PACK-INC*

Input: set of *n* items, assignment and incompatibility costs.

Initialization

1. generate at random an initial solution *s*;
2. set $s^\star = s$ and $f^\star = f(s)$;
3. set *Iter* = 0 (iteration counter).

While a stopping criterion is not satisfied, do

1. update the iteration counter: set *Iter* = *Iter* + 1;
2. determine the set *C* containing the *q* % most costly items [according to Eq. (14.16)];
3. generate the set *B* of non tabu candidate neighbor solutions obtained from *s* by modifying the container of an item $j \in C$ (exception: *B* can contain tabu solutions if their values are lower than $f^\star$);
4. set $s' = arg \min_{s'' \in B} f(s'')$; suppose that s' is generated from *s* by performing the move($j, C_t, C_{t'}$);
5. update the record: if $f(s' < f^\star$, set $f^\star = f(s')$ and $s^\star = s'$;
6. update the tabu status: do not reinsert *j* in C_t until iteration *Iter* + *tab*;
7. update the current solution: set $s = s'$.

Output: solution $s^\star$ with value $f^\star$.

Finally, the diversity $d(Z)$ of set *Z* of solutions is defined as the average distance between two solutions in *Z*:

$$d(Z) = \frac{\sum\limits_{s \in Z} d(s, Z - \{s\})}{|Z|} \qquad (14.21)$$

The value of δ is determined empirically at the start of the search, and thus at a moment where the diversity of the visited solutions is potentially high. Starting with $Z = \emptyset$, from the first time *h* (parameter fixed to 50) iterations without improvements of $s^\star$ (best solution visited during the search) have elapsed, at each cycle of *h* iterations, insert the best solution encountered during the last completed cycle into *Z*. Then, when $|Z| = z$ (parameter fixed to 10), set $\delta = d(Z)$. Then, at each $z \cdot h$ iterations, a new set *Z* of solutions is generated in the same way, and its diversity $d(Z)$ is computed. If $d(Z) < \delta$ (resp. $d(Z) \geq \delta$), the tabu duration *tab* [previously computed with Eq. (14.18)] of each new move is multiplied (resp. divided) by 5. All the ingredients are now available to formulate Algorithm 5.

14.6.2   *Adaptive Memory Method*

Using the same notation as in Sect. 14.3.3, a population *P* of ten solutions is used in [37]. To initialize *P*, ten solutions are generated and improved by tabu search during 1,000 iterations. The intensification operator is also the above tabu search, but used during 10,000 iterations.

The recombination operator is similar to the one proposed in [36], which is a generalization of X-GH. At each generation, an offspring solution $s_{(off)} = \{C_1^{(off)}, \ldots, C_k^{(off)}\}$ is constructed class by class from P. Suppose that the classes $C_1^{(off)}, \ldots, C_{t-1}^{(off)}$ have already been constructed from the set P of parent solutions, and that the parent solution $s_{r'}$ (with $r' \in \{1, \ldots, 10\}$) has provided the items of the class $C_{t-1}^{(off)}$. In addition, S denotes the set of previously assigned items (which are in $C_1^{(off)} \cup \ldots \cup C_{t-1}^{(off)}$). At that moment, it is required to build $C_t^{(off)}$. The items composing $C_t^{(off)}$ are provided by the solution $s_r = \{C_1^{(r)}, \ldots, C_k^{(r)}\}$ from P (with $r \neq r'$, so that the same parent solution cannot consecutively provide two classes) such that $|C_t^{(r)} - S|$ is maximal (break ties randomly). We therefore set $C_t^{(off)} = C_t^{(r)} - S$. At the end of this process, the non-already assigned items are successively inserted in $s_{(off)}$ in a greedy fashion.

The population update mechanism is based on the technique proposed in [36]. Let s be the solution provided by tabu search at the end of a generation. Let $s_{(worst)}$ be the worst solution of P and let $s_{(old)}$ be the oldest solution of P. If s is not worse than $s_{(worst)}$, then s replaces $s_{(worst)}$ in P, otherwise s replaces $s_{(old)}$. In the latter case, given that s cannot improve the quality of P, it is at least able to give it fresh blood (i.e., a bit of diversity) by replacing an old solution.

14.6.3   Variations of the Problem

The above model has the advantage of being able to reasonably account for two types of specific situations: (1) forbidding specific containers for certain items; (2) dealing with precedence constraints.

On the one hand, if one does not want to load item j in container t, it is relevant to set $a(j, t) = L$ (where L is an arbitrarily large number). As a result, the used solution method is likely to avoid assigning container t to item j (otherwise, the large cost L will be incurred in the objective function).

On the other hand, the model also allows to account for a precedence constraint of type (j, j'). To do it, we set $c(j, j') = L$, and we create k artificial items $j_1, j_2, \ldots, j_k$ associated with item j, as well as k artificial items $j'_1, j'_2, \ldots, j'_k$ associated with item j'. An artificial container k_0 is also introduced. All the associated assignment and incompatibility costs for these artificial items are zero, except the following ones:

(A) $a(j_i, t) = a(j'_i, t) = L$ for $i \neq t$;
(B) $c(j_p; j_q) = c(j'_p; j'_q) = L$ if $p \neq q$;
(C) $a(l, k_0) = L$ for each non artificial item l;
(D) $c(j; j_i) = c(j'; j'_i) = L$ for $i \in \{1, \ldots, k\}$;
(E) $c(j_p; j'_q) = L$ if $p > q$.

As a result, a solution in which the precedence constraint (j, j') is not satisfied is strongly penalized by the objective function. To better understand this, let us consider one by one the above constraints. For every $i \in \{1, \ldots, k\}$, the (A) constraints favor solutions where j_i and j'_i are in container i. The impact of the (B) constraints is as follows: at most one of the items of type j_i (say j_c) and at most one of the items of type j_i' (say j_d) must be assigned to container k_0 (otherwise the penalty incurred in the objective function will be high). Adding the (C) and (D) constraints means that items j and j' will be assigned to containers different from k_0 (say c and d, respectively) and the corresponding items j_c and j_d' are both assigned to container k_0. Finally the (E) constraints avoid c being larger than d, and this prevents the violation of precedence constraint (j, j'). The above technique can obviously be generalized if several precedence constraints have to be considered: even if each precedence constraint involves the creation of $2 \cdot k$ artificial items, the two metaheuristics proposed in [37] remain competitive because they can be used for instances with up to 10,000 vertices within a reasonable amount of computing time.

14.7   Conclusion

In this chapter, various NP-hard packing problems on identical containers have been investigated. The models and methods presented in this chapter build bridges between graph theory and packing/loading problems. It was showed, on the one hand that graphs are powerful modeling tools, and, on the other hand, that graph coloring metaheuristics can be very efficiently adapted to specific packing problems. The success of the best performing metaheuristics relies mainly on four factors:

- an efficient representation of a solution of the considered problem (e.g., fixing one of the problem's dimensions in advance, in order to minimize the number of conflicts or constraint violations);
- using an auxiliary objective functions different from the given objective function associated with the problem (e.g., minimizing the number of decision variables which have not received a value);
- using an aggressive local search as an intensification procedure (e.g., a tabu algorithm focusing on conflicts and using a type of move which eliminates at least one conflict at each iteration, even if it creates other conflicts in other components of the solution);
- using an information exchange system which accounts for the structural properties of the problem (e.g., given that two colorings are equivalent if the colors are renamed, the attachment of a vertex to a specific color is therefore not a relevant information to handle and transmit).

References

1. Aarts, E.H.I., Laarhoven, P.J.M.: Statistical cooling: a general approach to combinatorial optimization problems. Philips J. Res. **40**, 193–226 (1985)
2. Bloechliger, I., Zufferey, N.: A graph coloring heuristic using partial solutions and a reactive tabu scheme. Comput. Oper. Res. **35**, 960–975 (2008)

3. Blum, C., Roli, A.: Metaheuristics in combinatorial optimization: overview and conceptual comparison. ACM Comput. Surv. **35**(3), 268–308 (2003)
4. Bortfeldt, A., Gehring, H., Mack, D.: A parallel tabu search algorithm for solving the container loading problem. Parallel Comput. **29**(5), 641–662 (2003)
5. Calegari, P., Coray, C., Hertz, A., Kobler, D., Kuonen, P.: A taxonomy of evolutionary algorithms in combinatorial optimization. J. Heuristics **5**, 145–158 (1999)
6. Dorigo, M., Stuetzle, T.: The ant colony optimization metaheuristic: algorithms, applications, and advances. In: Glover, F., Kochenberger, G. (eds.) Handbook of Metaheuristics. International Series in Operations Research & Management Science, vol. 57, pp. 251–285. Springer US (2003)
7. Faigle, U., Kern, W.: Some convergence results for probabilistic tabu search. ORSA J. Comput. **4**, 32–37 (1992)
8. Galinier, P., Hao, J.K.: Hybrid evolutionary algorithms for graph coloring. J. Comb. Optim. **3**(4), 379–397 (1999)
9. Galinier, P., Hertz, A., Zufferey, N.: An adaptive memory algorithm for the graph coloring problem. Discret. Appl. Math. **156**, 267–279 (2008)
10. Garey, M., Johnson, D.S.: Computer and Intractability: A Guide to the Theory of NP-Completeness. Freeman, San Francisco (1979)
11. Gehring, H., Bortfeldt, A.: A parallel genetic algorithm for solving the container loading problem. Int. Trans. Oper. Res. **9**(4), 497–511 (2002)
12. Gendreau, M., Potvin, J.-Y.: Handbook of Metaheuristics. International Series in Operations Research & Management Science, vol. 146. Springer, Berlin (2010)
13. Gendreau, M., Iori, M., Laporte, G., Martello, S.: A tabu search algorithm for a routing and container loading problem. Transp. Sci. **40**(3), 342–350 (2006)
14. Glover, F.: Future paths for integer programming and linkage to artificial intelligence. Comput. Oper. Res. **13**, 533–549 (1986)
15. Glover, F., Hanafi, S.: Tabu search and finite convergence. Discret. Appl. Math. **119**(1–2), 3–36 (2002)
16. Grefenstette, J.J.: Incorporating problem specific knowledge into genetic algorithms. In: Genetic Algorithms and Simulated Annealing, pp. 42–60. Morgan Kaufmann, Los Altos (1987)
17. Hajek, B.: Cooling schedules for optimal annealing. Math. Oper. Res. **13**, 311–329 (1988)
18. Hertz, A., de Werra, D.: Using tabu search techniques for graph coloring. Computing **39**, 345–351 (1987)
19. Hertz, A., Zufferey, N.: Vertex coloring using ant colonies. In: Artificial Ants. Iste & Wiley, London (2010)
20. Hertz, A., Plumettaz, M., Zufferey, N.: Variable space search for graph coloring. Discret. Appl. Math. **156**, 2551–2560 (2008)
21. Lu, Z., Hao, J.-K.: A memetic algorithm for graph coloring. Eur. J. Oper. Res. **203**, 241–250 (2010)
22. Mack, D., Bortfeldt, A., Gehring, H.: A parallel hybrid local search algorithm for the container loading problem. Int. Trans. Oper. Res. **11**(5), 511–533 (2004)
23. Malaguti, E., Toth, P.: A survey on vertex coloring problems. Int. Trans. Oper. Res. **17**(1), 1–34 (2010)
24. Malaguti, E., Monaci, M., Toth, P.: A metaheuristic approach for the vertex coloring problem. INFORMS J. Comput. **20**(2), 302–316 (2008)
25. Meuwly, F.-X., Ries, B., Zufferey, N.: Solution methods for a scheduling problem with incompatibility and precedence constraints. Algorithmic Oper. Res. **5**(2), 75–85 (2010)
26. Mladenovic, N., Hansen, P.: Variable neighborhood search. Comput. Oper. Res. **24**, 1097–1100 (1997)
27. Moura, A., Oliveira, J.F.: A GRASP approach to the container-loading problem. Intell. Syst. **20**(4), 50–57 (2005)
28. Osman, I.H., Laporte, G.: Metaheuristics: a bibliography. Ann. Oper. Res. **63**, 513–623 (1996)

29. Pisinger, D.: Heuristics for the container loading problem. Eur. J. Oper. Res. **141**(2), 382–392 (2002)
30. Plumettaz, M., Schindl, D., Zufferey, N.: Ant local search and its efficient adaptation to graph colouring. J. Oper. Res. Soc. **61**, 819–826 (2010)
31. Taillard, E.D., Gambardella, L.M., Gendreau, M., Potvin, J.-Y.: Adaptive memory programming: a unified view of metaheuristics. Eur. J. Oper. Res. **135**, 1–16 (2001)
32. Talbi, E.-G.: A taxonomy of hybrid metaheuristics. J. Heuristics **8**(5), 541–564 (2002)
33. Zhang, D., Peng, Y., Leung, S.C.H.: A heuristic block-loading algorithm based on multi-layer search for the container loading problem. Comput. Oper. Res. **39**(10), 2267–2276 (2012)
34. Zufferey, N.: Metaheuristics: some principles for an efficient design. Comput. Technol. Appl. **3**(6), 446–462 (2012)
35. Zufferey, N.: Models and methods in graph coloration for various production problems. In: Metaheuristics for Production Scheduling. Hermès – Lavoisier, Paris (2013)
36. Zufferey, N., Amstutz, P., Giaccari, P.: Graph colouring approaches for a satellite range scheduling problem. J. Sched. **11**(4), 263–277 (2008)
37. Zufferey, N., Labarthe, O., Schindl, D.: Heuristics for a project management problem with incompatibility and assignment costs. Comput. Optim. Appl. **51**, 1231–1252 (2012)

Index

© Springer International Publishing Switzerland 2015
G. Fasano, J.D. Pintér (eds,), *Optimized Packings with Applications*, Springer
Optimization and Its Applications 105, DOI 10.1007/978-3-319-18899-7